# Solar Radiation Data

# Solar Energy R&D in the European Community

Series F:

# Solar Radiation Data

Volume 2

Publication arrangements: D. NICOLAY

# Solar Energy R&D
# in the European Community

## Series F       Volume 2

# Solar Radiation Data

Proceedings of the
EC Contractors' Meeting held in
Brussels, 18-19 October 1982

edited by

W. PALZ
Commission of the European Communities

D. REIDEL PUBLISHING COMPANY

Dordrecht, Holland / Boston, U.S.A. / London, England

for the Commission of the European Communities

**Library of Congress Cataloging in Publication Data**
Main entry under title:

Solar radiation data.

    (Solar energy R&D in the European community. Series F ; v. 2)
      1. Solar radiation—Europe—Congresses.   2. Solar energy—
Europe—Congresses.   I. Palz, Wolfgang.   II. Commission of the
European Communities.   III. Series.
QC911.82.E85S64    1983      551.5′271′094        83-2954
ISBN 978-94-009-7114-1     ISBN 978-94-009-7112-7 (eBook)
DOI 10.1007/978-94-009-7112-7

---

Organization of the Contractors meeting by
Commission of the European Communities
Directorate-General Science, Research and Development, Brussels

Publication arrangements by
Commission of the European Communities
Directorate-General Information Market and Innovation, Luxembourg

EUR 8333
Copyright © 1983, ECSC, EEC, EAEC, Brussels and Luxembourg
Softcover reprint of the hardcover 1st edition 1983

Published by D. Reidel Publishing Company
P.O. Box 17, 3300 AA Dordrecht, Holland

Sold and distributed in the U.S.A. and Canada
by Kluwer Boston Inc.,
190 Old Derby Street, Hingham, MA 02043, U.S.A.

In all other countries, sold and distributed
by Kluwer Academic Publishers Group,
P.O. Box 322, 3300 AH Dordrecht, Holland

D. Reidel Publishing Company is a member of the Kluwer Group

# P R E F A C E

This book gives a comprehensive overview of activities currently under way to produce, collect and compile radiation data as needed for the various types of solar energy applications in Europe. Contributions have been made by all contractors of the Commission of the European Communities, in particular the Meteorological Offices of the EC member countries. They all reported on their work at a meeting which was held in October 1982 in Brussels and of which these are the proceedings.

The Commission' work in this area follows a detailed strategy which was published earlier as part of the proceedings of Volume 1, Series F. Series F is especially devoted to publications on the European Communities' work on solar radiation data. Other volumes within Series F are in preparation and will deal with :

- solar radiation data on tilted planes;
- solar radiation data derived from meteorological satellite observations.

In addition, two new atlases are being prepared, one showing - for the area of the European Community - maps for solar radiation on titled planes of various inclinations and orientations, and the other showing - for the whole of Europe and the Eastern part of the Mediterranean - radiation data for horizontal planes. In the latter there will also be a statistical analysis section. Both atlases will be published in the course of 1983.

The present book contains many interesting new results for the European solar energy community and I am sure it will contribute to closing an important gap in the literature of the sun as an energy resource.

W. PALZ

C O N T E N T S


Preface

ACTION 1

CALIBRATION AND CHARACTERISTICS OF RADIOMETERS

Action leader's progress report

J.L. PLAZY, Agence francaise pour la maîtrise de
l'énergie, Sophia Antipolis, Valbonne, France

Report of action participant :

The implications for calibration and field use of
non-ideal cosine behaviour in Kipp and Zonen CM-5
pyranometers

# ACTION 1 - CALIBRATION AND CHARACTERISTICS OF RADIOMETRES

**Action Leader** :   J.-L. PLAZY
                      Agence française pour la maîtrise de l'énergie
                      SOPHIA-ANTIPOLIS
                      F-06565 VALBONNE CEDEX

**Participants**  : - Direction de la météorologie
                      Contract Nr ESF-005 F
                      P. GREGOIRE

                    - University College of Cardiff
                      Contract N₁ ESF 026 UK
                      J. McGREGOR

**Taks** : Intercomparison of pyranometers
           Characterization of pyranometers
           Handbook on the use of pyranometers

## SUMMARY

The first C.E.C substandard pyranometers comparison was held at Carpentras in June 1981. To check the stability of the method a second one was done in January 1982. The results show a larger deviation in winter than in summer but in general for a given pyranometer the calibration factor still within ± 1 % between these two periods. In fact to obtain better results it should be necessary to modelise with a bigger accuracy the main types of pyranometers available on the market.

To do it the work of the University College of Cardiff was a good approach but demonstrates that even if they are of the same type each pyranometer is itself an individual case.

If the correction method used during the Carpentras comparisons can allow to reach an accuracy of about 2 % in daily amount of radiation, it is not sufficient for more accurate requirements New progress in pyranometry will occur if more attention is given to developp a better instrumentation.

An other way to improve the accuracy of pyranometers measurements is to follow the recommandation included in the booklet on the use of pyranometer which is under drawing up and which will be published at the end of the programme by the C.E.C.

# 1 - GENERAL TASKS OF THE ACTION 1

The general tasks of the action 1 referred to the strategy paper is to check the homogeneïty of radiation measurements in the various countries of the European Communities and if possible to try to improve it by a better knowledge of the used instrumentation and by advisoring the users with some recomendations on pyranometry technique.

Two institues are involved in this action with a contract : the Direction de la Météorologie and the University College of Cardiff. But the general tasks of this action implies a cooperation between all the meteorological services of the European Communities and a collaboration with the action 4.1 contractors. It is the reason why the action 1 meetings are organised with the action 4.1 ones.

# 2 - RESULTS OF THE PYRANOMETERS COMPARISONS

In the action 1 frame the pyranometers of the various meteorological services used as substandard to calibrate field instruments were compared in June 1981 and January 1982 in Carpentras during a several days period. During the same time two pyranometers were involved in a round robin calibration.

## 2.1 - Results of the Carpentras Comparison

As announced in the previous Action 1 report published in the volume 1 of series F proceedings a second comparison was organised in winter period in Carpentras, from 19th January to first of February 1982, with the same methodology yet described to display the seasonal effect on the calibration of pyranometer by a comparison method.

Among the fourteen pyranometers compared in January, eight participated at the first session in June 1981. The italian group of pyranometers participated for the first time.

The results of the two sessions are summarized in the table I and on the graph II. Some comments can be done about them :

- if the comparisons would be not performed the spread of radiation measurements in the E.C countries could reach ± 8 %.

- after the comparisons and if the recommanded calibration factors are used the spread can be reduced to ± 2 %

- the characterisation and the modelisation of the substandard pyranometer should be improved if a better accuracy for the radiometric measurements is required.

| COUNTRY | SERIAL NUMBER | CALI FACTOR BEFORE C. at 15° C | CALI FACTOR JUNE 81 at 15° C | CALI FACTOR JEN.82 at 15° C | 81/82 RATIO |
|---|---|---|---|---|---|
| BELGIUM | 19682 F3 | 98.9 * | 94.4 | NP | — |
|  | 6217 A | 118.6 | 119.9 | NP | — |
| DENMARK | 76-3327 | 131.8 | 124.9 | NP | — |
|  | 800143 | 48.1 | 50 | NP | — |
| FRANCE | 16542F3 | 92.5 | 90.8 | 91.0 | 0.998 |
|  | 16540 F3 | 93.8 | 92.1 | NP | — |
|  | 800082 | 56.5 | 55.8 | 56.8 | 0.982 |
|  | 79-6617 | 116.4 | 115.7 | 115.1 | 1.005 |
|  | 79-5546 | 110.2 | 111.1 | NP | — |
| F.R.G. | 79-0057 | 57.5 | 58.2 | 58.2 | 1 000 |
| GREECE | 14899 F3 | 92.5* | 89.1 | 89.9 | 0.9911 |
| IRELAND | 69-0187 | 108.1 | 109.0 | 111.2 | 0.9802 |
| ITALY | 78-4691 | 117.0* | NP | 109.3 | — |
|  | 78-4707 | 120.0* | NP | 112.7 | — |
|  | 78-4634 | 120.0* | NP | 110.5 | — |
| NETHERLANDS | 1564 | 113.2 | 114.0 | 117.8 | 0.968 |
|  | 18006 F3 | 82,7 | NP | 82.0 | — |
| U.K | 2508 | 122,5 | 123,2 | 124,9 | 0.986 |

* calibration factor given by the manufacturer

TABLE I : RESULTS OF THE C.E.C SUBSTANDARD PYRANOMETERS COMPARISONS

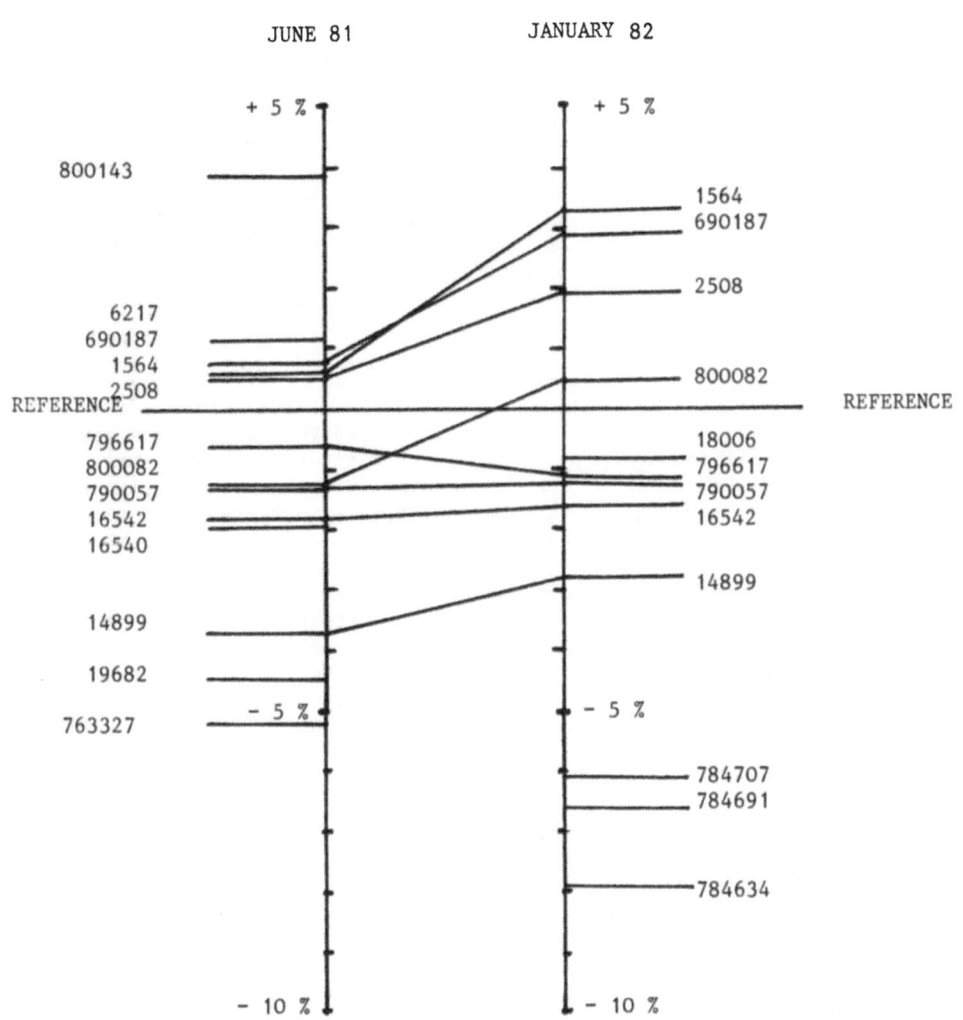

FIGURE II - DEVIATION OF THE SUBSTANDARD PYRANOMETERS USED WITH
THEIR INITIAL CALIBRATION FACTOR AGAINST THE REFERENCE

- the threshold of 2 % accuracy seems to be difficult to be diminished by usual methodology. New instrumentation development can perhaps bring new hopes in the pyranometry.

2.2 - Round robin calibration

During the years 1981 and 1982 two pyranometers, a Kipp and Zonen CM5 and an Eppley PSP, were successively calibrated by three meteorological services using their own method (U.K, FRG, FRANCE).

The results of this round robin comparison are given in table III. The maximal deviation is 2.2 % for the KIPP and ZONEN CM5 pyranometer and 1.5 % for the P.S.P one. This result can be considered as a good one according to the various calibration methods used : outdoor directly against the sun, outdoor by comparison with a substandard pyranometer, indoor by comparison with a substandard inside an integrating sphere and under a light parallel beam.

## 3 - PYRANOMETER CHARACTERISATION

The University College of Cardiff studied the characteristics of eight KIPP and ZONEN pyranometers manufactured in 1969, 1975, 1977 and 1979. For each of them the cosine response was measured with a spatial goniometer when the radiation is parallel then perpendicular to the line of the thermojunctions.

These experiments confirm the other studies on this topic showing that the sensitivity of old CM5 pyranometers increase at low angle of sun elevation giving a U shape at the sensitivity curves, and that the sensitivity of new CM5 pyranometers which have a raised thermopile decrease lightly with low angle of elevation giving a n shape. But these experiments indicate substantial variations of those shapes from a pyranometer to another even if they are of the same type.

Using the cosine response obtained by laboratory measurements the University College calculated the ratio between each pyranometer and a "perfect pyranometer" for various latitudes between 30 and 60 degrees and declinations and for various inclinations. These ratios are calculated on daily total of radiation.

This ratio was established too for a standard overcast sky using a distribution described by :

$$N \ (0) = N \ (0) \quad \frac{1 + b \cos \theta}{1 + b}$$

where O is the zenith angle and b a coefficient of linearity.

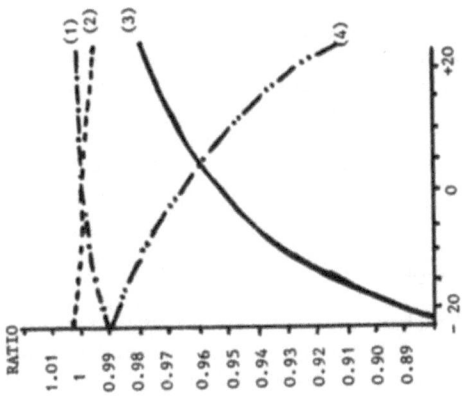

Fig. IV – Ratio between daily total radiation obtained with a field pyranometer and a perfect one versus the declination of the sun and for an horizontal and vertical position during a clear sky day

| b | - 0.3 | 0 | 1 | 2 | 3 |
|---|---|---|---|---|---|
| 796411 | 1.004 | 1.005 | 1.006 | 1.006 | 1.006 |
| 752957 | 0.910 | 0.965 | 0.970 | 0.972 | 0.945 |

Table V – Ratio between daily total radiation obtained with a field pyranometer and a perfect one an overcast sky conditions versus the coefficient of linearity b

| PYRANOMETER | COUNTRIES | METHODS | CALIBRATION FACTOR at 10°C | MAXIMUM DEVIATION |
|---|---|---|---|---|
| EPPLEY 16540 | FRANCE | OUTDOOR CAL. SUN-DISC | 93.8 | |
| | | SUBSTANDARD COMPARISON CARPENTRAS | 92.1 | |
| | F.R.G. | INDOOR CAL. DIRECT PARALLEL BEAM | 92.3 | 1.5 % |
| | | OUTDOOR CAL. BY COMPARISON | 92.2 | |
| | U. K. | INDOOR CAL. INTEGRATING SPHERE | 92.4 | |
| | | OUTDOOR CAL. BY COMPARISON | 93.4 | |
| KIPP and ZONEN 79-5546 | FRANCE | OUTDOOR CAL. SUN-DISC | 110.2 | |
| | | SUBSTANDARD COMPARISON CARPENTRAS | 111.1 | |
| | F.R.G. | INDOOR CAL. DIRECT PARALLEL BEAM | 111.7 | 2.2 % |
| | | OUTDOOR CAL. COMPARISON | 110.2 | |
| | U.K. | INDOOR CAL. INTEGRATING SPHERE | 109.3 | |
| | | OUTDOOR CAL. BY COMPARISON | 109.5 | |

Table III – Round robin calibration

The figure IV and table V give an exemple of level of the error which can be done on clear sky and an overcast sky conditions with a good pyranometer with a very small cosine effect (79.6411) and with a bad one with a large cosine effect (75.2957). In the first case the error do not exeed 1 % in the other case it can reach 10 % according to the season.

Among the conclusions of a such study it seems to be a prime necessity in a first step to invite the buyers of pyranometers to select them and to send back to the manufacturer the pyranometer which have too bad cosine response, by exemple more than 3° for a zenith angle of 30 degrees, in a second step to intend an action nearby the manufacturers to obtain a guarantee on the pyranometers characteristics.

## 4 - BOOKLET ON THE USING OF THE PYRANOMETER

The previous studies show clearly than the accuracy of radiometric measurements is not only a function of the intrinsic characteristics of the pyranometer but too a function of the way of using them. It is the reason why a draft of booklet giving recommandation on the using of pyranometers has been written and discussed by the action 1 and 4.1 experts and contractors. This booklet contains five chapters:

- description of pyranometers,
- calibration modes,
- use of pyranometers,
- associated measuring equipments,
- accuracy of radiometric measurements.

It will be achieved before the end of the second Solar energy R & D programme. At the present time, the draft in english version should be improved to take in account the discussions about it which took place in Copenhagen during the last action 1 and 4.1 meeting in September 1982.

## 5 - CONCLUSIONS

The action 1 preliminary results opens the eyes of the pyranometers users on the difficulties to do correctly radiometric measurements and on the maximum of accuracy which can be reached by such techniques Regular compaigns of comparison of primary or secondary standards prompt the specialized laboratories to maintain the radiometric scale with all the necessary attention and can help the less well equiped among them to maintain a sufficient level of accuracy.

For future work such comparisons could be repeated. By the same time studies on new instrumentation could be developped to increase the accuracy of the measurements. A classification of the pyranometer, more detailed than the W.M.O. one, could be established on condition that the manufacturer could guarantee a sufficient reproductibility of the pyranometers characteristics they sell.

# THE IMPLICATIONS FOR CALIBRATION AND FIELD USE OF NON-IDEAL COSINE
## BEHAVIOUR IN KIPP AND ZONEN CM-5 PYRANOMETERS

J. McGREGOR
Solar Energy Unit, University College, Cardiff

## Introduction

In an earlier report[1] the preliminary results of a laboratory
comparison of the cosine behaviour of a sample of seven Kipp and Zonen
CM-5 pyranometers were presented.  The comparison which was made
using the Spatial Goniometer at the National Institute of Agricultural
Engineering at Silsoe, Bedfordshire, demonstrated that Kipp and Zonen
CM-5 pyranometers can show substantial deviations from a true cosine
response and that the shape of the functional dependence with zenith
angle varies according to thermopile geometry.  For example the early
CM-5 instruments which have a low thermopile, display a typically
u-shaped cosine response with instrument sensitivity increasing at low
angles of incidence (such a behaviour is also typical of the earlier
CM-2 pyranometers).  However for the later CM-5 instruments which
have a raised thermopile the cosine response is typically n-shaped
showing decreasing sensitivity at low angles of incidence.

Another important finding which emerged from the preliminary analysis
was that substantial variations in cosine response occurred from
instrument to instrument of the same type and geometry, and it was
realised that this would have important consequences for pyranometer
operations.  Whilst the variation in height of the thermopile was
found to be the reason for the difference between n- and u-type
pyranometers, variations in the quality of the detector surface was
thought to be responsible for the differences in cosine behaviour for
instruments of the same type.

By utilizing the laboratory measured cosine response data a detailed
computer study has been undertaken, and the implications of the
deviations from true cosine behaviour have been assessed for the various
laboratories (i.e. integrating sphere, normal incidence and outdoor
comparison) as well as for the field operation of pyranometers.  Two
types of sky condition have been considered, the first representing
overcast sky conditions (which are also typical of the conditions found
in integrating spheres), and the second representing clear sky conditions.
For clear sky conditions the errors have been assessed as a function
of latitude and season.

An extension of this computer model has been made to investigate the
importance of poor instrument cosine response to measurements made by
pyranometers on tilted surfaces.  Throughout this computer study
reported here, it has been assumed that all instruments have the same
calibration for a direct-beam from zenith.

In this report we present the results of the laboratory and computer

investigations and draw up a list of conclusions and recommendations
that have been drawn from this study.

## Laboratory results

Most absorbing surfaces show some degree of deviation from a true cosine
behaviour and generally this becomes an increasingly important problem
for radiation from low incidence angles.  For thermopile pyranometers
with hemispherical glass domes the problem is further complicated
by internal reflections from the inner surface of the domes.  This
effect is also more important at lower incidence angles, when the
reflected radiation produces a distinct focussing effect on the
detector surface.

Yet another complication arises for the case of Kipp and Zonen CM-5
pyranometers.  These pyranometers are based on an original design of
Moll and Gorczynski and feature a thermopile of rectangular construction
which consists of fourteen manganin-constantan thermoelements linked up
via two parallel lines of eight copper mounting posts.  The line of hot
junctions is parallel to, and lies between the two lines of cold
junctions (2).  Clearly such a design implies rectangular symmetry and
this combined with the effect of focussing on the detector results in an
azimuthal dependence as well as the zenith angle dependence in
sensitivity of the pyranometer.

Such an azimuthal dependence is clearly demonstrated by comparing the
cosine plots shown in figures 1 and 2.  Figure 1 gives the cosine
response when the pyranometer is aligned such that the incident
radiation is parallel to the line of the thermojunctions, whilst figure 2
represents the case when the radiation falls perpendicular to the line
of the thermojunctions.  Apart from revealing azimuthal dependence
these plots also demonstrate the typically u-shaped behaviour of the
1969 - manufactured pyranometer whilst the 1975-1979 manufactured pyranometers show a n-shaped behaviour.  The plots reveal important variations
from one instrument to another.

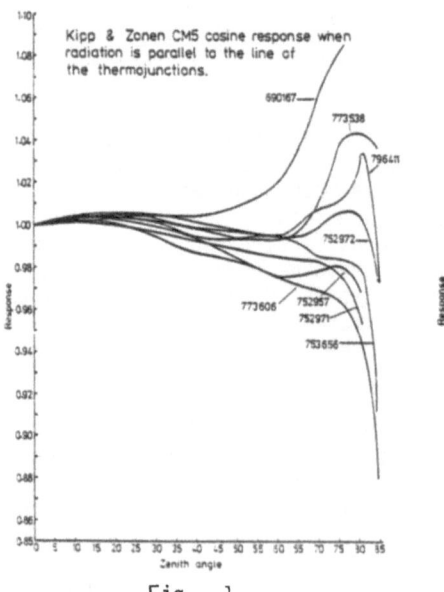

Fig. 1

Fig. 2

- 10 -

## Results from the Computer Study

### a) A diffuse sky

For the distribution of radiation in the diffuse sky we have assumed the 'Standard overcast sky' distribution to be described by

$$N(\theta) = N(o)\frac{(1 + b\cos\theta)}{1 + b}$$

where $\theta$ is the zenith angle of the radiation and b is a coefficient of linearity (it should be noted that Steven and Unsworth found b to equal 1.23 for average overcast sky conditions in Britain; for an isotropic sky b = o)

Using this description of the diffuse distribution the global irradiation is computed numerically using

$$G = \iint_{\theta\,z} N(\theta, z)\ d\theta\ dz$$

and for the case of $G_{\cos\ error}$ by

$$G_{\cos\ error} = \iint_{\theta\,z} N(\theta, z)\,\omega\,(\theta, z)\ d\theta\ dz$$

where $\omega(\theta, z)$ represents the weighting due to the deviation from a true cosine response and is a quantity that is measured in the laboratory.

Figure 3 shows plots of $G_{\cos\ error}/G$ as a function of b for each of the

instruments under test. This is a particularly revealing plot for it immediately provides an assessment of the absolute and the relative error which one would expect to encounter between an indoor integrating sphere and an indoor normal incidence calibration.

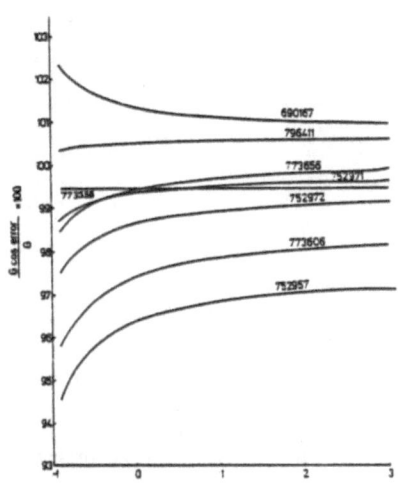

Fig. 3

### b) Direct radiation

The error in the estimation of direct radiation by any pyranometer at any instant can be made directly from a polar diagram of the instruments cosine response or for certain circumstances from the plots given in figures 1 and 2. These plots give the absolute error for any instrument

calibrated by normal incidence, whilst the relative errors between any
two pyranometers can be assessed by comparing the curves of the two pyranometers in question.   Whilst this method gives an accurate assessment of
error for any given angle of incidence and azimuth it does not predict the
performance over a full day.   In order to assess such performance a
numerical integration was made and this was weighted by the cosine behaviour
of the instrument.

The results of these integrations are shown in figure 4 and demonstrate
the latitudinal and seasonal dependence of the resulting error which arises
from the lack of true cosine behaviour in the instrumentation.   A
striking point which emerges from the plots is that u- and n-type pyranometers shown a fundamentally different behaviour and this highlights the
dangers of comparing or calibrating these two types of instruments against
each other.

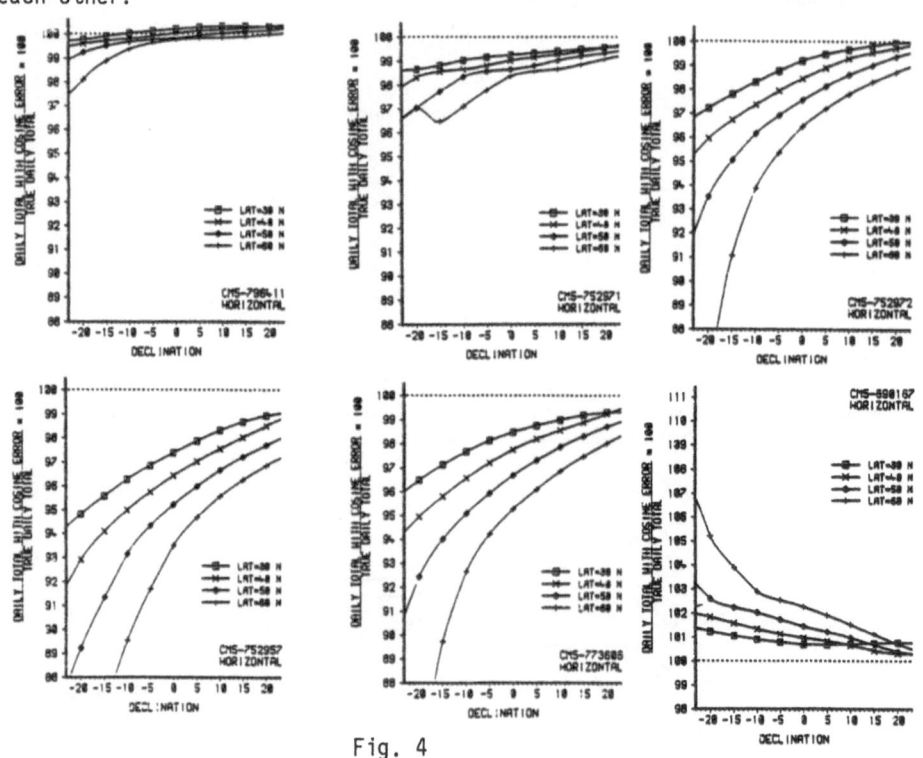

Fig. 4

c) Pyranometers on tilted surfaces

In this section we investigate the effect of tilting the pyranometers,
though it should be stressed that these results only include errors due to
the optical considerations of the cosine problem, they do not include any
contribution from the change in the thermal environment within the instrument when it is tilted.

The results are presented in figure 5 and show that purely on the basis of
cosine error substantial errors can occur in tilted pyranometers,
particularly for vertical surfaces during the summer months.

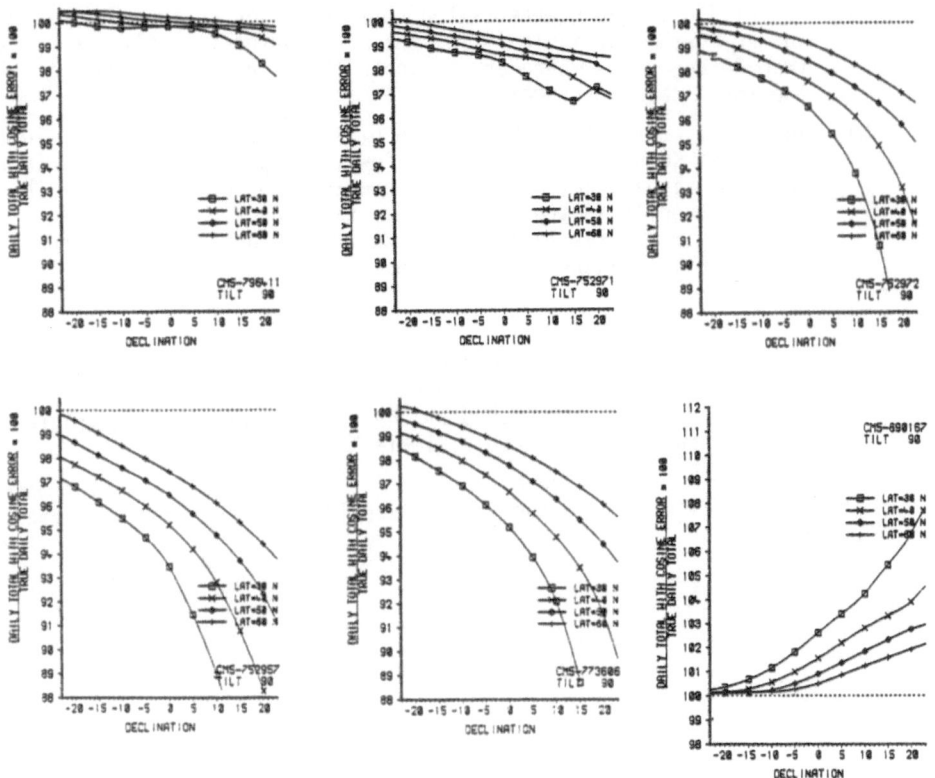

Fig. 5.

## Conclusions and Recommendations

The following conclusions and recommendations have been drawn from this study.

1) A poor cosine behaviour is genrally attributed to poor quality control of the detector surface.

2) u-type cosine plots are appropriate to low thermopile CM-5 instruments (which include CM-2 and early CM-5 instruments) Whilst n-type cosine plots are obtained for high thermopile instruments.

3) Neither indoor nor outdoor calibrations should be made by comparing an instrument exhibiting u-type behaviour with one exhibiting n-type behaviour.

4) The calibration of pyranometers in a diffuse sphere is unlikely to result in calibration errors in excess of 3.5% due to poor cosine response.

5) Outdoor calibration of pyranometers should be avoided in the winter months.   If this recommendation is followed, errors in excess of 3% are unlikely to be caused by poor instrument cosine response.

6) The method of normal incidence calibration is the most appropriate for instruments that are tilted at an angle of 45% and errors for instruments operated at this orientation are small, generally less than 1%.

7) Errors for the 90 degree tilted surface can be serious during the summer months particularly for the more southerly latitudes (less than 30 degrees), where errors up to 10-15% have been found for the worst case instruments.

8) The manufacturers' recommendation of aligning the thermojunctions East-West should be followed for low-thermopile-type pyranometers. Though a significant overall improvement can be obtained for the high thermopile instruments if the thermojunctions are aligned North-South. This fact is particularly important for instruments which exhibit a generally poor cosine behaviour.

## References

1) McGregor J. (1980) Some preliminary results from a comparative study of Kipp and Zonen CM5 pyranometers.
Proceedings of a workshop on the accuracy of pyranometers for radiation measurements, Brussels, 18-19 September 1980.

2) Bener P. (1950) The operation of the Moll-Gorcynski Solarimeter (U.K Met. Office Translation 1288) Arch. Met. Geoph. Bioklim., Vien, B2 pp 188-249.

ACTION 2

REFERENCE YEARS


Action leader's progress report
    S. EIDORFF, Thermal Insulation Laboratory, Technical
    University of Denmark, Lyngby, Denmark



Reports of action participants :

  - Short reference years with adjusted measured data

  - Establishment of short reference years for calculation of
    annual solar heat gain or energy consumption in residential
    and commercial buildings

  - SRY production by means of stochastic models

  - Short reference year, SRY

Action 2 - <u>REFERENCE YEARS</u>

Action budget  (CEC contribution) :  178.6 KUC

Duration of project :  until 1983-06-30

Action Leader :  S. EIDORFF
                 Thermal Insulation Laboratory
                 Technical University of Denmark
                 Building 118
                 DK - 2800  Lyngby,  Denmark

Participants:

- Technical University of Denmark
  Contract Nr. ESF-029-DK(G)
  H. LUND, S. EIDORFF

- Delft University of Technology
  Contract Nr. ESF-010-80-N(B)
  A.W. BOEKE, A. VAN PAASSEN, H. LIEM

- Analysis and Development of Energy Systems
  Contract Nr. ESF-007-I(S)
  C. MUSTACCHI, V. CENA, F. HAGHIGHAT

- Technological Institute, Denmark
  Contract Nr. ESF-030-DK(G)
  L. HALLGREEN

- Institut Royal Météorologique de Belgique
  Contract Nr. ESF-003-B(G)
  R. DOGNIAUX, M. LEMOINE

Task :

Development and production of a library of data
tapes;  various sites inside the European Community.

Summary

The aim of the project is to develop 4 operational methods for generating Short Reference Years (SRY) from existing basic ideas or procedures.  These methods will be evaluated, and the best method will be used
for generation of SRYs for a number of locations within EEC.

Three of the methods generate SRYs from deterministic and stochastic
properties extracted from the long term weather data set, and another method by adjusting the distributions of the weather parameters in selected
short periods in accordance with the distributions of the long term weather data set.

The evaluation will be done by comparing simulation runs with the
different SRYs and the weather data set from which the SRYs have been
derived.

The four methods are ready and SRYs generated for three locations.  A
lot of simulation runs with the SRY and a variety of systems within the
field of application of SRYs has been performed in the evaluation process.
Some examples are shown in this paper.  The full set will appear in the
final report.

The simulated systems and the mathematical evaluation procedure are
described briefly in this paper.

The evaluation has been concluded:

The method by S. Eidorff performed generally best in the test cases,
but method no. 2 by A.H.C. van Paassen et al was actually better in many
cases, so the former method is selected on a narrow margin.

Further could be concluded:

The SRYs are shown to be well suited for heating and cooling loads
and lesser suited for extreme temperatures and solar systems with long
storage times.

The quality of the generated SRYs varies much, quite randomly, and
it may therefore be necessary to test and document the performance of a
new SRY for a certain location before releasing the data for public use.

For the Test Reference Year Subtask is reported that the production
of the TRYs is completed, but some questions about the distribution and
formats are still to be clarified.

## 1. Task of Action 2 and the participants

The main task of Action 2 is, according to the Strategy paper for Project F, to develope and produce a library of data tapes for various sites inside the European Community.

The data tapes will hold two different forms of data sets: Test Reference Years, TRY, and Short Reference Years, SRY.

The selection of the best method to produce TRYs was finished in 1979. The production of data tapes with TRYs for the various locations is performed by Mr. R. Dogniaux and Ms. M. Lemoine. This work is completed and the actual distribution can start as soon as some questions about the distribution and the format are clarified.

The Short Reference Year project consists of 3 phases:
- Development and refinement of 4 different methods.
- Evaluation of these methods and selection of the best and
- Production of SRYs for the same European locations as for the TRYs.

The first two phases are completed, except for the final report.

The objective of Short Reference Years is to conserve computer time in order to make it feasible to perform hourly simulation on a micro computer or to test several design alternatives by simulation.

SRYs are generated for a certain location from an existing weather data set. A good SRY will give the same results with a simulation program as if the original weather data was used.

The four SRY generating methods which were developed or refined within this research programme are described in more detail in the annexes.

Three of the methods, by A.H.C. van Paassen et al, by C. Mustacchi et al and by L. Hallgreen generate the weather parameters by statistical processes after an analysis of the original weather data set.

The fourth method, by S. Eidorff, extracts typical days from the weather data set and adjust their hourly values in such a way that the distributions of the hourly values in the SRY are in accordance with the distributions in the weather data set.

It was agreed in the group that all methods shall comprise a fully automatic procedure for complete long term weather data sets and be so well documented that special treatment of certain data sets could be designed.

## 2. Evaluation of the SRY generating methods

### 2.1 Method

The evaluation of the candidate methods was based on simulation runs with solar energy systems and building thermal loads. The results of simulation runs on a certain system or building with the different SRYs were compared with a simulation run with the long term weather data set from which the SRYs were generated.

It is important that SRY must be representative for a wide variety of applications, because they will typically be used as input data for simulations, including both the building and the solar energy system.

In order to select the best method, on a high confidence level if possible, these simulations were performed on an appreciable number of systems and buildings with weather data sets from several locations.

The simulation program for the solar energy systems is designed by Prof. Mustacchi. The program for the building thermal load simulations is BA4 with the necessary enhancements for use with SRYs designed by H. Lund.

The simulated domestic hot water solar systems:
- Collector area in $m^2$ per 100 l daily usage at $45^{\circ}C$: (2,4,6).

- Storage tank capacity per m$^2$ collector area: (25 1, 50 1, 75 1)
- Total 9 systems.
- Water inlet temperature for the locations in Denmark, Holland and Belgium: 10$^o$C; for Carpentras: 15$^o$C.

For each system 5 observations are calculated:
- Total make up fuel consumption per year and per season.

The simulated rooms:
- The rooms have 1 outer wall with 1 window, floor area 20 m$^2$, and is placed in a typical office building.
- 2 rooms (heavy and lightweight).
- 3 orientations (north, east and south).
- 2 control strategies (const. room temperature at 21$^o$C, and heating only if temperature is below 21$^o$C).
- Total 12 rooms.

For each room the following 10 observations are calculated:
- Total heating per year and per season.
- Total cooling resp. number of degree-hours (T - 24$^o$C) above 24$^o$C per year and per season.

The number of observations for each location or weather data set is 165. As three locations were used (Copenhagen, Uccle or De Bilt, Carpentras) the total number of observations is 495 for each of the evaluated 4 SRY generating methods.

The mathematical evaluation of the SRY generating methods:
- A number of observations about each method are made.
- An observation is the absolute difference between a value of a certain characteristic obtained with the SRY and the long term weather data set from which the SRY has been derived.
- The said characteristic could f.inst. be the total cooling energy in the summer season necessary to maintain an indoor temperature of 21$^o$C in the lightweigh room with window facing East in Uccle.
- The observations are grouped in classes in order to compare them in a meaningful way, by normalization.
- The value number $V_i$ of a SRY method is the weighted average of all observations. The classes are weighted so that a yearly observation means twice as much as a seasonal observation to the value number. A method with a lesser $V_i$ performes better in this test.

Mathematically expressed:

$$V_i = \frac{\sum_{j=1}^{60} W_j \sum_{k=1}^{N_j} \frac{O_{ijk}}{\frac{1}{4N_j} \sum_{i=1}^{4} \sum_{k=1}^{N_j} O_{ijk}}}{\sum_{j=1}^{60} N_j W_j}$$

Where:
$W_j$ is the weighting factor.
$O$ is the value of an observation

Index i is the SRY method, i = 1 to 4,
Index j is the class, j = 1 to 60,
Index k is the characteristic of an observation, k = 1 to $N_j$,

It was pointed out by some of the participants in the group that small deviations in seasonal values which inherently will be small f.inst. heating in summer time get too much weight in this formula.
An alternative formula is therefore also calculated:

$$V_i = \frac{\sum\limits_{j=1}^{60} W_j \sum\limits_{k=1}^{N_j} \dfrac{O_{ijk}}{\dfrac{1}{4N_J} \sum\limits_{i=1}^{4} \sum\limits_{k=1}^{N_J} O_{iJk}}}{\sum\limits_{j=1}^{60} N_j W_j}$$

Where index J means the class number for the yearly observation corresponding to j.
In this way insignificant deviations will only participate in the evaluation by their value compared with the yearly deviations.

## 2.2  Results

The results of all 495 x 4 observations will appear in the final report. Here is only shown a few results as an example.
In the tables on the following pages are shown simulation results for the three different locations and for the four seasons plus the entire year. The rows with the city names show the actual simulation result with the weather data set from the said location. On the rows just below is shown the deviations in simulation results obtained by using the 4 different SRYs.
The actual values of the simulation results with the weather data sets shown are those obtained by S. Eidorff. However, the devations shown are those obtained by respective authors as the difference between the value for the weather data set and for the SRY.
The abbreviations for the SRYs are:
- SE:  S. Eidorff
- LH:  L. Hallgreen
- VP:  A.H.C. van Paassen et al
- CM:  C. Mustacchi et al

Table I

|  |  |  | WIN | SPR | SUM | AUT | YEAR |
|---|---|---|---|---|---|---|---|
| **Heating load (kWh)** | Lightweight | Copenhagen | 1249 | 591 | 65 | 540 | 2444 |
|  |  | SE-SRY dev. | -4 | -12 | -12 | -20 | -34 |
|  |  | LH-SRY dev. | -16 | 67 | 3 | -142 | -117 |
|  |  | VP-SRY dev. | 84 | -34 | 3 | -57 | 10 |
|  |  | CM-SRY dev. | 12 | 19 | -53 | 65 | 57 |
|  |  | Uccle/De Bilt | 1038 | 470 | 65 | 450 | 2034 |
|  |  | SE-SRY dev. | -2 | 24 | -31 | -19 | -16 |
|  |  | LH-SRY dev. | 8 | 25 | -33 | -44 | -71 |
|  |  | VP-SRY dev. | 1 | -4 | -12 | -61 | -65 |
|  |  | CM-SRY dev. | -181 | -18 | 272 | 64 | 143 |
|  |  | Carpentras | 668 | 278 | 0 | 150 | 1096 |
|  |  | SE-SRY dev. | 39 | -64 | 0 | 8 | -9 |
|  |  | LH-SRY dev. | 96 | -59 | 0 | 64 | 73 |
|  |  | VP-SRY dev. | 31 | -22 | 10 | 10 | 37 |
|  |  | CM-SRY dev. | -36 | -138 | 0 | -17 | -180 |
|  | Heavy | Copenhagen | 1241 | 549 | 25 | 499 | 2313 |
|  |  | SE-SRY dev. | -12 | 2 | -8 | -52 | -56 |
|  |  | LH-SRY dev. | -16 | 84 | 19 | -153 | -95 |
|  |  | VP-SRY dev. | 87 | -23 | -1 | -81 | -2 |
|  |  | CM-SRY dev. | -3 | 53 | -25 | 46 | 82 |
|  |  | Uccle/De Bilt | 1032 | 431 | 29 | 421 | 1913 |
|  |  | SE-SRY dev. | -13 | 40 | -28 | -55 | -45 |
|  |  | LH-SRY dev. | 7 | 39 | -15 | -50 | -58 |
|  |  | VP-SRY dev. | 4 | 15 | -28 | -82 | -78 |
|  |  | CM-SRY dev. | -201 | -20 | 299 | 67 | 152 |
|  |  | Carpentras | 621 | 224 | 0 | 95 | 940 |
|  |  | SE-SRY dev. | 60 | -83 | 0 | -15 | -30 |
|  |  | LH-SRY dev. | 121 | -44 | 0 | 47 | 96 |
|  |  | VP-SRY dev. | 48 | -41 | 4 | -13 | 6 |
|  |  | CM-SRY dev. | -72 | -122 | 0 | -47 | -277 |

Table I shows the heating load for a lightweight resp. heavy office room facing south with no cooling, and heating only if the room temperature is below 21°C.

Table II

| | | | WIN | SPR | SUM | AUT | YEAR |
|---|---|---|---|---|---|---|---|
| **Degree hours, T-24°C (h·k)** | **Lightweight** | Copenhagen | 37 | 512 | 2058 | 573 | 3179 |
| | | SE-SRY dev. | -11 | 12 | -3 | -120 | -120 |
| | | LH-SRY dev. | -27 | -193 | 75 | -282 | -425 |
| | | VP-SRY dev. | -10 | -141 | -105 | 162 | -107 |
| | | CM-SRY dev. | -15 | -348 | 1175 | -295 | 495 |
| | | Uccle/De Bilt | 32 | 561 | 2219 | 813 | 3625 |
| | | SE-SRY dev. | 2 | 30 | -485 | -229 | -699 |
| | | LH-SRY dev. | 24 | -368 | -161 | -675 | -1188 |
| | | VP-SRY dev. | -2 | 161 | 332 | 127 | 602 |
| | | CM-SRY dev. | 20 | -78 | -1112 | -396 | -1568 |
| | | Carpentras | 526 | 1221 | 6859 | 3550 | 12165 |
| | | SE-SRY dev. | -94 | 207 | -721 | 307 | -346 |
| | | LH-SRY dev. | -224 | 47 | -685 | -1489 | -2338 |
| | | VP-SRY dev. | -164 | -247 | -413 | 79 | -795 |
| | | CM-SRY dev. | 333 | 727 | 422 | 1009 | 2448 |
| | **Heavy** | Copenhagen | 0 | 83 | 1128 | 177 | 1388 |
| | | SE-SRY dev. | 0 | -54 | -332 | -84 | -477 |
| | | LH SRY dev. | 0 | -80 | -114 | -74 | -276 |
| | | VP-SRY dev. | 0 | -54 | -382 | -18 | -460 |
| | | CM-SRY dev. | 0 | -680 | 709 | -130 | 498 |
| | | Uccle/De Bilt | 2 | 137 | 1323 | 350 | 1812 |
| | | SE-SRY dev. | -2 | -83 | -690 | -233 | -1013 |
| | | LH-SRY dev. | 4 | -135 | -292 | -326 | -758 |
| | | VP-SRY dev. | 0 | 7 | 26 | 20 | 48 |
| | | CM-SRY dev. | 0 | -32 | -675 | -203 | -783 |
| | | Carpentras | 71 | 381 | 6563 | 2567 | 9582 |
| | | SE-SRY dev. | -26 | -156 | -1713 | 924 | -1002 |
| | | LH-SRY dev. | -45 | -76 | -1034 | -876 | -2054 |
| | | VP-SRY dev. | -24 | -303 | -1529 | -85 | -1979 |
| | | CM-SRY dev. | 40 | 140 | 73 | 653 | 870 |

Table II shows the degree hours above 24°C for a lightweight resp. heavy office room facing south with no cooling, and heating only if the room temperature is below 21°C.

Table III

| | | | WIN | SPR | SUM | AUT | YEAR |
|---|---|---|---|---|---|---|---|
| | | Copenhagen | 1050 | 512 | 293 | 733 | 2590 |
| | | SE-SRY dev. | -50 | -62 | -64 | -20 | -190 |
| | | LH-SRY dev. | 0 | -32 | -50 | 92 | 10 |
| | | VP-SRY dev. | 30 | -19 | -19 | -14 | -20 |
| | | CM-SRY dev. | -44 | 160 | - 5 | 98 | 210 |
| | Tank | Uccle/De Bilt | 1050 | 549 | 322 | 699 | 2620 |
| | | SE-SRY dev. | -10 | 44 | -33 | 5 | 0 |
| | 50 1 | LH-SRY dev. | -137 | -22 | -4 | 154 | -10 |
| | (MJ) | VP-SRY dev. | -20 | 36 | 5 | -9 | 10 |
| | | CM-SRY dev. | -102 | 9 | 78 | 81 | 70 |
| | | Carpentras | 468 | 132 | 20 | 139 | 759 |
| | | SE-SRY dev. | 20 | -118 | -7 | 3 | -101 |
| | | LH-SRY dev. | 190 | 52 | -16 | 145 | 371 |
| | | VP-SRY dev. | 14 | 2 | -5 | -13 | -4 |
| $2\ m^2$ | | CM-SRY dev. | -296 | - 40 | 41 | -21 | -317 |
| COLLECTOR | | Copenhagen | 1030 | 459 | 237 | 694 | 2420 |
| | | SE-SRY dev. | -92 | -5 | -106 | -14 | -220 |
| | | LH-SRY dev. | 0 | -19 | -71 | 89 | 0 |
| | | VP-SRY dev. | 20 | -43 | -25 | -40 | -90 |
| | | CM-SRY dev. | - 38 | 169 | -57 | 131 | 210 |
| | Tank | Uccle/De Bilt | 1040 | 501 | 265 | 655 | 2460 |
| | | SE-SRY dev. | -65 | 97 | -77 | 13 | -30 |
| | 150 1 | LH-SRY dev. | -135 | 9 | -2 | 157 | 30 |
| | (MJ) | VP-SRY dev. | -30 | 29 | 0 | -23 | -20 |
| | | CM-SRY dev. | -99 | -15 | 59 | 95 | 40 |
| | | Carpentras | 411 | 70 | 3 | 84 | 568 |
| | | SE-SRY dev. | 39 | -69 | -3 | -25 | -58 |
| | | LH-SRY dev. | 227 | 65 | -3 | 105 | 394 |
| | | VP-SRY dev. | -10 | -36 | -3 | -41 | -90 |
| | | CM-SRY dev. | -293 | -28 | 13 | -25 | -334 |

Table III shows the necessary fuel make up in a domestic hot water solar system in order to prepare 100 1 water a day at 45°C. Collector area 2 $m^2$ and tank capacities 50 and 150 1.

Table IV

| | | | WIN | SPR | SUM | AUT | YEAR |
|---|---|---|---|---|---|---|---|
| COLLECTOR 6 m$^2$ | Tank 150 l (MJ) | Copenhagen | 761 | 171 | 34 | 372 | 1340 |
| | | SE-SRY | -146 | 36 | -25 | 26 | -110 |
| | | LH-SRY | -26 | 16 | 13 | 131 | 130 |
| | | VP-SRY | 39 | -68 | -14 | -109 | -160 |
| | | CM-SRY | -122 | 20 | -18 | 44 | -80 |
| | | Uccle/De Bilt | 790 | 200 | 50 | 357 | 1400 |
| | | SE-SRY | -131 | 158 | -49 | 34 | 10 |
| | | LH-SRY | -292 | -55 | 1 | 99 | -250 |
| | | VP-SRY | -5 | 52 | 5 | -59 | -10 |
| | | CM-SRY | -243 | -68 | -37 | 90 | -257 |
| | | Carpentras | 132 | 0 | 0 | 14 | 146 |
| | | SE-SRY | 56 | 0 | 0 | -14 | 42 |
| | | LH-SRY | 113 | 2 | 0 | 3 | 117 |
| | | VP-SRY | 8 | 0 | 0 | -7 | -1 |
| | | CM-SRY | -133 | -12 | 0 | 0 | -146 |
| | Tank 450 l (MJ) | Copenhagen | 705 | 94 | 6 | 284 | 1090 |
| | | SE-SRY | -170 | 51 | -6 | -169 | -295 |
| | | LH-SRY | 39 | 76 | -6 | -16 | 90 |
| | | VP-SRY | 4 | -112 | -3 | -185 | -300 |
| | | CM-SRY | -87 | -2 | -2 | 19 | -80 |
| | | Uccle/De Bilt | 745 | 132 | 10 | 259 | 1160 |
| | | SE-SRY | -138 | 87 | -10 | -144 | -209 |
| | | LH-SRY | -247 | -15 | 16 | -28 | -277 |
| | | VP-SRY | -54 | -16 | -11 | -103 | -180 |
| | | CM-SRY | -224 | -71 | -12 | 126 | -182 |
| | | Carpentras | 69 | 0 | 0 | 0 | 69 |
| | | SE-SRY | 124 | 0 | 0 | 0 | 124 |
| | | LH-SRY | 79 | 0 | 0 | 0 | 79 |
| | | VP-SRY | -52 | 0 | 0 | 0 | -52 |
| | | CM-SRY | -67 | 0 | 0 | 0 | -67 |

Table IV shows the necessary fuel make up in a domestic hot water
solar system in order to prepare 100 l water a day at 45°C.  Collector area 6 m$^2$ and tank capacities 150 and 450 l.

The final result of the evaluation, the values of $V_i$ as described above, are:

| No. | SRY method | Original formula | Alternative formula |
|-----|------------|------------------|---------------------|
| 1. | S. Eidorff | 0.604 | 0.366 |
| 2. | A.H.C. van Paassen et al | 0.637 | 0.413 |
| 3. | L. Hallgreen | 1.170 | 0.809 |
| 4. | C. Mustacchi et al | 1.189 | 1.128 |

## 2.4 Analysis of results and comments

After studying all the simulation results and produced SRYs for the three locations the following observations are made:
- The quality of the produced SRYs varies surprisingly much, from one location to the other with the same SRY-method.
- The deviations of the SRY simulation results from weather data results are very small for heating and cooling loads, but rather larger for the extreme temperatures and the solar gain.
- While simulating solar systems with large tanks, appreciable quantities of energy are moved from one season to the following seasons. This seems to be a general problem for most of the SRYs.
- The expresson "Method no. 1 is better than no. 2" can not be stated on a confidence level of 90%. It can only be stated that no. 1 performed best in the test cases here, and most likely is the best.

## 3. Conclusion of the SRY evaluation

The full wording of the conclusion is not yet confirmed by the group. There is a general agreement of the following:

"The evaluation of the SRY methods based on 495 elementary test cases indicates that S. Eidorffs method most likely is the best. However, in many of the test cases method no. 2 by A.H.C. van Paassen et al was better, so the former method is elected on a narrow margin".

Extension of this has been proposed by mr. van Paassen.

Based on the simulation results the following may be added in the conclusion:

The SRYs are shown to be well suited for heating and cooling loads and lesser suited for extreme temperatures and solar systems with long storage times.

The quality of the generated SRYs varies much, quite randomly, and it may therefore be necessary to test and document the performance of a new SRY for a certain location before releasing the data for public use.

# 4. Production of Test Reference Years, TRY

Until now Mr. Dogniaux has selected TRY's for 27 stations, and hourly data have been compiled. Further, a number of stations in Italy are in production.

In Copenhagen we are now bringing the TRY's to a common format, and are performing the splitting of global radiation.

In Berlin a common research between Herman-Rietschel Institut, prof. H. Esdorn, Institut für Theoretische Meteorologie, prof. H. Fortak, and dr.-ing. Axel Jahn, has led to a climatic zoning of BRD and selection of TRY for 11 locations. This work will be finished this year, and it will most likely be advantageous to use these stations instead of the previously chosen 4 TRY's for BRD.

A common problem in many countries will be the distribution of TRY's. The Meteorological Services, but DWD, have agreed to distribute the TRY, but many of them have in recent years introduced systems of fees for delivery of meteorological data which will make a TRY, containing nearly 100,000 single values, so expensive that widespread use of TRY's will be impossible. Individual negotiations therefore have to be carried out with each Meteorological Service separately in order to find a suitable way.

Such ways can therefore, besides through the Met. Service, be e.g.: Distribution through a national Building Research Institution, a Solar Energy Society, or through a commercial software house.

Distribution can be either as: 1) A complete tape containing 30-35 stations for all EEC countries, 2) Tapes for each country, 3) Tapes, cassettetapes, floppy disks with one station only.

1) might give serious difficulties regarding different fee policies for the Met. Services, but shall be maintained in order to keep the image of a concerted EEC-action.

2) will in most cases fulfill the demands from consultants and other TRY-users.

3) is the only way to ensure that also smaller consulting enterprises with in-house mini- or desk-top computers can utilize the TRY's.

Table V.  Production and distribution of TRY's

| Country | B | D | DK | Eir | F | It | NL | UK | Gr |
|---|---|---|---|---|---|---|---|---|---|
| TRY's produced | 3 | 4(11) | 1 | 6 | 6 | | 3 | 4 | |
| Met. Serv. accept of distr. | + | (-) | + | + | + | | + | + | |
| Est. price, ECU | 107 | (high) | 60 | | | | ? | 185 | |
| for no. of stations | 27 | | 1 | | | | | 27 | |
| Other channels known | | + | | | | | | | |
| Reserv. against other channels | | | No | | ? | | | | |
| Climatic zones developed | | + | NA | | | | | | |
| Splitted radiation | 1 | | + | | | | | | |
| Method for split available not yet available | 2 | | | 6 | 2 4 | | 3 | 4 | |

# SHORT REFERENCE YEARS WITH ADJUSTED MEASURED DATA

Authors           :  S. EIDORFF, H. LUND

Contract number   :  ESF-029-DK(G)

Duration          :  24 months          1 July 1981 - 30 June 1983

Total budget      :  D.Kr. 781 000      CEC contribution: D.Kr. 549 000

Head of project   :  S. Eidorff, Thermal Insulation Laboratory

Contractor        :  Technical University of Denmark

Address           :  Thermal Insulation Laboratory
                     Technical University of Denmark
                     Building 118
                     DK - 2800 LYNGBY, Denmark

## Summary

Under this contract is developed an inexpensive and straight forward method to generate SRYs consisting of almost original measured weather data, which compare favourably with SRYs consisting of artificial data generated by statistical methods.

The method is a two step procedure.  In step 1 sequences of typical days are selected on daily values so that warm and cold periods of typical length are established.  In step 2 the hourly values in the SRY are adjusted so that the distributions of the weather parameters in the SRY are in accordance with the distributions in the original long term weather data set from which the SRY has been derived.

In the evaluation within Action 2 this SRY generating method performed best out of the four participating methods.

## 1.1 Introduction

A Short Reference Year, SRY, is a Test Reference Year consisting of appreciably fewer days than 365. The objective of such a year is to conserve computer time in order to make it feasible to perform simulation on a micro computer or to test several design alternatives by simulation.

Most SRY methods generate the weather parameters artificially by means of statistical methods which preserve some properties of the original long term weather data such as auto-correlation, cross correlation, means etc.

This alternative method uses the original measured weather data set, from which periods of a fixed length, e.g. 7 days, are selected. The periods should be as typical as possible for the month or season they are to present and consist of cold and warm periods of typical length.

The next step is to compare the distribution of the hourly weather parameters in the selected periods with the distributions in the original long term weather data set. The hourly values in the selected periods are then adjusted, as little as possible, to make the distributions identical.

## 1.2 Description of the method

Both the selection of typical periods in step 1 and the distribution adjustment in step 2 are performed on a monthly basis.

For the evaluation of the SRY generating methods in Action 2, it was decided to compare SRYs with two weeks per season. Consequently, the final months in this SRY have different lengths: the month in the middle of a season consists of six days while the other months consist of four days.

For a 6-days month are selected 2 periods of 3 consecutive days from the long term weather data set, so that one period is within the first half part of the month and the second in the remaining part. For a 4-days month is selected periods of 2 consecutive days respectively.

After studying the typical length of warm and cold periods in the original weather data set for Copenhagen, it was decided to have an ascending sequence of daily solar radiation in every second month in the SRY and a decending sequence in the intermediate months.

It is further decided to keep this "wavelength" for all SRYs regardless of the location, because it is believed to be more important to have control of the interference between the seasonal variations and the warm/cold periods than having the correct length of the warm and cold periods.

The selection of the typical periods for the SRY is based on the daily sum of direct normal radiation.

The ideal values of this parameter for the various days are determined from the long term weather data set, so that the distribution of the daily values in the SRY is as close to the distribution in the weather data set as possible and, concurrently, so that the values are in ascending resp. decending order.

The other weather parameters, e.g. temperature and diffuse radiation, are not taken into consideration in the selection step. They will not, however, be much out of range due to the natural connection between the weather parameters. In the completely prepared SRY the typical values of these will be ensured by step 2 in the procedure, the adjustment step.

In step 2 the hourly values of all weather parameters to be present in the SRY are adjusted so that the distributions, month by month, in the SRY are in accordance with the respective distributions in the long term weather data set.

In practice this is done by sorting all hourly values of e.g. outdoor temperature within each month in ascending order, both in the SRY and the weather data set. The values in the SRY are then substituted by the proper fractile values in the weather data set. After that the SRY is sorted in cronological order.

For the radiation data, only direct normal and diffuse radiation are adjusted. The hourly values of global radiation are then calculated from the adjusted values of the other parameters.

## 1.3  Results
The performance of a SRY may be determined by comparing simulation results carried out with the SRY and with the weather data set from which the SRY has been derived. These results should be similar.

In the evaluation process of the SRY generating methods in Action 2, a lot of simulation runs have taken place in order to cover a broad field of application of SRYs and to minimize the random errors.

All these simulation runs will be published in the final report from Action 2. Here is shown only a few results as an example.

In the tables below are shown simulation results for three different locations and for the four seasons plus the entire year. The rows with the city names show the actual simulation result with the weather data set from the said location. On the row just below is shown the deviations in simulation results obtained by using the SRY generated for that location.

Table I

|  |  | WIN | SPR | SUM | AUT | YEAR |
|---|---|---|---|---|---|---|
| Heating | Copenhagen | 1241 | 549 | 25 | 499 | 2313 |
|  | SRY dev. | -12 | 2 | -8 | -52 | -56 |
| load | Uccle | 1032 | 431 | 29 | 421 | 1913 |
| (kWh) | SRY dev. | -13 | 40 | -28 | -55 | -45 |
|  | Carpentras | 621 | 224 | 0 | 95 | 940 |
|  | SRY dev. | 60 | -83 | 0 | -15 | -30 |
| Degree | Copenhagen | 0 | 83 | 1128 | 177 | 1388 |
|  | SRY dev. | 0 | -54 | -332 | -84 | -477 |
| hours, | Uccle | 2 | 137 | 1323 | 350 | 1812 |
| $T - 24^{\circ}C$ | SRY dev. | -2 | -83 | -690 | -233 | -1013 |
| (h·K) | Carpentras | 71 | 381 | 6563 | 2567 | 9582 |
|  | SRY dev. | -26 | -156 | -1713 | 924 | -1002 |

Table I shows the heating load and degree hours above $24^{\circ}C$ for a heavy office room facing south with no cooling, and heating only if the room temperature is below $21^{\circ}C$.

Table II

| | | WIN | SPR | SUM | AUT | YEAR |
|---|---|---|---|---|---|---|
| **Tank**<br>**50 ℓ**<br>**(MJ)** | Copenhagen | 1050 | 512 | 293 | 733 | 2590 |
| | SRY dev. | -50 | -62 | -64 | -20 | -190 |
| | Uccle | 1050 | 549 | 322 | 699 | 2620 |
| | SRY dev. | -10 | 44 | -33 | 5 | 0 |
| | Carpentras | 468 | 132 | 20 | 139 | 759 |
| | SRY dev. | 20 | -118 | -7 | 3 | -101 |
| **Tank**<br>**150 ℓ**<br>**(MJ)** | Copenhagen | 1030 | 459 | 267 | 694 | 2420 |
| | SRY dev. | -92 | -5 | -106 | -14 | -220 |
| | Uccle | 1040 | 501 | 265 | 655 | 2460 |
| | SRY dev. | -65 | 97 | -77 | 13 | -30 |
| | Carpentras | 411 | 70 | 3 | 84 | 568 |
| | SRY dev. | 39 | -69 | -3 | -25 | -58 |

Table II shows the necessary fuel make up in a domestic hot water solar system in order to prepare 100 ℓ water a day at 45°C. Collector area 2 m² and tank capacity 50 ℓ resp. 150 ℓ.

## 1.4 Analysis of results and comments

After studying all the simulation results and the produced SRYs for the three locations the following observations are made:

- The quality of the produced SRYs varies surprisingly much. This is also true for the other SRY methods in Action 2.

- The obtained simulation results show too big sensitivity for accumulating components in the simulated system, i.e. heavy buildings and storage tanks. This is probably due to too fast and too large oscillations of the weather parameters.

- The extreme temperatures are generally too low. This could be caused by too bad correlation between radiation and temperature or too short warm periods.

- The ascending and decending sequences are not ensured good enough at the Carpentras SRY. This was derived from an only one year long weather data set, so it has obviously been difficult to select day sequences suiting several criteria.

- While simulating solar systems with large tanks, appreciable quantities of energy are moved from one season to the following seasons. This seems to be a general problem for most of the SRYs.

- The deviations of the SRY simulation results from the weather data results are very small for heating and cooling loads, but rather large for the extreme temperatures and the solar gain.

- The simulation results from the SRYs generated by this method are generally better than those from the SRYs generated by the other methods.

## 2. Conclusions

In this research project a new SRY generating method has been developed. It is very inexpensive and straight forward, and complies with the original aim of the project.

The characteristics of this method are mainly:

1. The weather data are original measured with only small modifications.

2. The total distribution of the hourly values, rather than some statistical values such as means and variances, is in accordance with the original long term weather data set.

3. The unavoidable interference in a SRY between the seasonal variations and the variations of warm/cold periods can be controlled.

4. Any weather parameter may be included in the SRY.

The SRYs are shown to be well suited for heating and cooling loads and lesser suited for extreme temperatures and solar systems with long storage times.

The quality of the generated SRYs varies much, quite randomly, and it may therefore be necessary to test and document the performance of a new SRY for a certain location before releasing the data for public use.

In the evaluation within Action 2 this SRY generating method performed best out of the four participating methods.

ESTABLISHMENT OF SHORT REFERENCE YEARS FOR CALCULATION OF ANNUAL SOLAR
HEAT GAIN OR ENERGY CONSUMPTION IN RESIDENTIAL AND COMMERCIAL BUILDINGS

Authors            : S.H. Liem, A.H.C. van Paassen

Contract number    : ESF - 010 - 80 N (B)

Duration           : 24 months 1 october 1980 - 1 october 1982

Total budget       : 366.500 FL   CEC contribution 117.280 FL

Head of the project : Dr.Ir. A.H.C. van Paassen
                     Department of Mechanical Engineering
                     Laboratory of Refrigeration and Indoor Climate
                     Technology

Contractor         : Delft University of Technology

Address            : Delft University of Technology
                     Julianalaan 134
                     2628 BL Delft, The Netherlands

## SUMMARY

This paper describes an automatic procedure to generate short series
of hourly weather data that can be considered as representative for the
outdoor climate during the year, because it possesses the same statistical
properties as the measured weather data from which it has been derived.
This short series is called a Short Reference Year (SRY).

The procedure involves the following steps: analysing the weather
data statistically, modelling the weather variations and generating new
hourly values of the weather variables - the SRY - by means of random
number generators. The procedure consists of program WEATHER for the
analysing and modelling of the weather data, program SRYEAR for the generation of the SRY and one or more subprograms to prepare the data set.
The advantage of this procedure is the ability to describe the climate
mathematically; this leads not only to a lot of additional statistical
information about the weather, that can be usefull for various kinds of
energy calculations, but also to a high degree of flexibility in changing
the length of the series without re-analysing the original weather data set.

SRY's of the Netherlands, Denmark and France are generated by this
procedure and afterwards they are tested according to the test procedure as
defined by the workgroup "Methods for Short Reference Years".

In general the heating and cooling loads, calculated with these SRY's
are within 6% of the multiyears'. The results of the hot water system
agree very well for the Netherlands and Denmark; the deviation for France
is rather high.

It is now possible to generate SRY's for various locations with this
automatic procedure; for energy calculations and hot water systems the
errors are acceptable low.

Work is going on to modify program SRYEAR for the generations of SRY's
which can be used for extreme values calculations or design conditions.

# 1. INTRODUCTION

Until now reference years consist of a complete year of hourly weather data, derived from measured meteorological data over a long period by means of some selection procedure. The approach used here is quite different; instead of selecting representative parts from the hourly measured data, new hourly data with the same statistical properties as the real data are generated. The reasons to choose for this approach are:
1. A mathematical characterization of the outdoor climate can be obtained.
2. The number of hourly data with the same statistical properties as the real climate can be reduced systematically.

# 2. THE THEORY USED IN THE PROCEDURE

The theory used in the procedure is based on the assumption that the global radiation is completely independent and the other weather variables are considered as more or less dependent on this global radiation (1,2). The daily course of the global radiation is approximated by a sine function with amplitude Aq. This Aq-value characterizes the daily type of the weather (clear, cloudy, etc.). For each type of weather there exists a certain shape of the mean daily course of each dependent variable. This shape can be described by a mathematical expression, in this case a polynomial equation. The independent part can be calculated by subtracting the average mean daily course from the measured hourly values. The residue (noise) is analysed and the statistical properties are calculated, see figure 1.

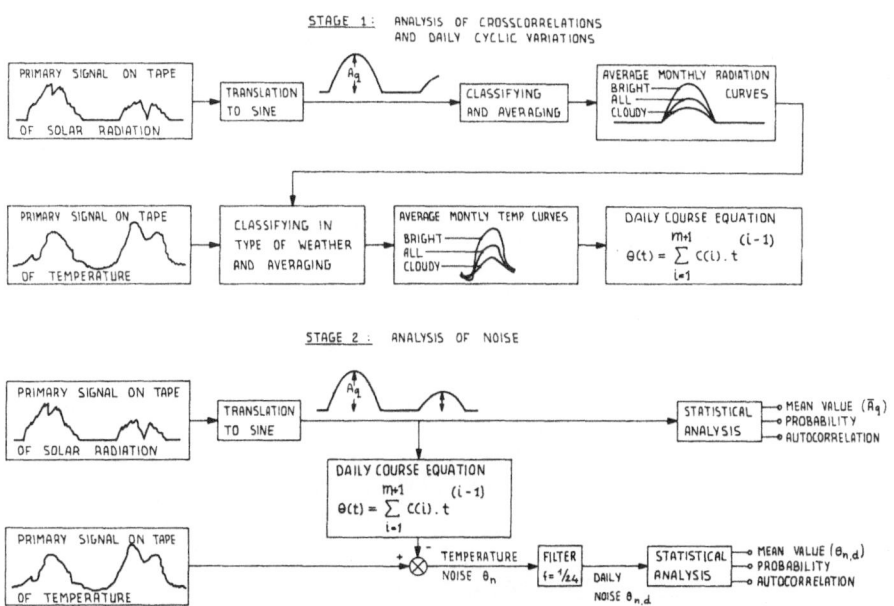

Figure 1. Procedure as used for the development of the weather model
(demonstrated for the solar radiation and temperature)

The mathematical model can now be constructed as follows: random number generators deliver uncorrelated random numbers, for each variable one series (Aq-values for the solar radiation and noise-values for the other variables). A special procedure scales and rearranges the random values of each series in such a way that this series gets the right distribution and autocorrelation. As said before, the solar radiation during one day is characterized by the value of the amplitude Aq. With the series of Aq-values, one value for each day, the daily types of the weather are specified. The daily course of the global radiation can be reconstructed by using the half sine function. The daily course of the other variables (temperature, humidity etc.) can now be calculated with the polynomial equations; the hourly values of these variables are obtained by super-imposing the daily course to the corresponding noise values of the generated series, see figure 2.

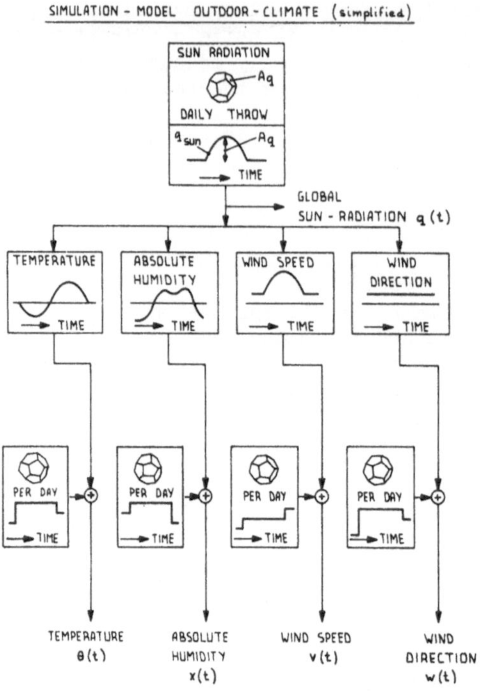

Fig. 2. The simplified weather model

## 3. THE AUTOMATIC PROCEDURE TO GENERATE SHORT REFERENCE YEARS (SRY's)

The automatic procedure to generate short reference years is realized by the two main programs: WEATHER, that analyses the measured weather data, calculates the statistical properties and derives the coefficients of the polynomial equations; and SRYEAR, that constructs the mathematical model and generates the new hourly values for the SRY. Besides these two programs this procedure contains one or more sub-programs to prepare the data set. By using delaying techniques of the computer system these programs can run simultaneously.

This automatic procedure has been used to generated short reference years of three weather stations:
a. The Netherlands - De Bilt (1961-1970)     SRYNL
b. Denmark - Varlose - Tastrup (1959-1973)   SRYDK
c. France - Carpentras (1970)                SRYF

## 4. TESTING THE GENERATED SHORT REFERENCE YEARS

Within the framework of the workinggroup "Methods for Short Reference Years" it was decided to test the generated short reference years with two test programs, an energy calculation program BA4 from Denmark and a domestic hot water system BENCHMARK from Italy. A more detailed description of the test procedure will be given in (3). The test parameters that will be used for the evaluation between the different methods are the errors of the seasonal and yearly total values of respectively the heatingload, cooling load and degree hours for program BA4 and the fossil fuel make up for program BENCHMARK. The degree hour is defined as the sum of the number of hours that the indoor temperature is above 24°C.

As an example Table I shows the results of program BA4 for a lightweight room with window facing south for the different weather stations.

Table II shows the results of program BENCHMARK for a flat plate collector of 2m² for the different weather stations.

The error Δ% is normalised to the yearly totals and is given by the equation:

$$\Delta\% = \left(\frac{XM-XS}{XMY}\right) * 100\%$$

XM = multiyear value
XS = short year value
XMY= multi year yearly total

Table I. Results of program BA4.

| Season | Strategy | Quantity | THE NETHERLANDS De Bilt 1961 - 1970 | | | DENMARK Varlose-Tastrup 1959 - 1973 | | | FRANCE Carpentras 1970 | | |
|---|---|---|---|---|---|---|---|---|---|---|---|
| | | unit | MY | SRY-NL | Δ% | MY | SRY-DK | Δ% | MY | SRY-F | Δ% |
| 1 | Heat ing only | Heat ing load (kWh) | 1163 | 1164 | 0. | 1250 | 1334 | -3.4 | 689 | 719 | -2.6 |
| 2 | | | 540 | 536 | 0.2 | 601 | 567 | 1.4 | 286 | 265 | 1.9 |
| 3 | | | 85 | 73 | 0.5 | 68 | 71 | -0.1 | 0 | 10 | -0.9 |
| 4 | | | 518 | 457 | 2.6 | 553 | 496 | 2.3 | 167 | 177 | -0.9 |
| Yearly | | | 2307 | 2242 | 2.8 | 2472 | 2482 | -0.4 | 1143 | 1180 | -3.2 |
| 1 | if T<21° | Degree hour (h°C) | 4 | 2 | 0.1 | 10 | 0 | 0.4 | 332 | 168 | 1.4 |
| 2 | | | 282 | 443 | -8.1 | 388 | 247 | 5.6 | 1188 | 941 | 2.1 |
| 3 | | | 1356 | 1688 | -16.8 | 1824 | 1719 | 4.1 | 6990 | 6577 | 3.6 |
| 4 | | | 337 | 464 | -6.4 | 313 | 475 | -6.4 | 3083 | 3162 | -0.7 |
| Yearly | | | 1979 | 2581 | -30.4 | 2535 | 2428 | 4.2 | 11594 | 10799 | 6.9 |
| 1 | Con stant Tem pera ture | Heat ing load (kWh) | 1175 | 1179 | -0.2 | 1266 | 1337 | -2.6 | 792 | 812 | -1.3 |
| 2 | | | 615 | 587 | 0.7 | 693 | 661 | 1.2 | 428 | 421 | 0.4 |
| 3 | | | 188 | 180 | 0.3 | 183 | 181 | 0.1 | 29 | 34 | 0.3 |
| 4 | | | 591 | 546 | 1.8 | 628 | 594 | 1.2 | 317 | 319 | -0.1 |
| Yearly | | | 2570 | 2514 | 2.2 | 2769 | 2786 | -0.6 | 1566 | 1594 | -1.8 |
| 1 | T=21° | Cool ing load (kWh) | 19 | 26 | -1.0 | 24 | 5 | 2.5 | 177 | 158 | 1.1 |
| 2 | | | 158 | 173 | -2.2 | 187 | 183 | 0.5 | 348 | 340 | 0.5 |
| 3 | | | 344 | 382 | -5.6 | 403 | 420 | -2.2 | 682 | 675 | 0.4 |
| 4 | | | 154 | 182 | -4.2 | 153 | 189 | -4.7 | 496 | 510 | -0.8 |
| Yearly | | | 674 | 760 | -12.8 | 767 | 793 | -3.4 | 1704 | 1679 | 1.5 |

Table I. Results of an energy calculation (BA4) for a light-weight room with south orientation for the Netherlands, Denmark and France for the multiyear and short series respectively.

Table II. Results of program BENCHMARK

| Season | Storage | Fossil fuel make up (MJ) with Collector area of 2 m² | | | | | | | | |
|---|---|---|---|---|---|---|---|---|---|---|
| | | THE NETHERLANDS De Bilt 1961 - 1970 | | | DENMARK Varlose-Tastrup 1959 - 1973 | | | FRANCE Carpentras 1970 | | |
| | (m³) | myear | SRYNL | Δ% | myear | SRYDK | Δ% | myear | SRYF | Δ% |
| 1 | 0.050 | 1060 | 1040 | 0.7 | 1080 | 1110 | -1.2 | 463 | 477 | -1.8 |
| 2 | | 567 | 603 | -1.3 | 499 | 480 | 0.8 | 134 | 136 | -0.3 |
| 3 | | 329 | 334 | -0.2 | 231 | 212 | 0.8 | 20 | 14 | 0.8 |
| 4 | | 729 | 720 | 0.3 | 709 | 695 | 0.6 | 147 | 134 | 1.7 |
| yearly | | 2690 | 2700 | -0.4 | 2520 | 2500 | 0.8 | 764 | 760 | 0.5 |
| 1 | 0.100 | 1050 | 1030 | 0.8 | 1070 | 1100 | -1.2 | 427 | 428 | -0.2 |
| 2 | | 536 | 569 | -1.3 | 461 | 430 | 1.3 | 94 | 73 | 3.3 |
| 3 | | 291 | 293 | -0.1 | 189 | 168 | 0.9 | 9 | 3 | 0.9 |
| 4 | | 703 | 682 | 0.8 | 683 | 651 | 1.3 | 114 | 80 | 5.3 |
| yearly | | 2580 | 2570 | 0.4 | 2410 | 2350 | 2.5 | 645 | 582 | 9.8 |
| 1 | 0.150 | 1050 | 1020 | 1.2 | 1070 | 1090 | -0.8 | 418 | 408 | 1.7 |
| 2 | | 525 | 554 | -1.1 | 446 | 403 | 1.8 | 74 | 38 | 6.1 |
| 3 | | 275 | 275 | 0. | 170 | 140 | 1.1 | 3 | 0 | 0.5 |
| 4 | | 693 | 670 | 0.9 | 673 | 633 | 1.7 | 99 | 58 | 6.9 |
| yearly | | 2540 | 2520 | 0.8 | 2360 | 2270 | 3.8 | 593 | 503 | 15.2 |

Table II. Results of a domestic hot water system calculation (BENCHMARK) with a collector area of 2 m² and different specific storages for the Netherlands, Denmark and France for the multi-year and the short series respectively.

## 5. DISCUSSION

As can be seen from Table I, the heating loads for all three stations are in good agreement with the multiyear results. The cooling load too is in good agreement with the multiyear results, except for The Netherlands, where it is too high. It seems that this is caused by the solar radiation. The agreement of the degree hours is not so well: the reason for this is the fact that the generated short reference year in this basic version gives mean average values, where as for the degree hours the extreme values are very important. The same difficulty has also been encountered by the other methods.

From Table II it can be seen that the results of the short reference year agree well with the multiyear results, except for France. A reason for this is the rather low fossil fuel make up required for this southern station with very much sun shine: a small absolute error brings about a high relative error.

## 6. CONCLUSIONS

With this automatic procedure it is possible to generate short reference years for various locations. The error for energy calculation programs and hot water system programs are acceptable.

The advantages of this approach is the ability to describe the climate in detail by mathematical expressions. This is very usefull for various kinds of energy calculations. An other advantage is the flexibility in changing the length of the series without re-analysing the original weather data set.

This basic version is not intended for extreme values calculations. Further investigation is going on to generate SRY's for this kind of calculations and design conditions.

## ACKNOWLEDGEMENT

This paper describes work at the Delft University of Technology supported by the Commission of the European Communities within the frame work of the Energy R and D programme.

## REFERENCES

1. Paassen, A.H.C. van. Indoor climate, outdoor climate and energy calculation.
   Delft University of Technology, 1981, WTHD 137.

2. Paassen, A.H.C. van; A.G. de Jong. The synthetical reference outdoor climate.
   Energy and Buildings, 2 (1979), pp. 151-161.

3. Liem, S.H.; A.H.C. van Paassen. Establishment of Short Reference Years for calculation of annual solar heat gain or energy consumption in residential and commercial buildings.
   Final report, Contract no. ESF-010-80 N(B), Delft, 1982.

# SRY PRODUCTION BY MEANS OF STOCHASTIC MODELS

Authors          : C. MUSTACCHI, V. CENA, M. ROCCHI

Contract number : ESF - 036-I(S)

Duration         : 12 months            1  January   - 31 December 1982

Total Budget     : Lit. 24,000,000      CEC contribution:  Lit. 12.000.000

Head of project : C. MUSTACCHI,    ADES s.r.l.

Contractor       : ADES s.r.l.


Address          : ADES s.r.l.
                   23, Via dei Giubbonari
                   00186  ROME

## Summary

The activities carried out consisted in performing statistical analysis on multiyears  meteorological datasets, calculating the coeffi —
cients of stochastic models and using these models to generate    complete SRY'S for three European  locations (Vaerlose - Tastrup, De Bild,
Carpentras). The computer code BENCH which simulates a domestic   hot
water system was developed and distributed to the other contractors in
the field. Both codes BENCH and BA4 (already available) were run using
as input either the original multiyear  datasets or the SRY'S generated by means of  our method. The results of these simulations   were
submitted to the Coordinator in charge of comparing them with similar
results for other contractors.
The stochastic models proved to be an efficient tool for the genera —
tion  of SRY'S. A reasonable human and computer effort is needed   to
calculate the models' coefficients, which are also very easy to store.

# 1. Introduction

The work carried out under the contract ESF-036-I(S) is the continuation of that carried out under a previous contract. The latter consisted in the following activities:

  a) statistical analysis (mean, average, higher moments, autocrrelation, crosscorrelation) on the multiyear meteorological datasets available for nine european locations (Vaerlose, Vaerlose-Tastrup Locarno, Uccle, Bracknell, Ispra, Valentia, Odeillo, De Bild)
  b) evaluation, for each parameter to be included in the Short Reference Year (SRY), of a satisfactory stochastic model, on the basis of the above statistical analysis
  c) determination for two locations (Vaerlose-Tastrup and De Bild, of the coefficients of the stochastic models
  d) production for the same two locations of complete SRY'S.

The work carried out under the present contract the following activities:

  e) iteration of activities a) through d) for an additional location (Carpentras) whose dataset became available
  f) analysis and development of the computer code BENCH, which simulates the behaviour of a sanitary hot water system
  g) test the three SRY'S by means of the BENCH and BA4 codes.

# 2. Methodology

Phase e) consisted in analysing the dataset available for Carpentras (France). The meteorological parameters included in the dataset were modelled by means of the following models:

| Variable | Model used |
|---|---|
| Hourly global radiation | Discrete AutoRegressive of order 1 DAR(1) |
| Hourly diffuse radiation | Linear relationship with hourly atmospheric transmittance (method of Orgill and Hollands) |
| Hourly wind velocity | Discrete AutoRegressive of order 2 DAR(2) |
| Hourly dry-bulb temperature | 5-term regression |
| Hourly relative humidity | Linear relationship between mean daily atmospheric transmittance + Discrete AutoRegressive of order 1 for the hourly residuals |

Phase f) consisted in the analysis and development of the computer code BENCH. This code was written to compare the performance of SRY'S, generated by different methods, with respect to full Multi-Year Datasets (MYD). The program can use either SRY or MYD data to estimate the incident solar radiation on a flat plate collector bank. The collected heat is to be used in a domestic hot water system. The program prints monthly and yearly values of thermal boad, collected energy and fossil fuel consumption.

After extensive testing and implementation of the modifications requested by other contractors in the same field, the BENCH code was released to all the contractors in charge of producing SRY'S.

Phase g) consisted in running the two available programs, BENCH and BA4, to test the different SRY'S. BENCH simulated hot water systems, BA4 simulated residential heating and cooling systems.

The test cases used by the BENCH code were as follows:
  - flat plate collector surface    2;4;6 m2
  - daily storage volume            25;50;75 liters/m$^2$ of collec tor

```
              - water mains temperature          10.0 °C (De Bild, Vaerlose - Tastrup)
                                                  15.0 °C (Carpentras)
```
The test cases used by the BA4 code were as follows:
```
    - type of room                   light; heavyweight
    - orientation                    South; North; East
    - control strategy               constant room temperature at  21 °C,
                                     with heating and cooling when needed;
                                     variable room temperature with heat-
                                     ing when below 21 °C and no cooling.
```

## 3.Results

The statistical analysis performed during phase e) (see for  example
the probability density function of atmospheric transmittance, fig. 1) allowed the calculation of all the coefficients in the stochastic models. By
means of these models, a complete SRY for Carpentras was generated.

The BENCH and BA4 programs were run, for each test case mentioned  in
section 2, using the original MYD and our SRY for all the three locations
considered (De Bild, Vaerlose - Tastrup, Carpentras). The results of   all
these simulations were submitted to the Coordinator, who was in charge  of
comparing them with similar results from other contractors.

## 4.Conclusions

Our procedure, which uses stochastic models, was used to generate
SRY'S for three European locations. The SRY'S were tested by means  of the
BA4 code (already available) and of the BENCH code, which we prepared    in
the framework of the present  contract.

The stochastic models proved to be an efficient tool for generating,
for a given location, any number of SRY'S. The human and computer workload
to calculate the coefficients of the models is reasonable, and the  models
require little memory to be stored.

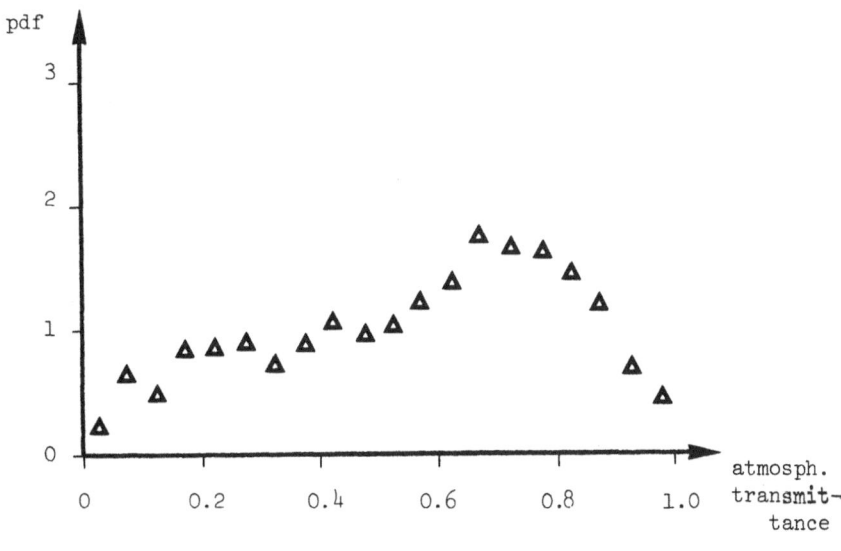

fig. 1

# SHORT REFERENCE YEAR, SRY

Authors           :  Lars Hallgreen

Contract number :  ESF-030-DK (G)

Duration          :  12 months                 1. november 1981 –
                                                31. october 1982.

Total budget      :  Dkr. 170.250              CEC contribution:
                                                Dkr. 85.125

Head of project :  Civ.ing. Lic. Techn. Lars Hallgreen

Contractor        :  Teknologisk Institut, Varme- og Installa-
                     tionsteknik

Adress            :  Gregersensvej, 2630  Tåstrup, Denmark

## Summary

This paper describes a proposal to a short Reference Year, selected for building thermal analysis and solar energy analysis. The approach used for development of the SRY climate model are briefly discussed. Furthermore the outline of a computer program for calculation of the SRY model parameter is illustrated and the results of parameter calculation for the Danish weather data is shown.

# 1. Introduction

The Short Reference Year (SRY) are selected for building thermal analysis and solar energy analysis. The SRY are synthetical generated weather data which concentrated the main meteorological characteristics of a year into 4 climate sequences each of 14'th days.

The SRY weather data can be used with most existing computer programs for building thermal analysis and solar energy analysis. Hereby, a reduction of 80 percent of the computer time requirements can be achieved. This reduction is due to the reduced number of time steps required to characterize an entire year.

For designing new buildings, passive solar energy houses or solar heating systems, engineers must perform scores of simulation analysis to evaluate the thermal consequences of each major design decision. The SRY weather data are very suitable for such analysis. For detailed analysis of the final construction or system it is recommanded to use the Test Reference Year (TRY).

The SRY weather data are generated by a mathematical climate model. This model are easy to implement as a subroutine and require limited memory compare with the SRY - or TRYweather data. So if small computer with limited memory are used it can be recommanded to implement the climate model and then calculate the weather data one day at the time. The synthetical weather data are not limited to a fixed number of days, but can be generated for an arbitrary number of days. Therefore, the implementation of the climate model gives a freedom to choose the number of days which suits the purpose best.

## 2. SRY - Climate model

The approach used for development of the SRY is based on the wellknown characteristic that every periodic time series can be described by an infinite number of sinus- and cosinus functions. Instead of using representative hourly measured meteorological data, the SRY generate new data, which gives a description of the main characteristics of the measured weather data.

Figure 1. shows the mathematical climate model for generation of the synthetical weather data. It is assumed that the cloud cover is completely independent, and the other climate variables are considered as more or less dependent of the daily mean cloud cover and the hourly cloud cover respectively.
The calculation procedure for generation of synthetical weather data is the following:

1.  Generation of a daily cloud cover profile by use of the Markow proces and calculation of the daily mean cloud cover.

2.  Calculation of sunrise, sunset and the solar attitude.

3.  The daily temperature profile (dry bulb and dew point temperature) are calculated as a sum of the yearly mean value, the yearly variation, the daily mean value and the daily variation.

4.  Calculation of the daily solar profile (normal- and global radiation).

It should be observed that the dry bulb temperature and the dew point temperature are generated by use of the same model.

## 3. Cloud cover model

There are especially two reason why the cloud cover are chosen as the force variable in the climate model:

1. There are a very strong correlation between the hourly cloud cover observation and the hourly solar radiation /1/.

2. Unlike solar radiation the cloud cover is observed every hour. Which means that the cloud cover is suitable as the force variable in a "continuous" model.

The cloud cover observations are made every hour by experienced observers who estimate the amount of cloud on a scale of 0 to 9.

Cloud cover:

0:   Clear sky
1:   Almost clear sky
2:   Slightly cloudy
3:   Slightly cloudy
4:   Half overcast
5:   Cloudy sky
6:   Cloudy sky
7:   Very cloudy sky
8:   Overcast sky
9:   No observation

The cloud cover model must have the capacity of reproducing the main statistical properties of the cloud cover. As basis for the cloud cover model a Markow process have been chosen.

## 4. Calculation of model parameters

The relatively simple mathematic structure of the SRY model causes the calculation of the model parameters to be systematized in such a way that these can easily be carried out by means of a computer program. In figure 2 the applied calculation procedure is schematically outlined.

The approach used for calculation of the model parameters is as previous mentioned, that the climate variations can be described by harmonic sequences as stated in the collection of formula figure 1. This means that by calculating the yearly variation the amplitude and the phase are determined by sinus function with the period of one year, which provides the best adaption to the measured weather data. Likewise by calculation of the daily variation, the amplitude and the phase are determined by the sinus functions with the period of 1 -, 1/2 - and 1/3 of a day, which provides the best adaption to be measured daily variations.

The amplitude and the phase for the yearly fluctuation are calculated recursively at intervals of 30 days. As a final expressions for the yearly amplitude and phase the mean value is applied.

The amplitude and phase for the daily fluctuation and the harmonic, are calculated at intervals of one days. The calculation is performed on the measured weather data minus the already determined yearly fluctuation and minus the yearly mean value. The amplitude and phase for the yearly variation and the daily variation respectively are determined by means of a filter, which use recursiv orthogonal decomposition.

The cloud cover factors are calculated as the ratio between measured global and normal solar radiation to determined global and normal solar radiation on a cloudless day respectively.

The dependence on the daily variation of the outdoor temperature and the daily mean cloud cover respectively the solar radiation and the hourly cloud cover are described by polynomial terms of the first or second degree. The coefficient of these polynomial are calculated by means of the IMSL routine RLFOR /5/.

Calculation of the daily mean variation of the outdoor temperature is carried out on the measured outdoor temperature minus the outdoor temperatur calculated by means of the SRY model with the measured cloud cover as independent variable. The regression model for the daily mean variation of the outdoor temperature as a function of the daily mean cloud cover is calculated by means of the SAS routine GLM /6/.

Tabel I, II and III shows the computer calculated SRY model parameters for the Danish weather data.

## 5. References.

/1/ Kimura K., Stephenson D.G., "Solar Radiation on cloudy days", ASHRAE Semianual Meeting, Jan. 1969.

/2/ Spliid H., Thyregod P., "Indledende undersøgelse af variationer i skydække", Delrapport, Teknologisk Institu 1981.

/3/ Kimura K., "Weather Information Telecommunication System for Computer Control of Air Conditioning", Second Symposium on the use of computers for environmental engineering related to Buildings, Paris 1974.

/4/ Hallgreen L., "A Short Reference Year for Denmark", Dansk VVS, 1979, No. 6.

/5/ "The IMSL Library", Customer Relations, GNB Building. 7500 Bellaire Boulevard, Houston, Texas.

/6/ "SAS user's Guide", SAS Institute Inc., 1979 Edition.

CLOUD COVER:

C(n) = "Markov chain model"

OUTDOOR AIR TEMPERATURE:

$T_A(n) = T_M + T_Y(N) + T_{DM}(N) + T_D(n)$

YEARLY MEAN TEMPERATURE: $T_M$

YEARLY TEMPERATURE VARIATION: $T_Y(N)$

$$T_Y(N) = \sum_{i=1}^{4} T_{Y,i} \sin(i\ W_Y\ N + P_{Y,i})$$

DAILY MEAN TEMPERATURE VARIATION: $T_{DM}(N)$

$T_{DM}(N) = A_0 + A_{1,1}\ T_{DM}(N-1) + A_{1,2}\ T_{DM}(N-2) + A_{2,1}\ C_{DM}(N)$

HOURLY TEMPERATURE VARIATION: $T_D(N)$

$T_D(n) = (B_0 + B_1\ C_{DM}(N))\ K_D$

$$K_D = \sum_{i=1}^{3} \frac{4\ \sin(W_D i\ (S_0 - S_M)/2)}{W_D^2\ i^2\ (2 - S_M + S_0)\ (S_M - S_0)} \sin(W_D i\ (n - (S_0 + S_M)/2))$$

$S_0$ = SUNRISE
$S_M$ = 3 PM

NORMAL SOLAR RADIATION:

$Q_N(n) = Q_{NO}(n)\ CC(n)$

CLOUD COVER FACTOR: $CC(n)$

$CC(n) = C_0 + C_1\ C(n)$

NORMAL RADIATION ON A CLOUDLESS DAY: $Q_{NO}(n)$

GLOBAL SOLAR RADIATION:

$Q_G(n) = Q_{GO}(n)\ CCF(n)$

CLOUD COVER FATOR: $CCF(n)$

$CCF(n) = D_0 + D_1 C(n) + D_2 C^2(n)$

GLOBAL RADIATION ON A CLOUDLESS DAY: $Q_{GO}(n)$

Fig. 1. SRY CLIMATE MODEL:

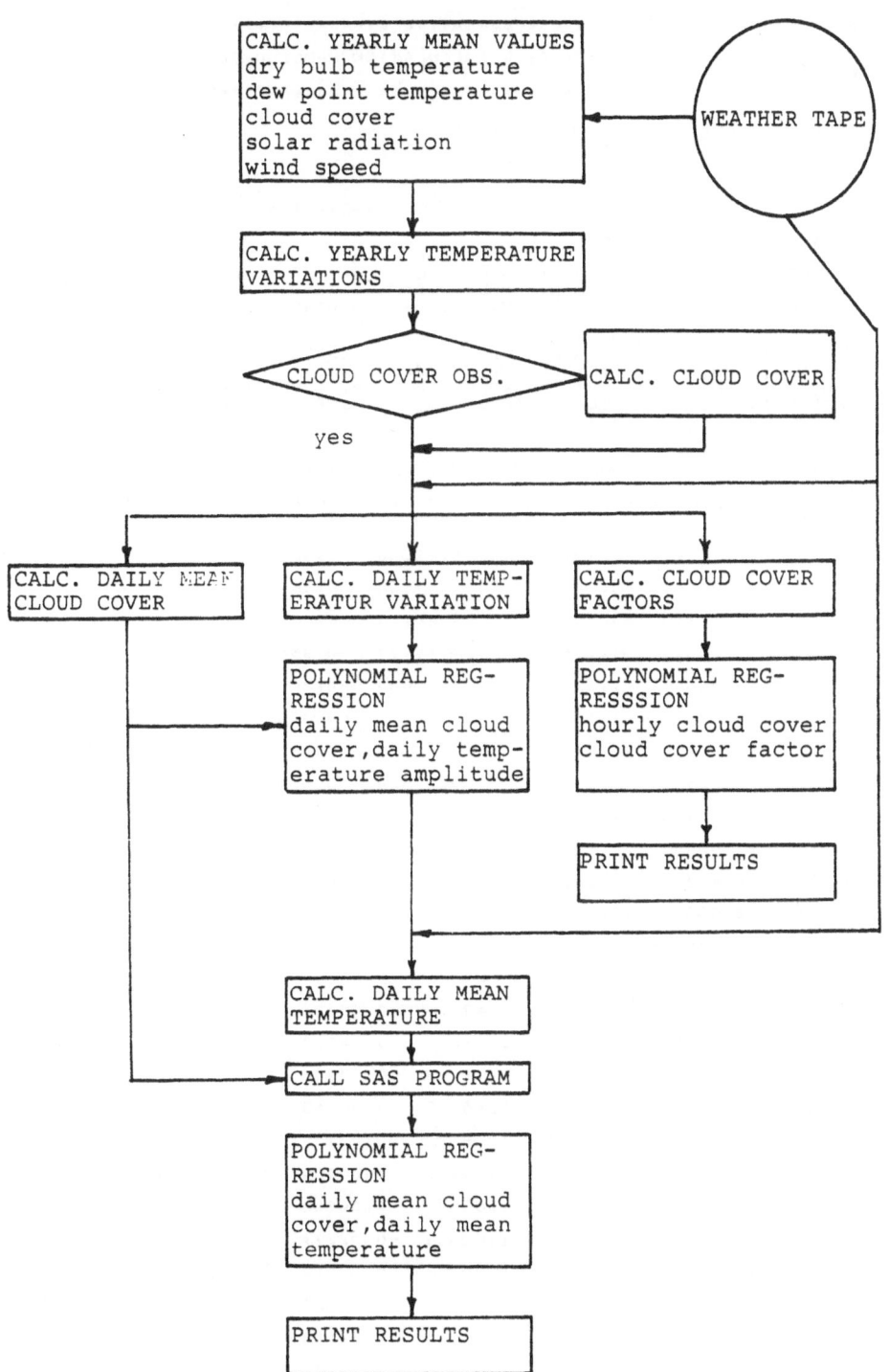

Fig. 2. Procedure for calculation of SRY model parameter.

```
*********************************************************************************
*                                                                               *
* RESULT OF THE SRY PARAMETER CALCULATION PROCEDURE                             *
*                                                                               *
*********************************************************************************

     CLIMATE ZONE:              DANMARK
     WEATHER DATA START:     59010101
     WEATHER DATA STOP :     73123124
     NUMBER OF OBSERVATION:    131495

     **************** *** YEARLY TEMPERATURE VARIATION   ********************

DRY BULB TEMPERATURE:
HARMONIC NO.:   1      YEARLY AMPLITUDE=      8.8760     YEARLY PHASE=      4.2781
HARMONIC NO.:   2      YEARLY AMPLITUDE=      0.7770     YEARLY PHASE=      2.8869
HARMONIC NO.:   3      YEARLY AMPLITUDE=      0.7385     YEARLY PHASE=      3.2978
HARMONIC NO.:   4      YEARLY AMPLITUDE=      0.6645     YEARLY PHASE=      3.6842

DEW POINT TEMPERATURE:
HARMONIC NO.:   1      YEARLY AMPLITUDE=      7.3397     YEARLY PHASE=      4.1487
HARMONIC NO.:   2      YEARLY AMPLITUDE=      0.8037     YEARLY PHASE=      2.8989
HARMONIC NO.:   3      YEARLY AMPLITUDE=      0.7399     YEARLY PHASE=      3.6556
HARMONIC NO.:   4      YEARLY AMPLITUDE=      0.7277     YEARLY PHASE=      3.6492


********              HOURLY TEMPERATURE VARIATION                 *********
        DRY BULB TEMPERATURE    DEW POINT TEMPERATURE
MONTH      B0          B1          B0          B1

   1    6.46844    -0.49711     5.82250    -0.38685

   2    8.31008    -0.74944     7.07538    -0.56939

   3   10.15199    -0.90477     4.28453    -0.14641

   4   13.46816    -1.17684     4.49475    -0.15610

   5   15.90048    -1.53294     4.86633    -0.24797

   6   16.69527    -1.64540     3.96243    -0.07334

   7   16.00136    -1.60349     3.81309    -0.10489

   8   16.28436    -1.69536     3.80538    -0.07910

   9   13.12756    -1.24116     3.95379    -0.02892

  10   11.98583    -1.18839     5.29093    -0.28636

  11    6.44298    -0.48902     5.08907    -0.25780

  12    5.63809    -0.39320     5.25770    -0.31635

**********************  CLOUD COVER FACTOR ***************************
MONTH    C0          C1          D0          D1          D2

   1    1.91579    -0.21984     1.47419    -0.08321    -0.00519

   2    1.50945    -0.17199     1.29306    -0.02912    -0.00988

   3    1.28306    -0.14853     1.18943    -0.02208    -0.00981

   4    1.18156    -0.13502     1.16895    -0.04143    -0.00677

   5    1.09250    -0.12413     1.09522    -0.03238    -0.00667

   6    1.03739    -0.12066     1.04740     0.00460    -0.01168

   7    0.96522    -0.11445     1.01983    -0.00258    -0.01303

   8    0.95167    -0.11133     1.03566    -0.02577    -0.00708

   9    1.00552    -0.11773     1.10358    -0.03414    -0.00669

  10    1.14678    -0.13747     1.15272    -0.03620    -0.00770

  11    1.39051    -0.15670     1.21693    -0.02643    -0.00989

  12    1.88025    -0.21797     1.52735    -0.13123    -0.00116
```

Table I. SRY parameter values.

ONE STEP TRANSISION MATRIX FOR CLOUD COVER MONTH  1/-  3

| FROM | 0 | 1 | 2 | 3 | 4 | 5 | 6 | 7 | 8 | 9 | TOTAL |
|---|---|---|---|---|---|---|---|---|---|---|---|
| 0 | 2491.0 | 305.0 | 64.0 | 53.0 | 19.0 | 17.0 | 14.0 | 11.0 | 5.0 | 23.0 | 3022.0 |
| 1 | 376.0 | 1308.0 | 274.0 | 127.0 | 54.0 | 30.0 | 35.0 | 19.0 | 8.0 | 2236.0 |
| 2 | 77.0 | 322.0 | 538.0 | 228.0 | 106.0 | 63.0 | 49.0 | 46.0 | 7.0 | 6.0 | 1442.0 |
| 3 | 27.0 | 156.0 | 251.0 | 415.0 | 203.0 | 165.0 | 105.0 | 59.0 | 13.0 | 5.0 | 1399.0 |
| 4 | 14.0 | 53.0 | 120.0 | 234.0 | 317.0 | 217.0 | 186.0 | 112.0 | 17.0 | 4.0 | 1274.0 |
| 5 | 2.0 | 38.0 | 69.0 | 121.0 | 217.0 | 378.0 | 341.0 | 231.0 | 37.0 | 6.0 | 1446.0 |
| 6 | 9.0 | 21.0 | 41.0 | 106.0 | 164.0 | 300.0 | 846.0 | 619.0 | 139.0 | 23.0 | 2288.0 |
| 7 | 6.0 | 19.0 | 47.0 | 76.0 | 124.0 | 197.0 | 523.0 | 2638.0 | 1016.0 | 33.0 | 4581.0 |
| 8 | 6.0 | 11.0 | 13.0 | 31.0 | 45.0 | 70.0 | 174.0 | 904.0 | 10657.0 | 400.0 | 12313.0 |
| 9 | 9.0 | 2.0 | 3.0 | 6.0 | 8.0 | 9.0 | 17.0 | 40.0 | 414.0 | 1672.0 | 2180.0 |

% PERCENT

| | 0 | 1 | 2 | 3 | 4 | 5 | 6 | 7 | 8 | 9 | MEAN |
|---|---|---|---|---|---|---|---|---|---|---|---|
| 0 | 82.4 | 10.1 | 2.6 | 1.8 | 0.6 | 0.6 | 0.5 | 0.4 | 0.2 | 0.8 | 0.4 |
| 1 | 16.8 | 58.5 | 12.3 | 5.7 | 2.4 | 1.3 | 1.6 | 0.8 | 0.2 | 0.4 | 1.4 |
| 2 | 5.3 | 22.3 | 37.3 | 15.8 | 7.4 | 4.4 | 3.4 | 3.2 | 0.5 | 0.4 | 2.5 |
| 3 | 1.9 | 11.2 | 17.9 | 29.7 | 14.5 | 11.8 | 7.5 | 4.2 | 0.9 | 0.4 | 3.4 |
| 4 | 1.1 | 4.2 | 9.4 | 18.4 | 24.9 | 17.0 | 14.6 | 8.8 | 1.3 | 0.3 | 4.3 |
| 5 | 0.6 | 2.6 | 4.8 | 8.4 | 15.0 | 26.1 | 23.6 | 16.0 | 2.6 | 0.4 | 5.1 |
| 6 | 0.4 | 0.9 | 1.8 | 4.6 | 8.0 | 13.1 | 37.0 | 27.1 | 6.1 | 1.0 | 5.9 |
| 7 | 0.1 | 0.4 | 1.0 | 1.6 | 2.5 | 4.0 | 10.7 | 58.1 | 23.8 | 0.7 | 6.6 |
| 8 | 0.0 | 0.1 | 0.1 | 0.3 | 0.4 | 0.6 | 1.4 | 7.3 | 86.6 | 3.2 | 7.9 |
| 9 | 0.4 | 0.1 | 0.1 | 0.3 | 0.4 | 0.4 | 0.8 | 1.5 | 19.0 | 76.7 | 8.6 |

ONE STEP TRANSISION MATRIX FOR CLOUD COVER MONTH  4/-  6

| FROM | 0 | 1 | 2 | 3 | 4 | 5 | 6 | 7 | 8 | 9 | TOTAL |
|---|---|---|---|---|---|---|---|---|---|---|---|
| 0 | 2155.0 | 407.0 | 67.0 | 36.0 | 10.0 | 6.0 | 8.0 | 1.0 | 3.0 | 13.0 | 2706.0 |
| 1 | 456.0 | 3282.0 | 529.0 | 197.0 | 65.0 | 39.0 | 22.0 | 12.0 | 5.0 | 1.0 | 4609.0 |
| 2 | 52.0 | 631.0 | 1328.0 | 481.0 | 191.0 | 82.0 | 49.0 | 25.0 | 3.0 | 2.0 | 2844.0 |
| 3 | 18.0 | 173.0 | 560.0 | 1071.0 | 500.0 | 235.0 | 121.0 | 36.0 | 5.0 | 0.0 | 2731.0 |
| 4 | 11.0 | 68.0 | 190.0 | 521.0 | 869.0 | 500.0 | 249.0 | 106.0 | 17.0 | 2.0 | 2541.0 |
| 5 | 5.0 | 20.0 | 90.0 | 246.0 | 492.0 | 1024.0 | 623.0 | 276.0 | 24.0 | 2.0 | 2608.0 |
| 6 | 1.0 | 16.0 | 36.0 | 109.0 | 253.0 | 593.0 | 1310.0 | 819.0 | 103.0 | 6.0 | 3248.0 |
| 7 | 2.0 | 9.0 | 19.0 | 57.0 | 125.0 | 273.0 | 707.0 | 2871.0 | 769.0 | 13.0 | 4802.0 |
| 8 | 1.0 | 3.0 | 7.0 | 14.0 | 32.0 | 53.0 | 155.0 | 701.0 | 5073.0 | 52.0 | 6091.0 |
| 9 | 3.0 | 1.0 | 0.0 | 1.0 | 2.0 | 3.0 | 3.0 | 14.0 | 70.0 | 208.0 | 305.0 |

% PERCENT

| | 0 | 1 | 2 | 3 | 4 | 5 | 6 | 7 | 8 | 9 | MEAN |
|---|---|---|---|---|---|---|---|---|---|---|---|
| 0 | 79.6 | 15.0 | 2.5 | 1.3 | 0.4 | 0.2 | 0.3 | 0.0 | 0.1 | 0.5 | 0.3 |
| 1 | 9.9 | 71.2 | 11.5 | 4.3 | 1.4 | 0.8 | 0.5 | 0.3 | 0.1 | 0.0 | 1.2 |
| 2 | 1.8 | 22.2 | 46.7 | 16.9 | 6.7 | 2.9 | 1.7 | 0.9 | 0.1 | 0.1 | 2.3 |
| 3 | 0.7 | 6.3 | 20.7 | 39.2 | 18.3 | 8.6 | 4.4 | 1.3 | 0.2 | 0.2 | 3.2 |
| 4 | 0.4 | 2.7 | 7.8 | 20.5 | 34.2 | 19.7 | 9.8 | 4.2 | 0.7 | 0.1 | 4.1 |
| 5 | 0.2 | 0.7 | 3.4 | 6.6 | 17.5 | 36.5 | 22.2 | 9.3 | 0.9 | 0.1 | 5.0 |
| 6 | 0.0 | 0.5 | 1.2 | 3.4 | 7.6 | 18.3 | 40.3 | 25.2 | 3.2 | 0.2 | 5.8 |
| 7 | 0.0 | 0.2 | 0.4 | 1.2 | 2.6 | 5.6 | 14.5 | 59.0 | 16.2 | 0.3 | 6.8 |
| 8 | 0.0 | 0.0 | 0.1 | 0.2 | 0.5 | 0.9 | 2.5 | 11.5 | 83.3 | 0.9 | 7.8 |
| 9 | 1.0 | 0.3 | 0.0 | 0.3 | 0.7 | 1.0 | 1.0 | 4.5 | 23.0 | 28.2 | 8.4 |

ONE STEP TRANSISION MATRIX FOR CLOUD COVER MONTH  7/-  9

| FROM | 0 | 1 | 2 | 3 | 4 | 5 | 6 | 7 | 8 | 9 | TOTAL |
|---|---|---|---|---|---|---|---|---|---|---|---|
| 0 | 2069.0 | 374.0 | 66.0 | 46.0 | 13.0 | 9.0 | 14.0 | 6.0 | 3.0 | 8.0 | 2608.0 |
| 1 | 434.0 | 3034.0 | 541.0 | 227.0 | 66.0 | 44.0 | 37.0 | 9.0 | 2.0 | 3.0 | 4420.0 |
| 2 | 67.0 | 689.0 | 1374.0 | 519.0 | 224.0 | 91.0 | 70.0 | 23.0 | 2.0 | 8.0 | 3096.0 |
| 3 | 20.0 | 66.0 | 229.0 | 585.0 | 526.0 | 249.0 | 141.0 | 65.0 | 5.0 | 3.0 | 3053.0 |
| 4 | 5.0 | 20.0 | 66.0 | 247.0 | 438.0 | 605.0 | 318.0 | 105.0 | 12.0 | 2.0 | 2866.0 |
| 5 | 0.0 | 36.0 | 122.0 | 247.0 | 631.0 | 1115.0 | 735.0 | 285.0 | 29.0 | 2.0 | 3207.0 |
| 6 | 4.0 | 13.0 | 61.0 | 139.0 | 287.0 | 754.0 | 1501.0 | 913.0 | 139.0 | 4.0 | 3875.0 |
| 7 | 1.0 | 8.0 | 29.0 | 59.0 | 126.0 | 285.0 | 819.0 | 3001.0 | 781.0 | 3.0 | 5112.0 |
| 8 | 0.0 | 2.0 | 4.0 | 17.0 | 24.0 | 54.0 | 171.0 | 697.0 | 3797.0 | 16.0 | 4784.0 |
| 9 | 3.0 | 3.0 | 5.0 | 2.0 | 2.0 | 2.0 | 9.0 | 7.0 | 13.0 | 57.0 | 108.0 |

% PERCENT

| | 0 | 1 | 2 | 3 | 4 | 5 | 6 | 7 | 8 | 9 | MEAN |
|---|---|---|---|---|---|---|---|---|---|---|---|
| 0 | 79.3 | 14.3 | 2.5 | 1.8 | 0.5 | 0.3 | 0.5 | 0.2 | 0.1 | 0.3 | 0.4 |
| 1 | 9.8 | 68.5 | 12.2 | 5.1 | 2.2 | 1.0 | 0.8 | 0.2 | 0.0 | 0.1 | 1.3 |
| 2 | 2.2 | 22.5 | 44.8 | 16.9 | 7.3 | 3.0 | 2.3 | 0.8 | 0.1 | 0.3 | 2.3 |
| 3 | 0.7 | 6.0 | 20.7 | 39.7 | 17.2 | 8.2 | 4.6 | 2.1 | 0.2 | 0.1 | 3.2 |
| 4 | 0.2 | 2.3 | 5.0 | 20.4 | 32.7 | 21.1 | 11.1 | 3.7 | 0.4 | 0.1 | 4.1 |
| 5 | 0.1 | 1.1 | 3.8 | 7.7 | 19.7 | 34.8 | 22.9 | 8.9 | 0.9 | 0.1 | 4.9 |
| 6 | 0.1 | 0.3 | 1.6 | 3.6 | 7.4 | 19.5 | 40.3 | 23.5 | 3.6 | 0.1 | 5.6 |
| 7 | 0.0 | 0.2 | 0.6 | 1.2 | 2.5 | 5.6 | 16.0 | 58.7 | 15.3 | 0.1 | 5.7 |
| 8 | 0.0 | 0.0 | 0.1 | 0.4 | 0.5 | 1.1 | 3.6 | 14.5 | 79.4 | 0.4 | 7.7 |
| 9 | 2.6 | 2.0 | 4.6 | 1.9 | 1.9 | 1.9 | 8.3 | 6.5 | 15.7 | 52.6 | 7.4 |

ONE STEP TRANSISION MATRIX FOR CLOUD COVER MONTH 10/- 12

| FROM | 0 | 1 | 2 | 3 | 4 | 5 | 6 | 7 | 8 | 9 | TOTAL |
|---|---|---|---|---|---|---|---|---|---|---|---|
| 0 | 1889.0 | 206.0 | 77.0 | 63.0 | 30.0 | 29.0 | 22.0 | 8.0 | 8.0 | 45.0 | 2437.0 |
| 1 | 357.0 | 1502.0 | 310.0 | 159.0 | 94.0 | 57.0 | 34.0 | 20.0 | 4.0 | 3.0 | 2536.0 |
| 2 | 73.0 | 414.0 | 652.0 | 271.0 | 134.0 | 83.0 | 85.0 | 48.0 | 9.0 | 10.0 | 1779.0 |
| 3 | 40.0 | 156.0 | 340.0 | 539.0 | 255.0 | 218.0 | 136.0 | 56.0 | 20.0 | 10.0 | 1760.0 |
| 4 | 17.0 | 63.0 | 147.0 | 287.0 | 413.0 | 284.0 | 253.0 | 90.0 | 26.0 | 7.0 | 1607.0 |
| 5 | 15.0 | 57.0 | 107.0 | 198.0 | 293.0 | 507.0 | 415.0 | 262.0 | 40.0 | 7.0 | 1893.0 |
| 6 | 7.0 | 28.0 | 79.0 | 130.0 | 200.0 | 394.0 | 971.0 | 770.0 | 143.0 | 16.0 | 2746.0 |
| 7 | 8.0 | 20.0 | 45.0 | 90.0 | 137.0 | 242.0 | 621.0 | 3297.0 | 1072.0 | 12.0 | 5550.0 |
| 8 | 6.0 | 8.0 | 13.0 | 37.0 | 44.0 | 81.0 | 191.0 | 974.0 | 10004.0 | 204.0 | 11562.0 |
| 9 | 23.0 | 1.0 | 1.0 | 0.0 | 7.0 | 5.0 | 16.0 | 25.0 | 235.0 | 872.0 | 1185.0 |

% PERCENT

| | 0 | 1 | 2 | 3 | 4 | 5 | 6 | 7 | 8 | 9 | MEAN |
|---|---|---|---|---|---|---|---|---|---|---|---|
| 0 | 77.5 | 10.9 | 3.2 | 2.6 | 1.2 | 1.2 | 0.9 | 0.3 | 0.3 | 1.8 | 0.6 |
| 1 | 14.1 | 59.2 | 12.2 | 6.1 | 3.7 | 2.2 | 1.3 | 0.5 | 0.2 | 0.1 | 1.4 |
| 2 | 4.1 | 23.3 | 36.6 | 15.2 | 7.5 | 4.6 | 4.6 | 2.7 | 0.5 | 0.6 | 2.5 |
| 3 | 2.2 | 8.9 | 19.4 | 30.3 | 14.5 | 12.2 | 7.6 | 3.1 | 1.1 | 0.6 | 3.4 |
| 4 | 1.1 | 5.2 | 9.1 | 17.9 | 25.7 | 17.7 | 15.7 | 5.5 | 1.6 | 0.4 | 4.2 |
| 5 | 0.8 | 3.0 | 5.6 | 10.3 | 15.4 | 29.7 | 21.9 | 13.8 | 2.1 | 0.6 | 4.9 |
| 6 | 0.3 | 1.0 | 2.9 | 5.0 | 7.3 | 14.3 | 35.4 | 23.3 | 5.2 | 0.6 | 5.8 |
| 7 | 0.1 | 0.4 | 0.8 | 1.7 | 2.5 | 4.4 | 11.2 | 59.4 | 19.3 | 0.2 | 6.8 |
| 8 | 0.1 | 0.1 | 0.1 | 0.3 | 0.4 | 0.7 | 1.7 | 8.4 | 86.5 | 1.8 | 7.8 |
| 9 | 1.9 | 0.1 | 0.1 | 0.0 | 0.6 | 0.4 | 1.4 | 2.1 | 19.8 | 73.6 | 8.5 |

Table II. SRY parameter values.

DRY BULB TEMPERATURE

| MONTH | $A_0$ | $A_{1,1}$ | $A_{1,2}$ | $A_{2,1}$ | R-SQUARE |
|---|---|---|---|---|---|
| 1 | -2.7113 | 0.7081 | 0.0152 | 0.5074 | 0.73 |
| 2 | -1.6897 | 0.9008 | -0.0901 | 0.29169 | 0.74 |
| 3 | -0.5650 | 0.9263 | -0.1173 | 0.0915 | 0.71 |
| 4 | -0.0993 | 0.9173 | -0.1200 | 0.0058 | 0.67 |
| 5 | 0.9203 | 0.8163 | -0.0681 | -0.2202 | 0.65 |
| 6 | 1.4525 | 0.7638 | -0.0790 | -0.3201 | 0.64 |
| 7 | 1.2813 | 0.8900 | -0.1966 | -0.2694 | 0.70 |
| 8 | 0.4979 | 0.8321 | -0.1358 | -0.1263 | 0.61 |
| 9 | -0.1522 | 0.8339 | -0.1246 | 0.0818 | 0.55 |
| 10 | -1.6411 | 0.7396 | -0.0642 | 0.3508 | 0.61 |
| 11 | -3.3918 | 0.7553 | -0.0615 | 0.4865 | 0.67 |
| 12 | -3.1935 | 0.8446 | -0.1081 | 0.4967 | 0.72 |

DEW POINT TEMPERATURE

| MONTH | $A_0$ | $A_{1,1}$ | $A_{1,2}$ | $A_{2,1}$ | R-SQUARE |
|---|---|---|---|---|---|
| 1 | -3.4878 | 0.6611 | 0.0117 | 0.6546 | 0.72 |
| 2 | -2.4756 | 0.8283 | -0.0835 | 0.4394 | 0.71 |
| 3 | -1.7900 | 0.7835 | -0.0729 | 0.3294 | 0.65 |
| 4 | -2.1026 | 0.7550 | -0.0332 | 0.4118 | 0.65 |
| 5 | -102291 | 0.7477 | -0.1177 | 0.2289 | 0.52 |
| 6 | -0.5045 | 0.7092 | -0.0217 | 0.1134 | 0.52 |
| 7 | -0.6897 | 0.7071 | -0.0342 | 0.1922 | 0.45 |
| 8 | -0.9771 | 0.6934 | -0.0272 | 0.2325 | 0.53 |
| 9 | -1.2617 | 0.6899 | -0.0049 | 0.3427 | 0.55 |
| 10 | -2.8098 | 0.5874 | 0.0539 | 0.5688 | 0.61 |
| 11 | -4.6403 | 0.6346 | 0.0048 | 0.6635 | 0.65 |
| 12 | -3.828 | 0.7444 | -0.0729 | 0.5913 | 0.69 |

Table III. SRY parameter values.

ACTION 3.1

GLOBAL IRRADIATION ON HORIZONTAL PLANE -

DEFINITION OF RADIATION CLIMATE ZONES

Action leader's progress report
    F. KASTEN, Deutscher Wetterdienst, Meteorologisches
    Observatorium Hamburg, Federal Republic of Germany

Report of action participant :

  - Heliothermic indexes for the evaluation of the potential
    of solar energy

## Action 3.1 - GLOBAL IRRADIATION ON HORIZONTAL PLANE.
## DEFINITION OF RADIATION CLIMATE ZONES


Action budget (CEC contribution) : 103.3 kUC

Duration of project : until 1983-06-30


Action Leader : F. KASTEN
                Deutscher Wetterdienst
                Meteorologisches Observatorium Hamburg
                Frahmredder 95
                D-2000 Hamburg 65


Participants :

- Deutscher Wetterdienst
  Contract No. ESF-004-80-D (B)
  F. KASTEN, H.J. GOLCHERT

- Institut Royal Météorologique de Belgique
  Contract No. ESF-003-B (G)
  R. DOGNIAUX, M. LEMOINE


Tasks :

- Improvement and extension of the first volume of the
  EC solar radiation atlas published in 1980.

- Definition of radiation climate zones in the EC.

## Summary

Preliminary maps of monthly means of daily global irradiation in Western Europe had been presented in the first volume of the European Solar Radiation Atlas published by the Commission of the European Communities. For the improvement of the Atlas, the data gaps between the radiometric stations were filled by making use of the daily sunshine duration data of the more numerous heliographic stations. The maps were extended beyond the borders of the region of the EC by collecting and processing the data from neighbouring countries and from the Mediterranean area including North Africa and the Near East.

For 310 European stations, uniform tables comprising the mean monthly means and the mean annual mean of daily global irradiation were established on the common data basis of the 10 year period 1966-1975. The annual course of daily global irradiation at each station was also shown on diagrams displaying the monthly means and their standard deviations. On the basis of the tabulated values, maps of the mean distributions of daily global irradiation in Europe in each month and on the annual average are being designed. The annual and the June maps are shown.

For the eastern part of the Mediterranean basin and the Near East area, tables and maps of monthly and annual means of daily sunshine duration were produced. By interpolation, daily sunshine duration was computed for each grid point on a $1^{\circ}$ latitude by $1^{\circ}$ longitude grid and tabulated.

Several possible methods of defining radiation climate zones were brought forward and discussed. At present, two methods are recommended. One method is for pure solar radiation applications. It defines global radiation zones by the number of months per year with mean daily global irradiation of at least 3 kWh·m$^{-2}$. A map displaying these zones is given, it resembles the mean annual distribution in a simplified manner. The other method defines zones by introducing a so-called heliothermic index which takes additional energy fluxes besides global radiation into account. This method is especially suited to solar thermal applications. A map of zones of equal heliothermic index is also given.

# 1. General tasks of the Action

## 1.1 Improvement and extension of the EC solar radiation atlas

Preliminary maps of monthly means of daily global irradiation in Western Europe had been presented in the first volume of the European Solar Radiation Atlas published by the Commission of the European Communities (1). The isolines on the maps have been drawn on the basis of 10 year records at 56 radiometric stationswithin the region of the European communities. Due to the relatively low density of this network, the maps derived from its data gave but a rather general impression of the large scale features of the global radiation distribution over Western Europe, and the isolines of irradiation also called isopyrs abruptly terminate at the borders of the EC-countries.

For the improvement of the atlas, the data gaps between the radiometric stations were to be filled by making use of the daily sunshine duration data of the more numerous heliographic stations. This subtask was to be performed by Deutscher Wetterdienst (DWD).

The extension of the maps beyond the borders of the EC-region was split into two subtasks: DWD collected and processed the data from neighbouring countries of the EC, and Institut Royal Météorologique de Belgique (IRMB) specialized on data acquisition from the Mediterranean area including North Africa and the Near East.

## 1.2 Definition of radiation climate zones in the EC

For the numerical simulation of the performance of solar energy systems or components, meteorological data sets are required as input. In order to minimize both the volume of data and the number of simulation runs, some kind of classification or division into zones of equal or similar radiation regime is desired.

A working group within Project F consisting of Mr. Dogniaux (IRMB), Mr. Bedel (Météorologie Nationale, France) and Mr. Kasten (DWD) was established with the aim to propose, discuss and possibly recommend methods of defining radiation climate zones.

# 2. Data collection and processing

## 2.1 Area of EC- and neighbouring countries

Europe is geographically located in the temperate zone and in the westerly wind belt of the general circulation of the atmosphere. As a meteorological consequence, solar radiation received at the earth surface is subjected not only to pronounced seasonal variations but to major year-to-year fluctuations. Therefore, many years means of data of the same time period from all stations are mandatory for drawing a true picture of the global radiation distribution in Europe. The period agreed upon earlier are the 10 years 1966-1975.

Additionally to the data base for the first edition of the atlas (1), simultaneous data of daily global irradiation G and daily sunshine duration S were supplied by the following EC-countries:
- Danmark: 3 radiometric stations (2 with G only),
- Italy: 15 radiometric stations,
- Greece: 1 radiometric station.

For each radiometric station, the Ångström relation

$$G/G_0 = a + b \cdot S/S_0 \qquad (1)$$

was established for each month; $G_0$ is daily extraterrestrial irradiation, $S_0$ is daily astronomical sunshine duration. The values of a and b for each station and each month were deter- mined by regression analysis and tabulated.

A second additional data source were the complete records of daily sunshine duration of the full period 1966-1975 ob- tained from the following EC-Countries:
- Ireland          : 11 heliographic stations,
- United Kingdom: 25 heliographic stations,
- Germany (F.R.): 67 heliographic stations,
- France          : 44 heliographic stations,
- Greece          :  8 heliographic stations.

These heliographic stations were associated as "satellites" to one or another appropriate radiometric station taking due account of the topographic and climatic conditions of the re- spective area as proposed by the National Meteorological Services who had supplied the data. A few heliographic sta- tions were associated to a radiometric station located in the adjacent country, for meteorological reasons.

With the regression coefficients a and b determined for the radiometric stations, the monthly means of daily sunshine duration S of the satellite heliographic stations were con- verted to monthly means of daily global irradiation G and then averaged over the 10 year period 1966-1975.

From outside the EC, many countries responded to our re- quest for supplying daily global irradiation and/or sunshine duration data of the 1966-1975 period, the following being selected for our purpose:
- Austria        :  5 radiometric stations,
- Czechoslovakia:  2 radiometric, 8 heliographic stations,
- Norway         :  1 radiometric stations,
- Poland         :  4 radiometric stations,
- Spain          :  1 radiometric (G only), 13 heliogr.stations,
- Sweden         :  5 radiometric, 4 heliographic stations,
- Switzerland    :  3 radiometric stations,
- Yugoslavia     : 11 radiometric stations.

For each radiometric station, the monthly Ångström regression coefficients a and b were computed and tabulated, and the dai- ly sunshine duration data of the heliographic stations of Czechoslovakia and Sweden converted to daily global irradia- tion. Since the Spanish data do not allow for an Ångström regression analysis, the coefficients a and b being necessary for the conversion of daily sunshine duration into daily glob- al irradiation were determined by the method of Dogniaux and Lemoine (2) who found a linear dependence of a and b on geo- graphical latitude.

Additionally, daily global irradiation data of the 1966- 1975 period were taken from the publications of the WMO World Radiation Data Center at Leningrad (3), namely:
- Bulgaria        : 1,          - Finland:  2,
- German Dem. Rep.: 4,          - Hungary:  4,
- Portugal        : 7,          - Romania   3.

A complete list of the number and types of stations of all countries the data of which were quality controlled and then processed to 10 year means of monthly means of daily global irradiation is compiled on Table I.

## 2.2 Mediterranean area
Data of daily sunshine duration and global irradiation were collected from the following Mediterranean countries besides those mentioned in section 2.1:
- Algeria,             - Cyprus,              - Egypt,
- Israel,              - Lebanon,             - Malta,
- Morocco              - Syria,               - Tunesia,
- Turkey.

Data from some other countries located in the latitude belt from 30°N to 44°N were also collected:
- Azores,              - Jordan,              -Madeira,
- USSR (South).

The data were either directly obtained from the National Meteorological Services or taken from the WMO Leningrad publications (3) and listed on uniform tables.

Since data records particularly of global radiation are scarce in several of these countries, the 10 year period 1966-1975 agreed upon for the European Solar Radiation Atlas could not be taken as common time basis for the analysis of the radiation regime in this area. In several cases, the data collection had to be put up with records of different periods of different lengths which are not felt to be serious shortcomings because the year-to-year fluctuations of global radiation in this area are less than in the temperate zone. Nevertheless, the global radiation distribution over the Mediterranean area derived from these data is recommended to be displayed separately from the global radiation maps based on the uniform 1966-1975 period.

## 3. State of the atlas
## 3.1 Europe
Collection and quality control of all 310 stations mentioned in section 2.1 have been finished. Computation and printout of the resulting tables, one for each station, of the mean monthly means have also been completed. The tables have the same format as in the first edition of the Atlas (1) except that values of global irradiation computed from sunshine duration are marked with a special label.

The results are also presented on diagrams, one for each station, displaying the mean annual course of daily global irradiation on the basis of the mean monthly means the standard deviations of which are also plotted in the diagrams. Examples are given in Fig. 1 and Fig. 2 which show the mean annual courses of daily global irradiation at two Italian cities, one in a northern industrial area, the other on Sicily in the far south. Furthermore, the annual courses of all stations of each country considered are summarized on one or two diagrams leaving out the standard deviation bars. Figs.3 and 4 show these summarizing diagrams for Italy as examples.

The mean monthly means of daily global irradiation are entered into geographical maps, one for each month and one for the annual mean, and interpolated by drawing isolines. Pure mathematical interpolation is not legitimate but may lead to gross errors in certain cases. Rather, due account has to be

taken of the geographic and climatic conditions of the area
such as coastlines, river valleys, mountains etc.

Three maps have been completed so far. As examples, the
maps of June and of the annual means are presented as Figs.
5 and 6. Copies of all maps are planned to be sent to the
National Meteorological Services for comments and suggestions
which will be used to improve the patterns of the isopyrs before the maps are delivered to the printer.

## 3.2 Mediterranean

Tables and maps of monthly and annual means of daily sunshine duration have been produced of
- the eastern part of the Mediterranean basin,
- the Near-East comprising Syria, Lebanon, Israel,and Jordan,
- Turkey and Cyprus separately.

By interpolation, monthly and annual means of daily sunshine duration have been computed for each point of a $1^{\circ}$ latitude by $1^{\circ}$ longitude grid stretching from $30^{\circ}$N to $44^{\circ}$N and
from $16^{\circ}$E to $46^{\circ}$E, and tabulated.

Complete sets of the tables and maps of sunshine duration
have been published (4). On the basis of this data material,
daily global irradiation values will be derived, compiled on
tables and maps, and published.

## 4. Recommendations on radiation climate zones

Several methods of defining radiation climates were proposed and discussed by the members of the working group. Three
of these proposals were selected for detailed consideration by
the working group.

The method of Bedel is based on relative daily sunshine
duration $S/S_0$ as parameter. Of each station, the mean numbers
of days per month with $S/S_0 < 0.2$, $> 0.8$, or between these
two values are determined for 4 selected months. By a statistical method to be defined, the stations are arranged into
classes which represent the radiation climate zones.

The method of Kasten relies on the annual mean of daily
global irradiation, $\bar{G}$, and the number of months per year, $N_3$,
with monthly mean $\geqslant 3$ kWh·m$^{-2}$. It was shown that

$$N_3 \approx 2 \, \bar{G}/\text{kWh·m}^{-2} \tag{2}$$

so that either one of the two quantities is sufficient. The
radiation climate of Europe was proposed to be divided into
4 zones whose irradiations may be designated as very high,
high, moderate, or low, respectively, corresponding to $N_3 \geqslant 8$,
$= 7$, $= 6$, or $= 5$. A map which is based on the first edition
of the European Solar Radiation Atlas (1) shows the division
of Western Europe into the four radiation climate zones thus
defined, see Fig. 7. Because of Eq.(2), the pattern resembles
the distribution of daily global irradiation in a simplified
manner.

The method of Dogniaux is based on a new quantity called
heliothermic index $t_h$ in $^{\circ}$C:

$$t_h = t_a + (G + L\downarrow)/h_c' \tag{3}$$

where     $t_a$ = air temperature, in $^{\circ}C$,

$G$ = global solar irradiance, in $W \cdot m^{-2}$,

$L_{\downarrow}^{*}$ = net terrestrial surface irradiance, in $W \cdot m^{-2}$,

$h_c'$ = heat transfer coefficient due to convection at wind velocity 0, in $W \cdot m^{-2} \cdot K^{-1}$, of the surface.

By making several simplifying assumptions, $t_h$ can be calculated solely from daily global irradiation and daily mean air temperature by the help of a set of equations. A map of the distribution of the heliothermic index based on the annual means of daily global irradiation as given in the first edition of the European Solar Radiation Atlas (1) and on corresponding annual mean air temperatures is shown as Fig. 8.

Since the data basis required for Bedel's method is presently not available from the European countries except France, only the remaining two methods are recommended for use within the Solar R&D Programme of the EC at the moment. The classification by Kasten defines global radiation zones which are useful for pure solar radiation applications; the heliothermic index of Dogniaux takes additional energy fluxes into account and therefore defines zones which are especially tuned to solar thermal applications, building climatology and the like.

References

(1) Commission of the European Communities: European solar radiation atlas, vol.I: Global radiation on horizontal surfaces, ed. by W. Palz. Grösschen-Verlag, Dortmund (1979).

(2) R. Dogniaux, M. Lemoine: Classification of radiation sites in terms of different indexes of atmospheric turbidity. Contribution to Action 3.2, Project F, CEC Solar Energy R&D Programme, Brussels (Oct 1982).

(3) USSR Chief Administration of the Hydrometeorological Service, A.I. Voeikov Main Geophysical Observatory: Solar radiation and radiation balance data (The world network), sponsored by World Meteorological Organization, Leningrad (since 1964).

(4) R. Dogniaux, M. Lemoine: Distribution du rayonnement solaire dans les régions limitrophes du bassin méditerranéen oriental et occidental, Fascicule I - Durée d'insolation. Inst.Roy.Mét.Belg. Misc. Série B no. 54 (1982).

| Country | G and S | G only | S only | total |
|---|---|---|---|---|
| Austria | 5 | | | 5 |
| Belgium | 4 | | | 4 |
| Bulgaria | | 1 | | 1 |
| Czechoslovakia | 2 | | 8 | 10 |
| Denmark | 2 | 2 | 5 | 9 |
| Finland | | 2 | | 2 |
| France | 7 | | 44 | 51 |
| German Dem.Rep. | | 4 | | 4 |
| Germany (F.R.) | 8 | 1 | 67 | 76 |
| Greece | 1 | | 8 | 9 |
| Hungary | | 4 | | 4 |
| Ireland | 3 | | 11 | 14 |
| Italy | 28 | | | 28 |
| Netherlands | 5 | | | 5 |
| Norway | 1 | | | 1 |
| Poland | 4 | | | 4 |
| Portugal | | 7 | | 7 |
| Romania | | 3 | | 3 |
| Spain | | 1 | 13 | 14 |
| Sweden | 5 | | 4 | 9 |
| Switzerland | 3 | | | 3 |
| United Kingdom | 11 | | 25 | 36 |
| Yugoslavia | 11 | | | 11 |
| | | | | |
| All countries | 100 | 25 | 185 | 310 |

Table I: Number of stations of European Countries
the data of which were processed to
10 year monthly means of daily global
irradiation and then entered into monthly
maps.
G: daily global irradiation available,
S: daily sunshine duration available.

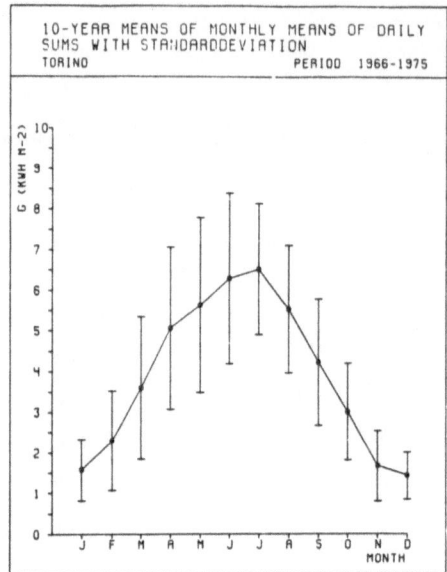

Fig. 1

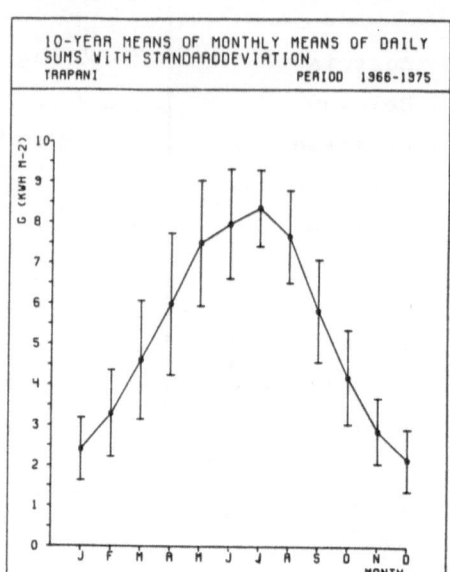

Fig. 2

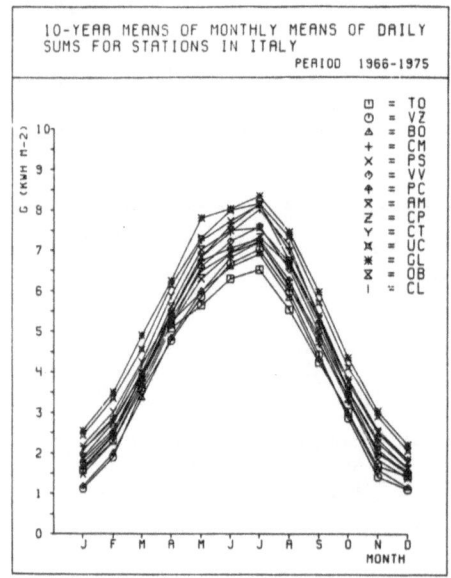

Fig. 3

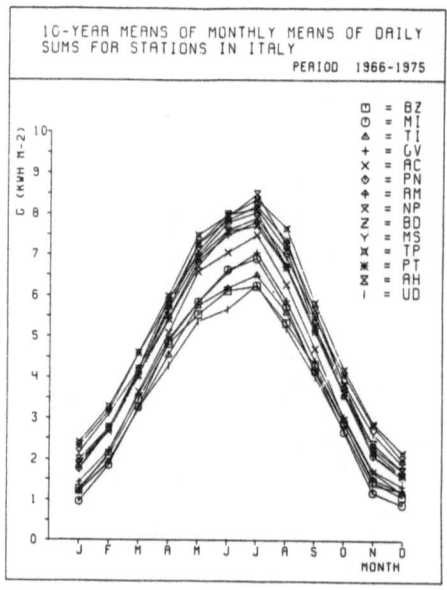

Fig. 4

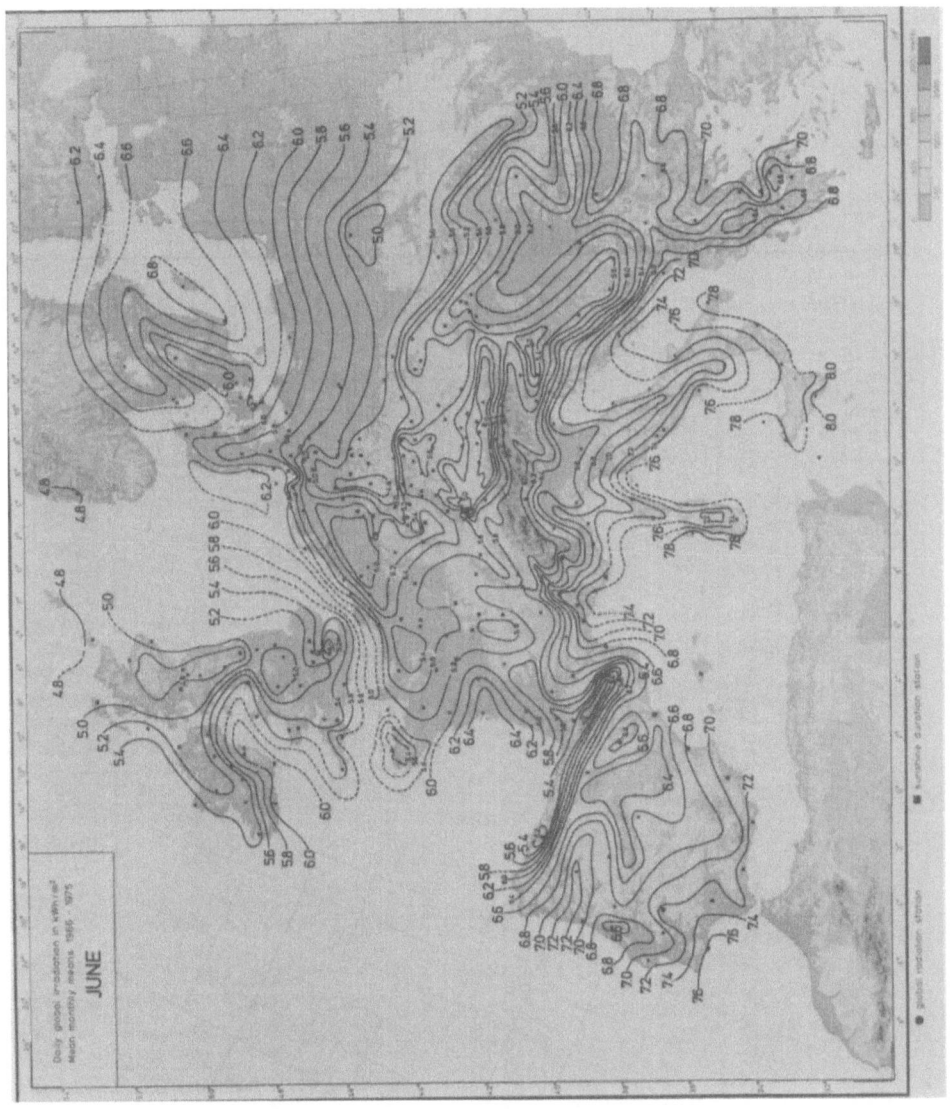

Fig. 5

- 59 -

Fig. 6

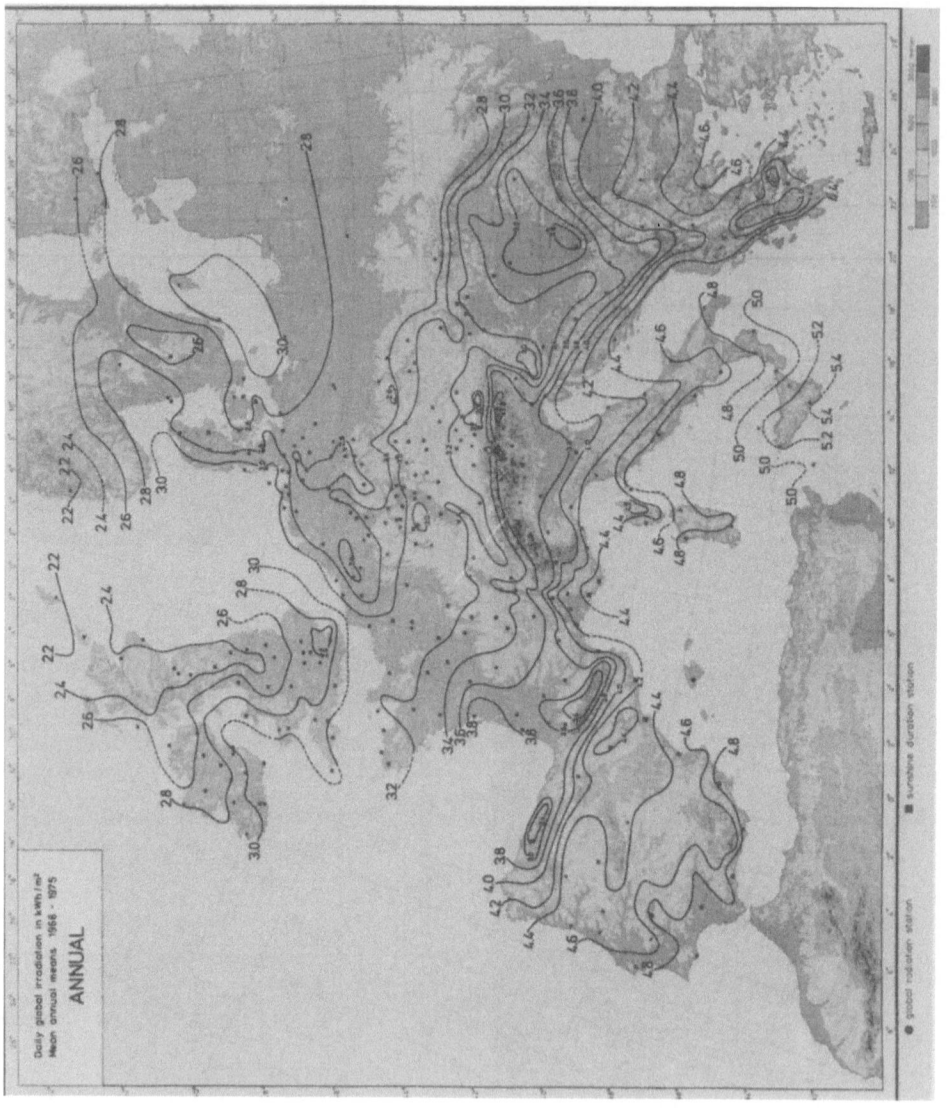

Daily global irradiation in kWh/m²
Mean annual means 1966 - 1975
ANNUAL

● global radiation station    ■ sunshine duration station

- 60 -

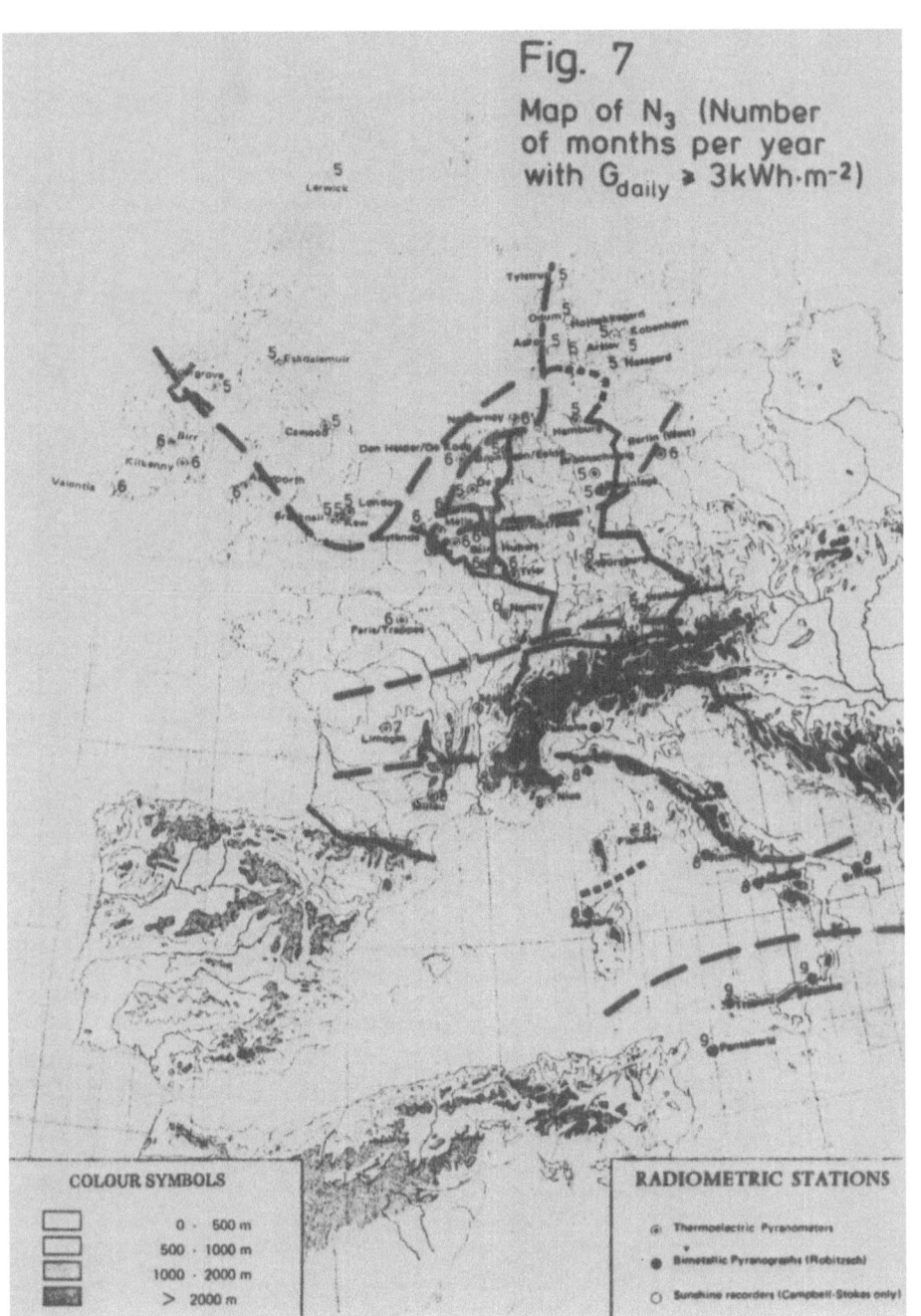

Fig. 7
Map of $N_3$ (Number of months per year with $G_{daily} \geqslant 3 \, kWh \cdot m^{-2}$)

COLOUR SYMBOLS

☐ 0 - 500 m
☐ 500 - 1000 m
☐ 1000 - 2000 m
▓ > 2000 m

RADIOMETRIC STATIONS

⊙ Thermoelectric Pyranometers
● Bimetallic Pyranographs (Robitzsch)
○ Sunshine recorders (Campbell-Stokes only)

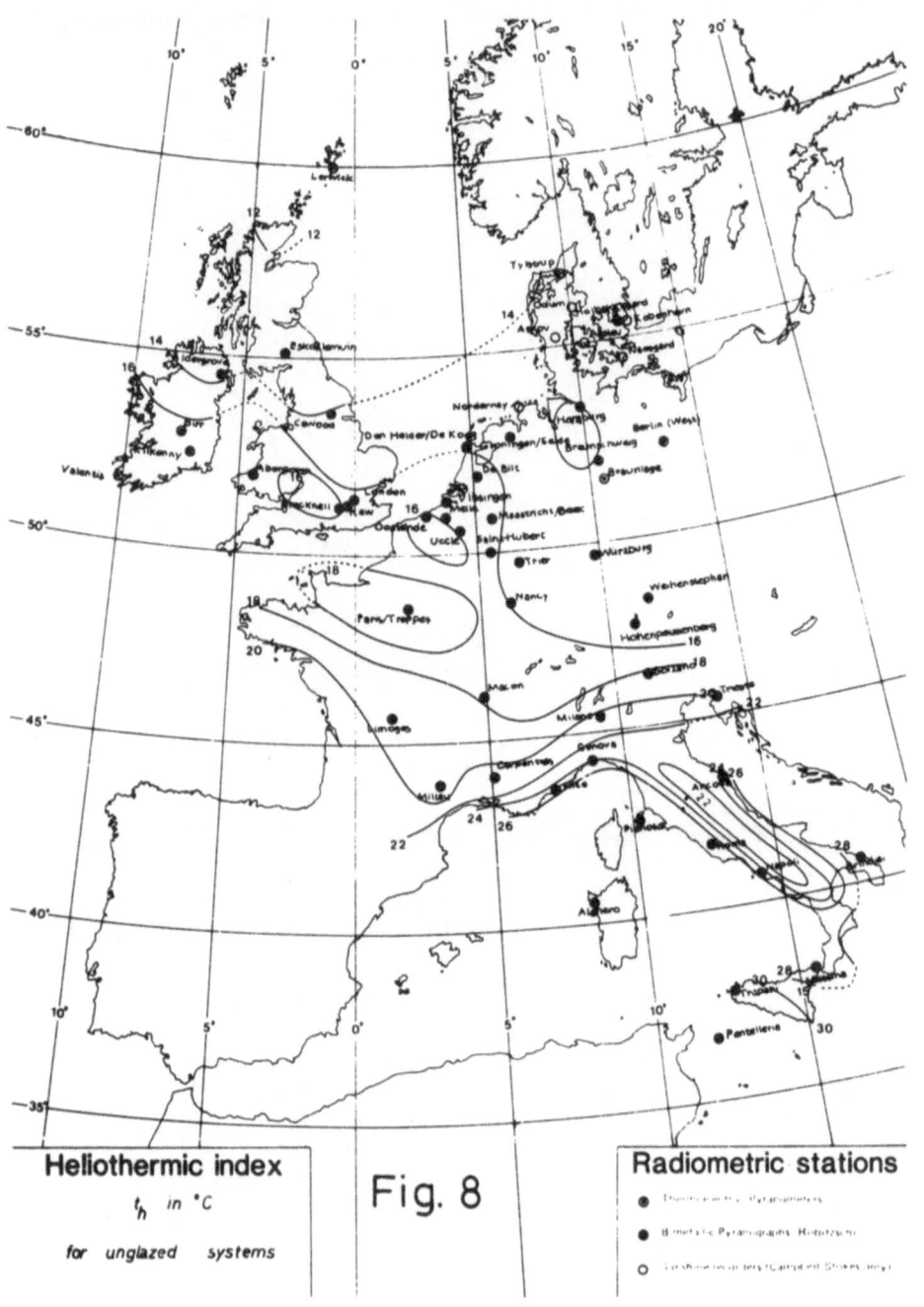

## Heliothermic index

$t_h$ in °C

for unglazed systems

# Fig. 8

## Radiometric stations

● Thermoelectric Pyranometers

● Bimetallic Pyranographs Robitzsch

○ Sunshine recorders (Campbell-Stokes only)

# HELIOTHERMIC INDEXES FOR THE EVALUATION
# OF THE POTENTIAL OF SOLAR ENERGY

R. DOGNIAUX
Institut Royal Météorologique de Belgique

## SUMMARY

The object of the present note is to propose and define simple indexes for charaterizing and differentiating, at the mesoscale range, climatic areas answering criteria suited for heliothermic applications.

The meteorological elements most liable to influence the behaviour of heliothermic systems are solar and atmospheric irradiances, air temperature and wind speed. It is quite natural to call on a simple function of these variables to define, as simply as possible, *heliothermic indexes* that can be applied to either unglazed systems or glass-shielded systems.

Two different proposals are presented as examples to comply with these two types of application.

# 1. INTRODUCTION

Studies of thermal exchanges through the partition walls of systems in contact with a meteorological environnement refer to the well known *equivalent temperature "air-sun"* defined by

$$t_e = t_a + \frac{\alpha Q\downarrow - \varepsilon L\uparrow}{h_c} \tag{1}$$

where :

$t_a$    denotes the air temperature,

$\alpha$    the solar absorptivity of the wall,

$\varepsilon$    the emissivity of the wall,

$Q\downarrow$    the total downward irradiance (solar plus atmospheric),

$L\uparrow$    the longwave irradiance emitted by the wall,

$h_c$    the outside surface coefficient of heat transfert of the surface in contact with air due to convection.

The *equivalent temperature* represents, in fact, the effective temperature reached by the ambient air in contact with an irradiated surface but not influenced by latent heat exchanges, of which the absorptivity and the surface conductance are known.

For our purpose, we shall particularize this concept by introducing some simplifying constraints related to the nature of the solar system under consideration.

# 2. HELIOTHERMIC INDEX FOR UNGLAZED SYSTEMS : $t_h$

Assuming a flat horizontal surface, behaving as a black body ( $\alpha = \varepsilon = 1$), shielded from the wind (wind speed = 0) and located at the ground surface without any heat exchange by conduction with the soil, equation (1) can be written.

$$t_h = t_a + \frac{G + L\updownarrow}{h'_c} \tag{2}$$

where :

$t_h$    denotes the relevant heliothermic index in °C,

$t_a$    the air temperature in °C,

$G$    the global solar irradiance on the surface in $W.m^{-2}$,

$L\updownarrow$   the longwave net irradiance (difference between downward and upward longwave radiation) in $W.m^{-2}$,

$h'_c$   the outside surface coefficient of heat transfer of the wall by convection for a <u>wind speed equal to zero</u>, in $W.m^{-2}.°C$.

## 3. HELIOTHERMIC INDEX FOR GLASS-SHIELDED SYSTEMS : $t'_h$

In the particular case of a system protected by a glass, the heliothermic index deduced from equation (1) takes the form :

$$t'_h = t_a + \frac{\alpha SG}{U} + \frac{\varepsilon L\updownarrow}{h_e}$$

where :

$t'_h$   denotes the relevant heliothermic index,

$t_a$   the air temperature in °C,

$S$   the solar factor of the screen (glass and frames),

$G$   the global solar irradiance on the surface in $W.m^{-2}$,

$U$   the areal coefficient of thermal transmittance in $W.m^{-2}K^{-1}$,

$\alpha$   the solar absorptivity of the system,

$\varepsilon$   its emissivity,

$L\updownarrow$   the longwave net irradiance in $W.m^{-2}$,

$h_e$   the total outside surface coefficient of heat transfer of the system in $W.m^{-2}K^{-1}$.

## 4. ANALYTICAL EXPRESSIONS OF THE HELIOTHERMIC INDEXES

### 4.1 CASE 1 : UNGLAZED SYSTEMS : INDEX $t_h$

$$t_h = t_a + \frac{G + L\updownarrow}{h'_c} \tag{2}$$

with      $L\updownarrow = L\downarrow - L\uparrow$

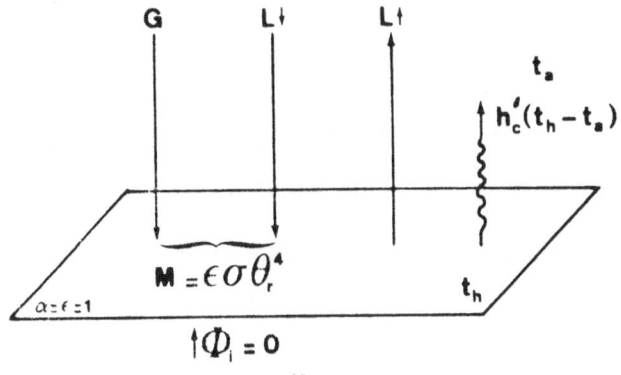

## Expression for h'$_c$

The superficial exchanges between a reference horizontal surface and an external environment which result from natural convection are governed by the outside surface coefficient of heat exchange. For a wind speed equal to zero, this coefficient can be written :

$$h'_c = ( 1.744 - 0.0035 \frac{\theta_r + t_a}{2} ) ( \theta_r - t_a )^{0.333} \ \text{W.m}^{-2}.\text{K}^{-1} \qquad (4)$$

where $\theta_r$ denotes the *superficial radiation temperature* reached by the surface exposed to the radiation fluxes, in absence of any other exchange of energy by convexion or thermal leakage, for example.

With the hypothesis of a surface similar to a black body, this radiation temperature can be deduced from the *radiant emittance* $M_e$ of the surface by application of the Stefan-Boltzmann law :

$$\theta_r = (\frac{1}{\sigma} M_e )^{0.25}$$

with $M_e = G + L\downarrow$    in $\text{W.m}^{-2}$

and

$$\sigma = 5.67 .10^{-8} \ \text{W.m}^{-2}.\text{K}^{-4}$$

Values of G  (global solar irradiance) can be found in tables, namely in the European Radiation Atlas, published by the European Communities.

Values of L↓ (downward longwave irradiance) can be derived, in the absence of observational actual data, by the DOGNIAUX's equation (7), using merely the ration $S/S_o$ of the effective to the theorical duration of sunshine and the ground level water pressure, both measured at the station :

$$L\downarrow= \sigma T_a^4 (0.6 + 0.056 \ p^{0.5}) + (1-S/S_o) [ \ \sigma T_a^4 (0.135 - 0.128 \ p^{0.5}) + 43 \ p^{0.5}$$
$$(7)$$

where $T_a$ is the air temperature in Kelvin.

If p is not readily available, an average annual value can be accepted, viz :

$$p = 10 \ \text{mb}$$

With this assumption, equation (7) becomes :

$$L\downarrow = 0.78 \ \sigma \ T_a^4 + (1 - S/S_o) (136 - 0.27 \ \sigma \ T_a^4 ) \qquad (8)$$

Values of L↑   (upward longwave irradiance). Since the upward longwave irradiance is a function of the effective temperature of the air in contact with the surface, $t_h$, $L\uparrow$ is expressed by

$$L\uparrow = \sigma(t_h + 273)^4 \qquad (9)$$

under the conditions of the abovementioned hypothesis.
Combining equations (2), (4), (8) and (9), we finally arrive to

$$T_h \ ( \sigma T_h^3 + h'_c ) = G + L\downarrow + h'_c T_a \qquad (10)$$

in which the temperatures are expressed in Kelvin.

From (10), the index $t_h$ is deduced by conversion into the Celsius scale

$$t_h = T_h - 273 \qquad (11)$$

## 4.2 CASE 2 : GLASS-SHIELDED SYSTEMS : INDEX $t'_h$

$$t'_h = t_a + \frac{\alpha SG}{U} + \frac{\varepsilon L\uparrow}{h_e} \qquad (3)$$

In this instance, equation (3) is applied to the following particular conditions :

- black body reference surface, 1 sq.m.area, under glazing and without any heat exchange with the soil;
- as regards the coefficients of the equation (3), the following values are introduced :

$$\alpha = \varepsilon = 1$$
$$S = 0.7$$
$$U = 7 \ W.m^{-2}.K^{-1}$$
$$h_e = 20 \ W.m^{-2}.K^{-1} \quad \text{(corresponding to a wind speed} \approx 3 \ m.s^{-1})$$

Equation (3) then becomes :

$$t'_h = t_a + 0.1 \ G + 0.05 \ L\uparrow \qquad (12)$$

with

$$L\downarrow = (1-S/So) (136-0.27 \ \sigma T_a^4) - 0.165 \ G - 0.12\sigma \ T_a^4 \qquad (13)$$

according to DOGNIAUX's equation proposal for $L\downarrow$ and $L\uparrow$

## 5. ESTIMATION OF THE HELIOTHERMIC INDEXES FOR THE E.C. COUNTRIES

Approximate distributions of the indexes $t_h$ and $t'_h$ relating to uncovered and glass-shielded systems respectively have been computed for the countries of the European Community.

The required data on global irradiances and durations of sunshine were taken from the E.C. European Radiation Atlas, while the air temperature data were inferred from the WMO Climatic Atlas of Europe.

The downward and upward components of the net irradiance have been computed by means of the relationships proposed by R. DOGNIAUX in "Estimation des composantes du bilan énergétique", I.R.M. Note préliminaire n° 17; 1981.

The values of the indexes, either calculated for the stations tabulated in the Radiation Atlas or interpolated for points evenly distributed at intervals of one degree in longitude and latitude, are given in appended tables.

Tentative corresponding maps have been drawn also at different scales :
- general maps covering the E.C. area, with indexes bearing on annual average values of the meteorological variables $t_a$, G and $S/S_o$
- detailed map for France with indexes bearing on annual average.

6. CONCLUSION

The purpose followed in presenting, as examples, two types of indexes particularized according to the heliosystems used for solar energy captation, is to provide to *the solar users community* with some simple indexes allowing to disclose and select on a large geographical scale (mesoscale) the area most suited for a specific application, and also to determine, for these homogeneous regions, the average meteorological parameters – and their relevant distributions – essential for the calculations of the efficiency and rentability of solar equipments.

On the other part, such evaluations of the distributions of the meteorological parameters can help *the meteorologist* in his attempt to classify, in terms of heliothermic characteristics, sites for which the variations of the climatic elements, taken individually or together, can be studied statistically in form of reference periods (reference years for example).

A selection of various indexes judiciously established to define radiation climate zones according to different fields of solar energy application (tower power plant, photovoltaïc power generation, photobiological and photochemical processes, biomass, solar energy in dwellings, agriculture and industry,...) could contribute to gain a better knowledge of the terrestrial solar potential and of its geographical distribution.

We trust that the Commission of the European Communities through its Directorate General XII for Research, Science and Education will initiate some concrete recommendations with respect to this proposal.

# ENVIRONMENTAL PARAMETERS AND HELIOTHERMIC INDEXES FOR SOME

## STATIONS OF THE EUROPEAN COMMUNITY

| STATIONS | $\phi$ | $\lambda$ | H | $t_a$ | G | $\sigma$ | $L{\downarrow}$ | $L{\updownarrow}$ | $h'_c$ | $t_h$ | $t'_h$ |
|---|---|---|---|---|---|---|---|---|---|---|---|
| IR BIRR | 53.05N | 07.54W | 72 | 9.2 | 115 | .28 | 317 | -35 | 3.83 | 16.1 | 18.9 |
| IR KILKENNY | 52.40N | 07.16W | 64 | 9.5 | 122 | .29 | 328 | -37 | 3.93 | 16.9 | 19.6 |
| IR VALENTIA | 51.56N | 10.15W | 20 | 10.0 | 118 | .28 | 310 | -37 | 3.84 | 17.0 | 20.0 |
| UK LERWICK | 60.08N | 01.11W | 82 | 7.5 | 90 | .22 | 305 | -26 | 3.46 | 12.6 | 15.2 |
| UK ESKDALEMUIR | 55.19N | 03.12W | 242 | 7.7 | 95 | .25 | 304 | -28 | 3.54 | 13.1 | 15.8 |
| UK ALDERGROVE | 54.39N | 06.13W | 68 | 7.5 | 105 | .28 | 302 | -31 | 3.74 | 13.8 | 16.5 |
| UK CAWOOD | 53.50N | 01.08W | 6 | 7.8 | 105 | .27 | 304 | -31 | 3.73 | 14.1 | 16.8 |
| UK ABERPORTH | 52.08N | 04.34W | 133 | 10.0 | 119 | .32 | 308 | -39 | 3.83 | 17.0 | 20.0 |
| UK LONDON | 51.31N | 00.07W | 77 | 10.1 | 102 | .32 | 309 | -36 | 3.47 | 15.4 | 18.5 |
| UK KEW | 51.28N | 00.19W | 5 | 10.2 | 108 | .33 | 309 | -37 | 3.59 | 16.0 | 19.1 |
| UK BRACKNELL | 51.23N | 00.47W | 73 | 10.0 | 111 | .31 | 309 | -37 | 3.68 | 16.2 | 19.3 |
| F TRAPPES | 48.46N | 02.01E | 168 | 9.7 | 129 | .37 | 306 | -42 | 3.96 | 17.5 | 20.5 |
| F NANCY | 48.41N | 06.13E | 217 | 9.5 | 129 | .35 | 306 | -41 | 4.01 | 17.4 | 20.4 |
| F MACON | 46.18N | 04.48E | 217 | 10.0 | 143 | .42 | 305 | -46 | 4.16 | 18.8 | 22.4 |
| F LIMOGES | 45.52N | 01.11E | 402 | 11.0 | 144 | .42 | 308 | -43 | 4.12 | 19.7 | 23.0 |
| F MILLAU | 44.07N | 03.01E | 720 | 7.8 | 163 | .47 | 296 | -49 | 4.50 | 18.6 | 21.7 |
| F CAPPENTRAS | 44.05N | 05.03E | 105 | 13.0 | 179 | .62 | 308 | -53 | 4.42 | 23.7 | 27.7 |
| F NICE | 43.39N | 07.12E | 10 | 10.0 | 177 | .61 | 297 | -59 | 4.50 | 21.1 | 24.7 |
| B OOSTENDE | 51.12N | 02.52E | 4 | 10.1 | 123 | .34 | 308 | -40 | 3.88 | 17.3 | 20.4 |
| B MELLE | 50.59N | 03.50E | 15 | 10.0 | 108 | .33 | 308 | -37 | 3.60 | 15.9 | 18.9 |
| B UCCLE | 50.48N | 04.21E | 105 | 10.0 | 111 | .32 | 308 | -37 | 3.67 | 16.2 | 19.2 |
| B SAINT-HUBERT | 50.02N | 05.24E | 563 | 7.7 | 117 | .32 | 301 | -35 | 3.92 | 15.2 | 17.7 |
| NL GRONINGEN-EELDE | 53.08N | 06.55E | 5 | 9.0 | 116 | .30 | 306 | -36 | 3.85 | 16.0 | 18.8 |
| NL DEN HELDER | 52.55N | 04.47E | 2 | 9.0 | 124 | .33 | 325 | -38 | 3.77 | 16.6 | 19.5 |
| NL DE BILT | 52.06N | 05.11E | 40 | 9.7 | 113 | .31 | 338 | -37 | 3.74 | 16.2 | 19.2 |
| NL VLISSINGEN | 51.27N | 03.36E | 22 | 10.0 | 120 | .32 | 328 | -39 | 3.85 | 17.0 | 20.1 |
| NL MAASTRICHT-BEEK | 50.55N | 05.46E | 116 | 9.3 | 115 | .30 | 327 | -36 | 3.81 | 16.1 | 19.0 |
| D NORDERNEY | 53.43N | 07.09E | 13 | 9.0 | 125 | .35 | 304 | -39 | 3.97 | 16.6 | 19.5 |
| D HAMBURG | 53.38N | 10.00E | 14 | 7.7 | 112 | .33 | 331 | -34 | 3.82 | 14.5 | 17.2 |
| D BERLIN(WEST) | 52.23N | 13.18E | 51 | 8.0 | 117 | .33 | 332 | -36 | 3.90 | 15.2 | 17.9 |
| D BRAUNSCHWEIG | 52.18N | 10.27E | 81 | 7.6 | 110 | .32 | 301 | -33 | 3.80 | 14.2 | 16.9 |
| D BRAUNLAGE | 51.43N | 10.37E | 601 | 7.6 | 112 | .32 | 301 | -34 | 3.84 | 14.4 | 17.1 |
| D WUERZBURG | 49.48N | 09.54E | 259 | 7.5 | 127 | .33 | 300 | -36 | 4.10 | 15.7 | 18.4 |
| D TRIER | 49.45N | 06.40E | 265 | 9.5 | 122 | .32 | 307 | -38 | 3.91 | 16.8 | 19.6 |
| D WETTENSTEPHAN | 48.24N | 11.44E | 467 | 4.5 | 134 | .35 | 291 | -34 | 4.34 | 13.7 | 16.2 |
| D HOHENPEISSENBERG | 47.48N | 11.01E | 975 | 4.5 | 138 | .41 | 238 | -37 | 4.35 | 14.0 | 16.4 |
| DK TYLSTRUP | 57.11N | 09.57E | 13 | 7.7 | 115 | .37 | 297 | -35 | 3.89 | 14.7 | 16.7 |
| DK JOUM | 56.19N | 10.05E | 61 | 7.3 | 115 | .36 | 299 | -36 | 3.90 | 14.4 | 17.1 |
| DK KOBENHAVN | 55.41N | 12.36E | 20 | 7.5 | 117 | .36 | 299 | -36 | 3.91 | 14.7 | 17.4 |
| DK HOJBAKKEGARD | 55.40N | 12.19E | 30 | 7.3 | 113 | .35 | 290 | -35 | 3.94 | 14.6 | 17.3 |
| DK ASKOV | 55.23N | 09.07E | 64 | 7.8 | 114 | .34 | 301 | -35 | 3.85 | 14.7 | 17.4 |
| DK ARSLEV | 55.18N | 10.27E | 43 | 7.8 | 117 | .35 | 301 | -36 | 3.90 | 14.9 | 17.7 |
| DK NAESGARD | 54.52N | 12.07E | 17 | 7.8 | 117 | .35 | 301 | -36 | 3.90 | 14.9 | 17.7 |
| I BOLZANO | 46.23N | 11.20E | 241 | 10.0 | 155 | .43 | 304 | -49 | 4.32 | 19.3 | 23.1 |
| I TRIESTE | 45.39N | 13.45E | 20 | 13.5 | 156 | .46 | 315 | -55 | 4.15 | 22.6 | 26.4 |
| I MILANO | 45.26N | 09.17E | 103 | 11.5 | 156 | .40 | 310 | -50 | 4.27 | 21.2 | 24.6 |
| I GENOVA | 44.25N | 08.51E | 3 | 14.0 | 167 | .49 | 315 | -58 | 4.26 | 23.9 | 27.5 |
| I ANCONA | 43.37N | 13.31E | 104 | 13.0 | 174 | .45 | 313 | -57 | 4.40 | 23.9 | 27.6 |
| I PIANOSA | 42.35N | 10.05E | 27 | 14.0 | 198 | .57 | 312 | -66 | 4.58 | 26.2 | 30.5 |
| I ROMA | 41.48N | 12.35E | 131 | 10.5 | 192 | .56 | 301 | -60 | 4.67 | 22.9 | 26.7 |
| I NAPOLI | 40.51N | 14.18E | 72 | 15.5 | 192 | .53 | 319 | -66 | 4.47 | 27.1 | 31.4 |
| I BRINDISI | 40.39N | 17.57E | 10 | 17.5 | 202 | .55 | 326 | -71 | 4.48 | 29.4 | 34.2 |
| I ALGHERO | 40.38N | 08.17E | 40 | 13.6 | 205 | .59 | 311 | -67 | 4.56 | 26.4 | 30.7 |
| I MESSINA | 38.12N | 15.33E | 57 | 10.0 | 203 | .56 | 299 | -62 | 4.82 | 27.1 | 27.2 |
| I TRAPANI | 37.55N | 12.30E | 14 | 17.5 | 219 | .61 | 324 | -75 | 4.63 | 30.7 | 35.6 |
| I PANTELLERIA | 36.45N | 11.58E | 170 | 20.0 | 212 | .57 | 335 | -76 | 4.47 | 32.2 | 37.4 |

## Symbols

$t_a$ = mean annual temperature in °C

$G$ = mean irradiance of global radiation

$\sigma$ = relative sunshine duration

$L\!\downarrow$ = mean irradiance of downward long wave radiation

$L\!\updownarrow$ = mean irradiance of net long wave radiation

$h'_c$ = external heat transfer coefficient for wind speed equal to zero

$t_h$ = heliothermic index for unglazed systems

$t'_h$ = heliothermic index for glass-shielded systems

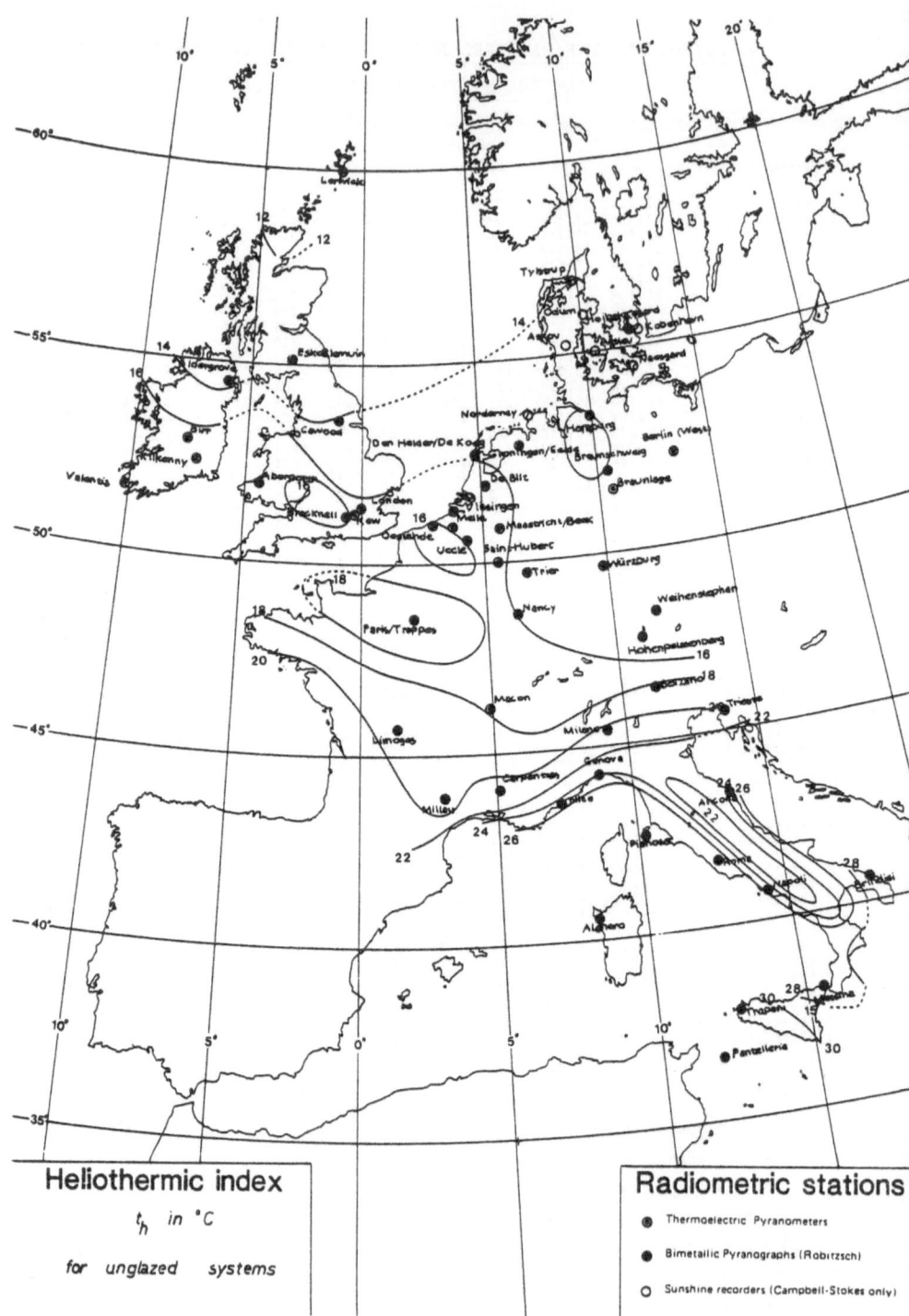

## Heliothermic index

$t_h$  in °C

*for  unglazed  systems*

## Radiometric stations

- ● Thermoelectric Pyranometers
- ● Bimetallic Pyranographs (Robitzsch)
- ○ Sunshine recorders (Campbell-Stokes only)

## Heliothermic index

$t'_h$ in °C

*for glass - shielded systems*

## Radiometric stations

● Thermoelectric Pyranometers

● Bimetallic Pyranographs (Robitzsch)

○ Sunshine recorders (Campbell-Stokes only)

F R A N C E

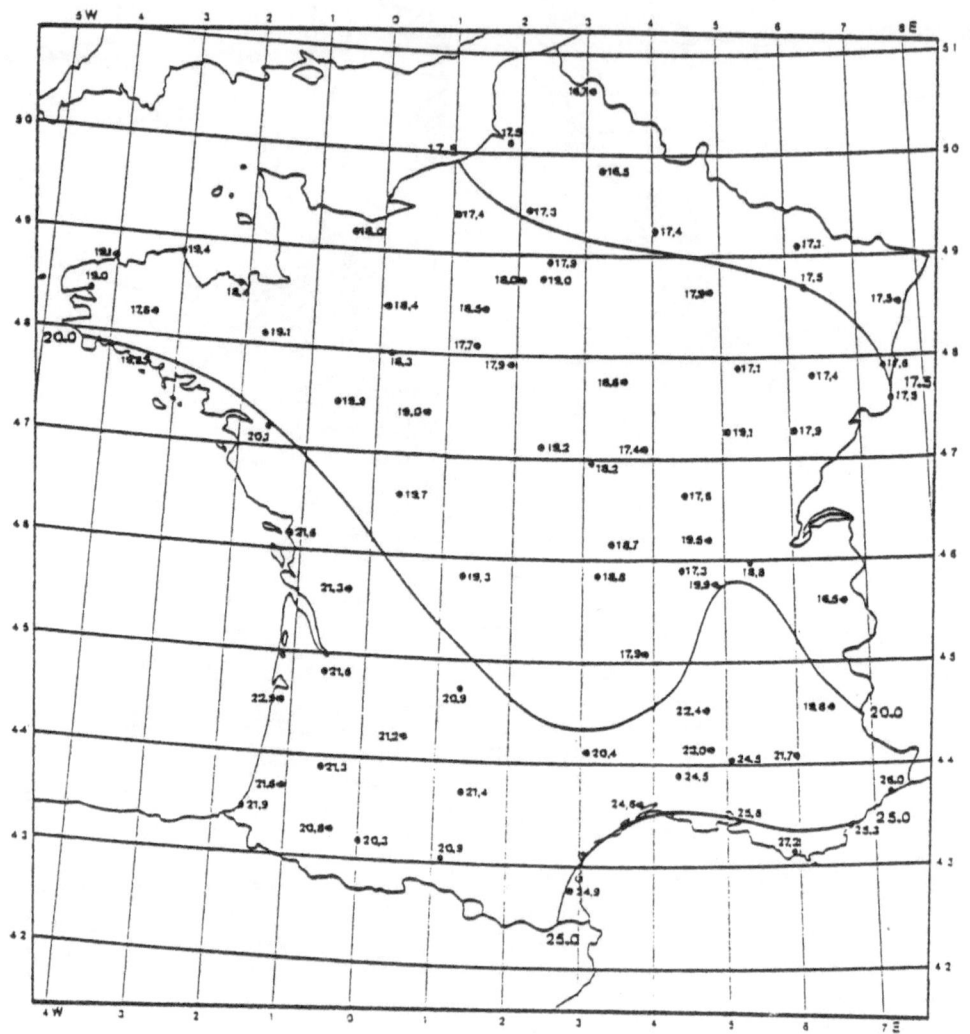

Heliometric index

Annual means

Unglazed systems

ACTION 3.2


## GLOBAL IRRADIANCE ON TILTED PLANES
## DIRECT IRRADIATION ON HORIZONTAL AND TILTED PLANES
## CALCULATION OF TURBIDITY FACTOR


Action leader's progress report
R. DOGNIAUX, Institut Royal Météorologique de Belgique


Reports of action participants :

- Classification of radiation sites in terms of different
  indices of atmospheric transparency

- Parameterization of radiation fluxes as function of solar
  elevation, cloudiness and turbidity

- Diffuse sky radiation on tilted surfaces

- Climatological values of solar irradiation on the horizontal and several inclined surfaces at De Bilt

- Development of methods - Analysis of solar radiation data
  of a high altitude station (1 580 m )

- A statistical analysis of the components of solar radiation on a vertical south-facing surface

- Development of a systematic method for the prediction of
  solar radiation on inclined planes

- Calculation method for the radiation data on inclined
  surfaces and calculated atlas data

- Empirical study of the angular distribution of sky
  radiance and of ground reflected radiation fluxes

- Estimation of hourly solar irradiation over the UK

ACTION 3.2. - GLOBAL IRRADIANCE ON TILTED PLANES
          - DIRECT IRRADIATION ON HORIZONTAL AND TILTED PLANES
          - CALCULATION OF TURBIDITY FACTOR


ACTION BUDGET (EC CONTRIBUTION) : 156 kUC


ACTION LEADER : R. DOGNIAUX
                INSTITUT ROYAL METEOROLOGIQUE DE BELGIQUE
                AVENUE CIRCULAIRE, 3
                B - 1180 BRUXELLES


PARTICIPANTS : - INSTITUT ROYAL METEOROLOGIQUE DE BELGIQUE
                 (CONTRACT N°ESF-003-B(G) : R. DOGNIAUX)
               - DEUTSCHER WETTERDIENST OBSERVATORIUM HAMBURG
                 (CONTRACT N°ESF-004-D(B) : F. KASTEN)
               - METEOROLOGIE NATIONALE - PARIS
                 (CONTRACT N°ESF-005-F(G) : J.A. BEDEL)
               - KONINKLIJK METEOROLOGISCH INSTITUUT
                 (CONTRACT N°ESF-006-NL(B) : W.H. SLOB)
                      SUBCONTRACTOR :
                   INSTITUT OF APPLIED PHYSICS
                   TNO-TH (TPD): G.J. VAN DEN BRINK AND J.K.M. VERDONSCHOT
               - CNRS LABORATOIRE D'ENERGETIQUE SOLAIRE
                 (CONTRACT N°ESF-009-F(G) : J.F. TRICAUD)
               - IRISH METEOROLOGICAL SERVICE
                 (CONTRACT N°ESF-018-EIR(G) : E.J. MURPHY
               - DEPT. OF BUILDING SCIENCE - UNIVERSITY OF SHEFFIELD
                 (CONTRACT N°ESF-021-UK(H) : J.K. PAGE)
               - LMT LICHTMESSTECHNIK GmbH - BERLIN
                 (CONTRACT N°ESF-020-D(B)/ESF-035-D(B) : J. KROCHMANN)
                      SUBCONTRACTOR :
                   SWISS METEOROLOGICAL INSTITUTE
                   (CONTRACT N°ESF-020-80-D) : P. VALKO
               - METEOROLOGICAL OFFICE - U.K.
                 (CONTRACT N°ESF-024-80-U.K.(H) : R. RAWLINS)

TASKS         : - CALCULATION OF GLOBAL IRRADIATION ON PLANES ORIENTED
                  SOUTH AT THE ANGLES OF LATITUDE, 30°, 60° and 90°,
                  EAST AND WEST, SOUTH-EAST AND SOUTH-WEST VERTICALS
                  (MONTHLY AND ANNUAL MEANS OF DAILY VALUES)
                - CALCULATION OF DIRECT IRRADIATION ON HORIZONTAL PLANES
                  (HOURLY VALUES)
                - CALCULATION OF THE TURBIDITY FACTOR

# 1. SUMMARY

Action 3.2., part of Project F of the Solar Energy R. & D.
Programme, deals with the estimation of the hourly and daily irradiations
on tilted planes based on a E.C. agreed method selected during the first
programme on solar radiation (1975-1979) and refined on the basis of an
improved statistical analysis of actual measurements of radiation on
inclined surfaces.

The E.C. Solar Radiation Prediction Method, - a description of
which and of the recommended computational programme suitable for its
application on a main frame, mini or micro computer will be published
very soon in the Series F of Solar Energy R. & D. in the European Community
Collection - requires, as input data, either the global irradiance or irradiation on a horizontal surface, or the duration of sunshine associated
to a parameter of turbidity characterizing the clarity of the atmosphere
at the site.

Consequently, in the framework of the action 3.2., some studies
were carried out to investigate the validity of the parameterization of
radiation fluxes in terms of cloudiness, turbidity, sky radiance distribution, ground reflectance (albedo), altitude of the station, ...

Different indices of atmospheric turbidity were submitted as
possible means of classification of sites for solar energy applications.

The final product of the action 3.2. will be the publication in
the course of 1983 of an atlas of isopyr curves for global radiation on
tilted planes including tables giving the ratios of inclined surface
irradiances to the horizontal ones for clear sky conditions in terms of
solar altitude, slope and azimuth of the surfaces, Linke turbidity factor
and tables giving the monthly means of daily global irradiations for
overcast sky conditions for different latitudes together with the slope
multiplication factor for their conversion for different tilted planes.

## 2. GENERAL TASK OF THE ACTION

For the convenience of the statement, the general description of the task will follow the scheme adopted in the previous action leader's reports prepared for the former contractors meeting. References will be made to the strategy paper published in volume 1 of Series F : concrete results and technical notes of the participants will be included as appendices and corresponding references are shown in brackets in the text.

ITEM 1. : APPLICATION OF THE E.C. SOLAR RADIATION PREDICTION METHOD FOR
         THE EVALUATION OF IRRADIATIONS ON TILTED PLANES ; PRODUCTION
         OF DATA FOR THE ATLAS.
         by J.K. PAGE : SHEF. UNIV.(G.B.) and J. KROCHMANN : LICHTMES.(G.)

The scientific basis of the E.C. solar radiation prediction method has been developped conjointly by J.K. Page and R.J. Flynn at the University of Sheffield and by J. Krochmann, R. Rattunde and S. Aydinli at the L.M.T. Lichtmesstechnik GmbH in Berlin. The model has been developped to provide monthly mean climatological values of the short wave solar irradiance and also the daily global irradiation to be calculated for slopes of any orientation for any site within the E.E.C. area. The model computes the monthly mean hourly values of the irradiance from the clear sky components and overcast sky components using a specially derived interpolative technique. It therefore provides simultaneously information about three conditions, clear day, overcast day and monthly means.
Two basis inputs are used : observed monthly mean daily hours of bright sunshine and an appropriate monthly mean value of the Linke turbidity factor selected to model the representative atmospheric clarity of the actual site for clear days.

The prediction models are irradiance models. They predict the instantaneous fluxes of the components of solar energy onto a surface per unit area. Values of irradiation are then obtained by the subsequent numerical integration of a series of irradiance predictions made at suitable time intervals over the required period.

The prediction method has been extensively checked against actual slope and horizontal surface radiation observations in the E.C. AREA. An improved instrumental situation in the Southern regions of the Alps (Italy) and Greece should yield new observations which will permit the method to be checked for these areas.

Concerning the modelling of irradiation for the average conditions of sunshine, the original Sheffield model and the original Berlin model used different techniques. Absence of observations of hourly sunshine produced at first some difficulties in producing an improved model on an hourly basis. These difficulties were nevertheless overcome by introducing the concept of *relative sunshine probability*. This concept defined as the ratio of the actual monthly mean hourly relative sunshine duration at a given solar altitude to the predicted relative sunshine duration value when the sun is vertically overhead can be expressed by a pragmatic mathematical function deduced from observations in several European stations (fig.1).

From this function, one then obtains the estimated mean monthly hourly relative sunshine durations for any given hour in a particular month. These values are then used as inputs in the E.C. computational method to split the radiation associated with each hour into clear and overcast fractions which are then treated separately respectively for the direct and diffuse components of the global radiation.

The concept of relative sunshine probability is an interesting one, but it leads to less accurate estimates at the hourly level than the use of daily relative sunshine duration. The present relative duration of sunshine method underestimates the direct beam contribution at low solar altitudes and overestimates the direct beam at high solar altitudes. Better results are achieved using a 4° solar altitude relative daily sunshine duration model with some slight correction for solar altitudes below about 10°. Other improvements include using a variable turbidity model which reflects the actual effects of solar altitude on Linke turbidity.

A Guide manual for practical utilization of the method including tables of predicted radiation data on inclined surface for selected
sites in the E.C. region (see fig. 2) and general tables of monthly means
of daily sums of global radiation for clear sky conditions for monthly
means and overcast sky conditions is now finalised for publication in the
Series F of the solar energy R. and D. collection (see for examples fig.
3 and 4).

Separates tables for the stations selected for the first atlas
will be published in a second volume, an example of such tables is given
in figure 5.

Another final product of this part of action 3.2. will be the
publication in the course of 1983 of maps of isopyrs of global radiation
on tilted planes on a monthly and annual basis.

ITEM 2. : IMPROVEMENT OF METHODS OF CALCULATION OF IRRADIATION ON AN
HOURLY BASIS INCLUDING THE ESTIMATION OF THE DIFFERENT
COMPONENTS.

Besides the studies reported in the preceding item, several
methods for the separation of the direct and diffuse components of
irradiation have been validated by other way (K N M I - NL; INS. OF APP.
PHYS. NL; MET OFFICE-UK; MET NAT - F).

2.1. CLIMATOLOGICAL VALUES OF SOLAR IRRADIATION ON THE HORIZONTAL AND
SEVERAL INCLINED SURFACES AT DE BILT (NL) by W.H. SLOB

For the horizontal surface climatological monthly means of daily
global, direct and diffuse irradiations were calculated from the available
data (80 years of sunshine duration, 20 years of global solar radiation
and 10 years of direct irradiation).

Monthly regression equations were calculated for the global, the
direct and the diffuse daily irradiation as a function of relative
sunshine. Then the average monthly relative sunshine calculated over the
80 years was put into these equations to obtain climatological values on
the horizontal surface.

The climatological direct and diffuse irradiation on the horizontal were the input parameters for the climatological values on the inclined surfaces. Assuming that the sunshine was evenly distributed over the day and that during sunshine the Linke turbidity had a constant monthly value, a multification factor could be calculated for any inclined surface. With this factor which is the ratio of the direct irradiation on the inclined and the direct irradiation on the horizontal surface, the climatological value of the direct on the inclined surfaces was calculated.

For the diffuse component, one first calculated the diffuse for an isotropic sky and then multiplied this value with an experimental factor which accounts for the anisotropy of the sky.

This factor was determined from the measurements on inclined surfaces in Cabauw and it depends on the orientation, the month of the year, and the relative sunshine duration.

For 11 different orientations, namely East 90°, East 45°, South-East 45°, South 90°, South 67.5°, South 45°, South 22.5°, South-East 45°, West 45° and West 90° the climatological monthly means of daily global irradiation were calculated.

## 2.2. CALCULATION OF GLOBAL IRRADIANCE ON INCLINED SURFACES

J.K.M. VERDONSCHOT AND G.J. VAN den BRINK - DELFT (NL)

Many empirical relations exist for calculating the global irradiance on an inclined surface depending on the availability of the components of radiation on an horizontal surface used as input data. The relative accuracy depends on the available climatological parameters.

The criterium for applying a certain relation depends on the available climatological parameters and the required information. In their method, the authors consider those situations that at least the global irradiance on the horizontal plane is available. In that case, empirical relations are needed; these are :

- the separation of solar irradiance into direct and diffuse components.
- the calculation of the diffuse irradiance on an inclined surface from the horizontal diffuse irradiance.

Several methods for the separation of the direct and diffuse irradiance components of the global radiation were validated. It appears that the clearness index gives comparable results as the measured duration of sunshine.

For the calculation of diffuse irradiance on inclined surfaces, some relations for different conditions of sky cloudiness were tested.

The results of the different described procedures for calculating the global irradiance on an inclined surface from the available climatological parameters were compared and analysed. For each calculation step, the most suited relation was applied. One may conclude that the yearly irradiance on a tilted plane can be calculated within a few percents. However, the hourly standard deviation is in the order of 10-20% dependant on the available climatological parameters.

## 2.3. ESTIMATION OF HOURLY SOLAR IRRADIATION OVER THE UNITED KINGDOM
by F. RAWLINS (MET. OFFICE - U.K.)

The aim is to provide hourly estimates of the components of solar irradiation -- global, diffuse and direct -- on horizontal surfaces averaged over a month (or longer period) from records of bright sunshine duration.A method was devised to determine global irradiation on an hourly basis using coefficients from the (daily) Angström equation and gave improved results compared to a method which ignored hourly sunshine statistics, although long term averages were simular. Little progress was achieved in the estimation of hourly diffuse irradiation for different geographical sites from sunshine records alone. Linke turbidities derived from hourly direct irradiation distributions at Kew and Bracknell were compared with those of actual measurements and were found to give similar seasonal variations. The ratios of average to 'clear' direct irradiations (calculated from the derived turbidities) were presented as a function of the bright sunshine fraction which then allows the average hourly direct irradiation to be determined from sunshine statistics providing that a representative turbidity is known. Maps of global irradiation could be prepared for regions where there are numerous sunshine recording stations but maps of direct and hence diffuse, irradiation require information concerning mean turbidities. The extension of schemes of this type, involving simple regression from sunshine fractions, is demeed to be an unpromising approach to the investigation of irradiation on vertical and inclined surfaces.

## 2.4. DIFFUSE SKY RADIATION ON TILTED SURFACES - J.A. BEDEL, J. JAN AND V. PERARNAUD (METEO NAT. F.).

The analysis of direct and global radiation on tilted surfaces for Trappes and Carpentras integrated on a period of 6 minutes has shown that the splitting up of the diffuse radiation from the sky into 2 components (DISO : isotropic + DIR : direct) was not allowable. Consequently a third component DZ is introduced (called zenithal) which can be either positive or negative. It is negative, in particular, for high values of the atmospheric turbidity.

Then the validity of the model was estimated with actual data for Trappes and Carpentras by computation of cumulative frequency curves. This method also allows to estimate the validity of the model based on a isotropic diffuse sky radiation. At last, statistic tables of the daily global irradiation on slopes were produced using the proposed method of calculation of diffuse sky radiation validated from actual global and direct irradiations measured at Trappes and Carpentras.

## 2.5. A STATISTICAL ANALYSIS OF THE COMPONENTS OF SOLAR RADIATION ON A VERTICAL SOUTH - FACING SURFACE by E.J. MURPHY (IRISH MET. OFFICE).

For the months of June and December 1981, the measured hourly values of the diffuse and ground reflected radiation components on a vertical South-facing surface at Valentia Observatory are compared with the theoretically expected values and are analysed with respect to solar elevation (E) and D/G (as representative of sunshine duration).

An asymmetry between morning and afternoon values was found for both components. The asymmetry in the diffuse component may be explained by the fact that a range of hills at a distance of 1.5 km South-East of the Observatory has a screening effect on the diffuse

radiation. A possible explanation for the asymmetry in the ground
reflected component is the fact that the ground to the South-East of
the pyranometers slopes upwards while it is level to the South-West.

Significant linear equations were found to fit the data for
June.

It was not possible to derive reliable equations for December
as there are not sufficient ranges of elevation available for analysis.
Data for an 18-month period will be included in the analysis for the
final report and any seasonal variation should become apparent.

ITEM 3. : PARAMETERIZATION OF RADIATION FLUX IN TERMS OF OTHER
METEOROLOGICAL PARAMETERS.

3.1     PARAMETERIZATION OF RADIATION FLUXES AS FUNCTION OF SOLAR
ELEVATION, CLOUDINESS AND TURBIDITY - F. KASTEN, H.J. GOLCHERT
and M. STOLLEY (DEUTSCHER WETTERDIENST, HAMBURG, D)

3.1.1. Global radiation dependent on total cloud amount

The parameterization formula for global radiation G as
function of total cloud amount N found earlier for Hamburg :

$$G(N)/G(o) = 1 - a(N/8)^b \tag{1}$$

is confirmed to be valid for 7 other stations by adoption of appropriated
sets of constants a and b.

3.1.2. Global radiation from cloudiness sky as function of turbidity

A simple parameterization formula is found for global
irradiance from cloudless sky, G(o), as function of Linke turbidity factor
$T_L$ :

$$G(o)/kW.m^{-2} = 0.83 \ \bar{I}_o \sin\gamma.\exp(-0.026T_L/\sin\gamma) \tag{2}$$

where $\bar{I}_o$ = 1367 $kW.m^{-2}$ is the solar constant and $\gamma$ , the solar
elevation.

Formulae (1) and (2) combined describe global irradiance
G as function of solar elevation $\gamma$, total cloud amount N and turbidity
factor $T_L$.

### 3.1.3. Dependence of atmospheric radiation on temperature and water vapor content

Eight different parameterization formulae for atmospheric radiation from cloudless sky as function of temperature and water vapor pressure are shown to fairly agree with 12 years records of hourly measurements in Hamburg. Further analysis including comparison to the results of rigorous radiative transfer calculations is required.

### 3.2. PARAMETERIZATION OF GLOBAL RADIATION AS FUNCTION OF SUNSHINE DURATION R. DOGNIAUX (I.R.M. - B.)

Systematic studies using the material collected from European countries for which the extension of the atlas is anticipated were achieved for correlating the global radiation with the relative duration of sunshine and for determining the variations of their coefficients on a geographical scale as function of the radiation climatic area.

The sum A + B of the coefficients of the well-known Ångström regression equation can be regarded as a parameter characterizing the average monthly or annual conditions of the transparency or clearness of the atmosphere for clear sky at a given station.

Let we call this quantity the *atmospheric transparency index* (ATI).

This index can itself be expressed as a sum of two terms :

$$ATI = \overline{(A + B)} + R$$

in which the first term A + B represents the variation of ATI for standardized conditions of atmospheric transparency termed "clear atmosphere". Such values follow from a first approximation of ATI by a linear regression equation in terms of the latitude (see p. 106). The second term R, which appears as the deviation from the standardized conditions, results both from errors of measurements and from specific local atmospheric turbidity conditions. This residual term can consequently be corrolated to the Linke turbidity factor (see p. 107).

The hereunder table summarizes the different indices allowing to characterize a site; by an appropriate choice of them, it is possible to estimate with a reasonable accuracy the mean daily sum of global irradiation of a site.

ITEM 4 :   STUDY OF THE ANGULAR DISTRIBUTION OF SKY RADIANCE AND GROUND
           REFLECTED RADIATION FLUXES.
           MEASUREMENTS OF IRRADIANCES AND RADIANCES OF THE UPPER (SKY)
           AND OF THE LOWER (GROUND) HEMISPHERES - P. VALKO

Complementary to methods using totals or averages of irradiation
over long time intervals, this project envisages an opposite procedure :
- measuring the instantaneous angular components of both radiance
  and irradiance over the whole 4π sr sphere;
- parameterizing the relationships between radiance and irradiance on the one hand and angular, atmospheric and environment specific quantities on the other;
- integrating the established relationships over hours and days
  to compute the irradiation of any exposed surface.

Photographs showing the instrumentation used for the study and
figures representing irradiance measurements and radiance measurements
with silicon - diode - scans have been collected and analyzed.

Two series of diagrams were prepared giving for different solar
hights and turbidity gradations the ratios between diffuse irradiance
on an inclined surface to that of a horizontal surface as function respectively of the solar surface azimuth angle and the slope of the receiving
surface on the one hand and as function of these two variables plus the
turbidity and the solar height on the other hand.

Series of polar diagrams were also analyzed with respect to the
distribution of radiance over the sky for three solar height groups
11-17°; 31-32°; 49-55° with, into each group, different ranges of turbidity.

All radiance distributions cover the spectral range of 300 -
1100 nm, the high spectral sensitivity of the silicon diode in the red
and near infrared regions having been rectified by using a blue filter.

Some important results have been obtained from the study of the
important material so far achieved.

| Class of turbidity according [1] | Atmospheric clearness | Atmospheric parameters β | Atmospheric parameters w | Site characteristics | Average linke turbidity factor $\overline{T}$ according [2] | Correction index R | Atmospheric transparency index ATI for $\phi = 45°$ (x) |
|---|---|---|---|---|---|---|---|
| 1 | Very clear | 0.05 | <2 | Dry air Mountain site Polar & desert Climates | <2.5 | 0.05 ± 0.02 | >0.79 |
| 2 | Clear | 0.05 0.10 | <7 <2.5 | Rural site Temperate climates | 3.25 ± 0.75 | 0.00 ± 0.02 | 0.775 ± 0.015 |
| 3 | Fairly Polluted | 0.10 0.20 | <7 <0.5 | Urban site Temperate climates | 4.5 ± 0.5 | -(0.04 ± 0.02) | 0.74 ± 0.02 |
| 4 | Strongly Polluted | 0.20 | >1 | Industrial site Tropical climates | >5.0 | <-0.05 | <0.72 |

(x) For other latitudes, ATI is given by equation (5)

### 3.3 ANALYSIS OF SOLAR RADIATION DATA ON A HIGH ALTITUDE STATION (1580m)

J.K. TRICAUD (LAB. D'ENERGETIQUE SOLAIRE ODEILLO F.)

All the equipments for the records of data on tenths of hour values basis are operative. Data for the period 1979-1982 are recorded on tapes and disks: they comprise global and diffuse irradiations on horizontal, vertical surfaces and on a South facing surface tilted according to the latitude of the station.The daily sums are analysed, for three selected months of 1982, in function of the duration of sunshine. Linear regressions equations have been carried out for the ratios of the different components to the global horizontal irradiation as functions of the duration of sunshine and of the sun's declination. The variations of the ratios of the ground reflected irradiation on a South facing vertical surface are also investigated.

It is contemplate, for the final report, to study the variations of the data as function of the Linke turbidity factor, of the declination of the sun and of the duration of sunshine.

The dependence of the diffuse irradiation with solar altitude will be also investigated.

Work and results may be summarized as follows :

1. Sky radiance was measured - using a system of four scanning devices - at 121 points of the sky dome within about two minutes in a great number of cases. The measuring campaigns were made 1979-1981 at four different sites, representing different climatic conditions.

2. Measured radiances have been normalized to the radiance of the zenith. Each data set composed by the 121 quasi-instantaneous relative values has been plotted against the polar coordinates (elevation above the horizon and azimuth relative to the sun's position) of the scanned points.

3. Using 182 polar diagrams as a representative selection of 551 (clear sky) cases, site-independent functional relationship could be found between normalized radiance on the one hand, and solar height, air turbidity, as well as the angular coordinates of the respective points in the sky on the other.

4. Air turbidity could be considered as function of the ratio of diffuse to global irradiance on horizontal surface and the solar height angle.

5. For each set of measurements radiance has been numerically integrated over the sky hemisphere and related to the diffuse irradiance on the horizontal surface. The latter was simultaneously measured using appropriate instruments. Thus, for each set of measurements an adjusting factor could be determined individually for equalizing the effect of different sensor chracteristics of the different devices.

6. Zenith radiance "calibrated" this way and related to the full view angle of the sky dome, as seen from the horizontal surface, has been used to find the functional link to horizontal surface diffuse irradiance, solar height and atmospheric turbidity. With other words, measured diffuse irradiances were compared with those generated by a sky having all over the same radiance as that measured in the zenith. The result is a highly correlated relationship to the solar height and - as expected - independency on turbidity.

The significance of these results is obvious :

(i) The angular distribution of clear sky radiance can, both absolutely and relatively, be determined in <u>real time</u> at any place where global and diffuse irradiance on the horizontal surface is known.

(ii) The sky component of the diffuse irradiance on a surface of arbitrary tilt and orientation can be determined in <u>real time</u> - by numerical integration of sky radiance over the sky hemisphere respective to that surface - at any place where global and diffuse irradiance on the horizontal surface is known.

First comparisons of the radiance distributions for overcast skies as depending on type of the cloud cover have also been made. The denser the clouds, the more the maximum of radiance is shifted towards the zenith.

The results presented for clear skies will be formulated analytically and compared with those gained by other authors.

At present the instruments scan sky radiance at 121 fixed points and in 77 pyranometers positions respectively. The development of the hardware and software of a new control program for stearing the devices in diffe- rent positions is almost completed and will be operational till the end of 1982. This new system will allow a flexible choice of the scanning sequence, thus measuring both radiance and irradiance in an angular grid of arbitrary configuration. Measurements over regions of the sky near the horizon and nearer to the sun will be the most important improvements. The multiparameter relationship for sky radiance will be extended for cloudy and overcast cases by using cloud parameters as determined by the digitalized information content of the fish-eye-photographs. The influence of type and amount of cloudiness on the global and diffuse irradiances upon vertical surfaces are parallely studied by using data from Locarno- Monti.

Proper separation of sky- and reflected components of the total diffuse irradiance on surfaces of arbitrary tilt and orientation will be achieved by integrating numerically the measured radiance distribution functions and subtracting the result from simultaneously measured irradiances. These computations will involve radiance and irradiance measurements over both the upper and the lower (10 m mast-) hemisphere.

SHORT DESCRIPTION OF INDIVIDUAL CONTRACTOR'S WORK

See Solar Radiation Data; Series F; Vol. 1; Solar Energy
R. and D. in the E.C. p. 38.

EXTENSIVE REPORTS AVAILABLE FROM THE AUTHORS

- Page J.K. : European Solar Radiation Atlas : solar radiation prediction
          methods and data for inclined surfaces.
- Krochmann J., Aydinli S., Rattunde R., :
          The final Berlin - Model on the availability of solar
          radiation and daylight.
- Verdonschot J.K.M., G.J. Van den Brink, W.H. Slob
          Climatology of solar irradiance on inclined surface IV.
          Part 1 : Measurements.
          Part 2 : Validation of calculation models.
          Part 3 : Climatology of solar irradiance on the horizontal
              and inclined surfaces in De Bilt.
- Dogniaux R. : Classification of radiation sites in terms of different
          indexes of atmospheric turbidity.
- Valko P.  : Empirical study of the angular distribution of sky radiance
          and of ground reflected radiation flux.

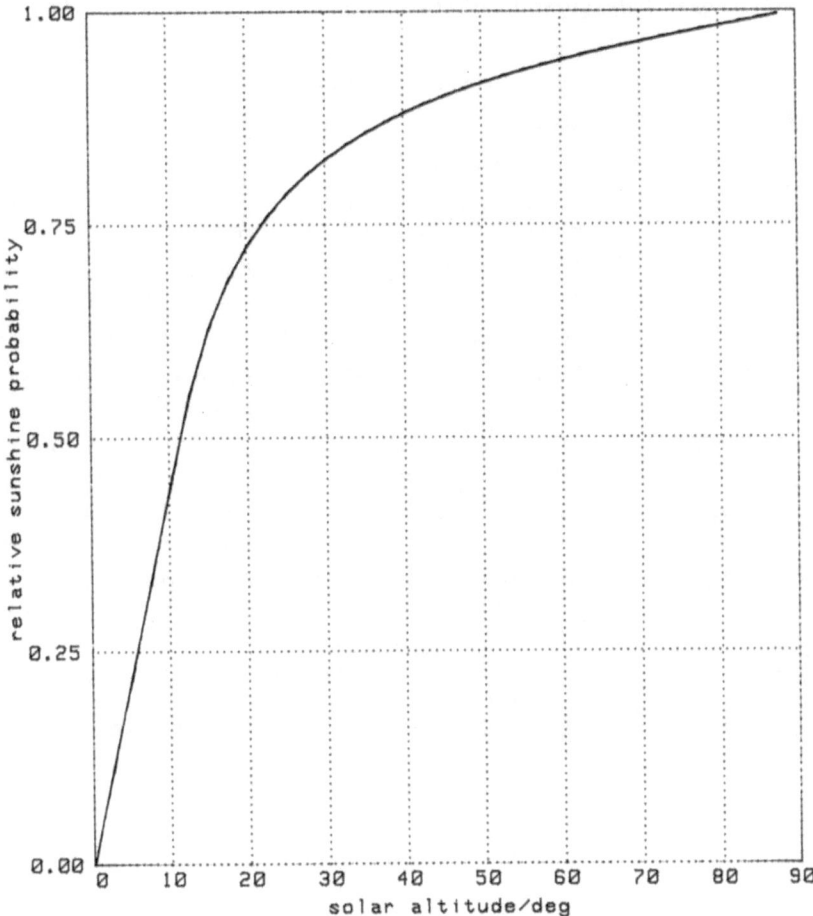

Fig. 1 - Relative sunshine probability as a function of solar altitude

Fig. 2 : Expected content of the guide manual for practical utilization of the EEC.

SOLAR RADIATION PREDICTION METHODS AND DATA FOR INCLINED SURFACES

1. PREFACE            To be prepared by the Commission

2. SYMBOLS & TERMINOLOGY

3. CHAPTER 1          Introduction

4. CHAPTER 2          Scientific Basis of the EEC Solar Radiation Prediction
                      Method for Inclined Surfaces

5. CHAPTER 3          Recommended computational Routines for Irradiance and
                      Irradiation Predictions for Inclined Surfaces

6. CHAPTER 4          Recommended Algorithms for Solar Energy Computations

7. CHAPTER 5          Tables of Predicted Radiation Data on Inclined
                      Surfaces for Selected Sites in the EEC Region

8. APPENDIX 1         The Determination of Turbidity Type from Ångström
                      Regression Equations

9. APPENDIX 2         Short Method for the Computation of Clear Sky Inclined
                      Surface to Horizontal Surface Diffuse Irradiance Ratio

10.APPENDIX 3         Example of Inclined Surface Calculations using the Desk
                      Computer Methodology with Standard Proforma to Structure
                      Calculations

11. APPENDIX 4        Bibliography of Publications Relating to CEC Contracts
                      with Institut fur Lichttechnik, Lichtmesstechnik Berlin
                      and the Department of Building Science, University of
                      Sheffield.

Department of Building Science,
University of Sheffield,
Western Bank,
SHEFFIELD        S10 2TN

Ratio of Inclined Surfaces Clear Sky Irradiance to Horizontal Irradiance in (%)

Linke Turbidity Factor : 2

| Angle of Slope | Solar altitude | Wall Solar Azimuth Angle | | | | | | | | | | | | |
|---|---|---|---|---|---|---|---|---|---|---|---|---|---|---|
| | | 0 | 15 | 30 | 45 | 60 | 75 | 90 | 105 | 120 | 135 | 150 | 165 | 180 |
| 15 | 5 | 130 | 129 | 126 | 120 | 114 | 107 | 100 | 94 | 93 | 94 | 94 | 94 | 94 |
| | 15 | 124 | 123 | 121 | 117 | 113 | 107 | 99 | 98 | 95 | 93 | 90 | 89 | 88 |
| | 30 | 119 | 119 | 117 | 115 | 111 | 106 | 99 | 98 | 97 | 95 | 94 | 93 | 92 |
| | 45 | 115 | 114 | 113 | 111 | 109 | 105 | 99 | 99 | 98 | 97 | 96 | 95 | 94 |
| | 60 | 111 | 110 | 110 | 108 | 106 | 103 | 99 | 99 | 99 | 99 | 98 | 97 | 97 |
| | 75 | 107 | 107 | 106 | 106 | 104 | 102 | 99 | 100 | 101 | 101 | 101 | 100 | 100 |
| | 90 | 105 | 105 | 105 | 105 | 105 | 105 | 105 | 105 | 105 | 105 | 105 | 105 | 105 |
| 30 | 5 | 154 | 152 | 146 | 136 | 123 | 110 | 98 | 92 | 95 | 96 | 96 | 96 | 97 |
| | 15 | 142 | 140 | 136 | 129 | 121 | 110 | 98 | 93 | 89 | 89 | 89 | 89 | 89 |
| | 30 | 133 | 132 | 129 | 124 | 118 | 109 | 97 | 94 | 91 | 87 | 84 | 82 | 81 |
| | 45 | 124 | 124 | 121 | 118 | 113 | 105 | 96 | 94 | 92 | 90 | 88 | 86 | 85 |
| | 60 | 116 | 116 | 114 | 112 | 108 | 102 | 96 | 95 | 95 | 94 | 92 | 91 | 90 |
| | 75 | 109 | 109 | 108 | 106 | 104 | 100 | 96 | 97 | 98 | 98 | 97 | 96 | 96 |
| | 90 | 104 | 104 | 104 | 104 | 104 | 104 | 104 | 104 | 104 | 104 | 104 | 104 | 104 |
| 45 | 5 | 171 | 168 | 159 | 145 | 127 | 110 | 96 | 91 | 94 | 94 | 95 | 95 | 96 |
| | 15 | 153 | 151 | 145 | 136 | 123 | 109 | 95 | 86 | 87 | 87 | 87 | 87 | 87 |
| | 30 | 141 | 139 | 135 | 128 | 119 | 107 | 93 | 87 | 81 | 79 | 79 | 78 | 78 |
| | 45 | 128 | 127 | 124 | 119 | 112 | 102 | 91 | 87 | 83 | 79 | 76 | 73 | 72 |
| | 60 | 116 | 115 | 113 | 110 | 105 | 98 | 90 | 88 | 86 | 84 | 82 | 80 | 79 |
| | 75 | 105 | 105 | 104 | 102 | 99 | 95 | 90 | 90 | 90 | 90 | 89 | 88 | 87 |
| | 90 | 99 | 99 | 99 | 99 | 99 | 99 | 99 | 99 | 99 | 99 | 99 | 99 | 99 |
| 60 | 5 | 179 | 175 | 164 | 147 | 125 | 107 | 93 | 88 | 89 | 89 | 90 | 91 | 91 |
| | 15 | 156 | 154 | 147 | 135 | 121 | 105 | 90 | 82 | 82 | 82 | 82 | 82 | 82 |
| | 30 | 141 | 139 | 134 | 126 | 115 | 101 | 87 | 77 | 74 | 74 | 73 | 73 | 72 |
| | 45 | 125 | 124 | 120 | 114 | 106 | 95 | 84 | 77 | 71 | 69 | 68 | 68 | 67 |
| | 60 | 110 | 109 | 107 | 102 | 97 | 90 | 82 | 78 | 75 | 71 | 68 | 66 | 65 |
| | 75 | 97 | 96 | 95 | 93 | 89 | 86 | 81 | 80 | 79 | 78 | 76 | 75 | 74 |
| | 90 | 89 | 89 | 89 | 89 | 89 | 89 | 89 | 89 | 89 | 89 | 89 | 89 | 89 |
| 75 | 5 | 178 | 174 | 161 | 142 | 118 | 99 | 87 | 83 | 82 | 81 | 82 | 83 | 84 |
| | 15 | 152 | 149 | 141 | 129 | 113 | 97 | 84 | 77 | 75 | 74 | 74 | 74 | 74 |
| | 30 | 134 | 132 | 127 | 117 | 105 | 92 | 79 | 70 | 67 | 66 | 66 | 65 | 65 |
| | 45 | 116 | 115 | 110 | 104 | 95 | 85 | 74 | 65 | 63 | 62 | 61 | 60 | 60 |
| | 60 | 99 | 98 | 95 | 90 | 85 | 78 | 72 | 66 | 61 | 60 | 60 | 59 | 58 |
| | 75 | 84 | 83 | 82 | 79 | 76 | 73 | 70 | 68 | 65 | 63 | 61 | 59 | 59 |
| | 90 | 74 | 74 | 74 | 74 | 74 | 74 | 74 | 74 | 74 | 74 | 74 | 74 | 74 |
| 90 | 5 | 167 | 163 | 151 | 130 | 106 | 88 | 80 | 75 | 72 | 71 | 72 | 72 | 73 |
| | 15 | 141 | 138 | 129 | 116 | 100 | 85 | 75 | 69 | 65 | 64 | 64 | 64 | 64 |
| | 30 | 121 | 119 | 113 | 104 | 92 | 79 | 69 | 63 | 58 | 57 | 56 | 56 | 55 |
| | 45 | 102 | 100 | 96 | 89 | 80 | 71 | 63 | 58 | 55 | 53 | 52 | 52 | 51 |
| | 60 | 83 | 82 | 79 | 75 | 69 | 64 | 60 | 57 | 54 | 52 | 52 | 51 | 50 |
| | 75 | 67 | 67 | 65 | 62 | 60 | 58 | 57 | 55 | 54 | 53 | 52 | 52 | 51 |
| | 90 | 57 | 57 | 57 | 57 | 57 | 57 | 57 | 57 | 57 | 57 | 57 | 57 | 57 |

Fig. 3 - Example of tables for the EEC Atlas on slopes (clear sky conditions) computed by S. Aydinly and R. Rattunde, from a new algorithm developed by J.K. Page.

Monthly means of daily sums of global radiation on overcast days for differrent lattitude in (W.h.M-2) (WRR)

| geog. Latitude | Jan | Feb | Mar | Apr | May | Jun | Jul | Aug | Sep | Oct | Nov | Dec | Ann |
|---|---|---|---|---|---|---|---|---|---|---|---|---|---|
| 30 | 798 | 965 | 1208 | 1411 | 1557 | 1614 | 1589 | 1475 | 1297 | 1063 | 849 | 747 | 1216 |
| 31 | 777 | 947 | 1194 | 1405 | 1558 | 1618 | 1592 | 1473 | 1286 | 1046 | 829 | 725 | 1205 |
| 32 | 756 | 929 | 1181 | 1398 | 1558 | 1622 | 1594 | 1470 | 1275 | 1028 | 809 | 703 | 1195 |
| 33 | 734 | 911 | 1167 | 1390 | 1559 | 1625 | 1596 | 1466 | 1263 | 1010 | 788 | 681 | 1184 |
| 34 | 713 | 892 | 1152 | 1382 | 1558 | 1627 | 1597 | 1463 | 1251 | 992 | 768 | 659 | 1173 |
| 35 | 691 | 873 | 1137 | 1374 | 1558 | 1630 | 1598 | 1458 | 1239 | 974 | 747 | 636 | 1161 |
| 36 | 669 | 854 | 1122 | 1365 | 1556 | 1631 | 1599 | 1454 | 1227 | 955 | 725 | 613 | 1149 |
| 37 | 647 | 835 | 1107 | 1356 | 1555 | 1632 | 1598 | 1448 | 1214 | 936 | 704 | 591 | 1137 |
| 38 | 625 | 815 | 1091 | 1353 | 1552 | 1633 | 1598 | 1443 | 1200 | 916 | 682 | 567 | 1125 |
| 39 | 602 | 795 | 1075 | 1344 | 1550 | 1633 | 1597 | 1437 | 1187 | 897 | 661 | 544 | 1112 |
| 40 | 579 | 775 | 1058 | 1335 | 1547 | 1633 | 1595 | 1430 | 1173 | 877 | 639 | 521 | 1098 |
| 41 | 556 | 755 | 1041 | 1326 | 1543 | 1632 | 1593 | 1423 | 1158 | 853 | 617 | 497 | 1085 |
| 42 | 533 | 734 | 1024 | 1316 | 1539 | 1636 | 1591 | 1416 | 1144 | 833 | 594 | 473 | 1071 |
| 43 | 510 | 714 | 1007 | 1306 | 1535 | 1638 | 1588 | 1408 | 1128 | 814 | 572 | 448 | 1057 |
| 44 | 487 | 693 | 989 | 1296 | 1530 | 1639 | 1584 | 1400 | 1113 | 795 | 549 | 427 | 1043 |
| 45 | 463 | 672 | 971 | 1285 | 1524 | 1640 | 1586 | 1392 | 1097 | 775 | 526 | 406 | 1030 |
| 46 | 436 | 650 | 953 | 1274 | 1519 | 1640 | 1585 | 1383 | 1081 | 755 | 503 | 385 | 1015 |
| 47 | 416 | 629 | 935 | 1262 | 1512 | 1639 | 1584 | 1373 | 1065 | 735 | 480 | 364 | 1001 |
| 48 | 395 | 607 | 916 | 1250 | 1506 | 1638 | 1582 | 1364 | 1048 | 714 | 457 | 343 | 987 |
| 49 | 374 | 585 | 897 | 1238 | 1506 | 1637 | 1579 | 1353 | 1031 | 693 | 433 | 322 | 973 |
| 50 | 354 | 563 | 877 | 1225 | 1501 | 1635 | 1576 | 1343 | 1014 | 673 | 407 | 301 | 958 |
| 51 | 333 | 541 | 858 | 1212 | 1496 | 1633 | 1572 | 1332 | 996 | 652 | 387 | 279 | 943 |
| 52 | 311 | 518 | 838 | 1198 | 1491 | 1630 | 1568 | 1320 | 979 | 630 | 366 | 258 | 928 |
| 53 | 290 | 496 | 818 | 1184 | 1485 | 1627 | 1564 | 1309 | 960 | 609 | 345 | 237 | 912 |
| 54 | 269 | 473 | 797 | 1170 | 1479 | 1623 | 1559 | 1297 | 942 | 587 | 324 | 215 | 897 |
| 55 | 248 | 450 | 777 | 1156 | 1472 | 1619 | 1554 | 1284 | 923 | 565 | 303 | 193 | 881 |
| 56 | 226 | 427 | 756 | 1141 | 1465 | 1624 | 1548 | 1271 | 904 | 544 | 282 | 170 | 865 |
| 57 | 205 | 404 | 735 | 1126 | 1457 | 1623 | 1542 | 1258 | 885 | 521 | 261 | 153 | 850 |
| 58 | 183 | 381 | 714 | 1110 | 1449 | 1622 | 1542 | 1250 | 865 | 499 | 240 | 135 | 835 |
| 59 | 160 | 358 | 692 | 1094 | 1441 | 1620 | 1539 | 1239 | 846 | 477 | 219 | 118 | 819 |
| 60 | 143 | 331 | 671 | 1078 | 1432 | 1617 | 1536 | 1228 | 826 | 454 | 197 | 100 | 804 |
| 61 | 125 | 310 | 649 | 1061 | 1429 | 1615 | 1532 | 1216 | 805 | 432 | 176 | 83 | 789 |
| 62 | 108 | 290 | 627 | 1045 | 1423 | 1618 | 1528 | 1203 | 785 | 409 | 152 | 61 | 773 |
| 63 | 90 | 269 | 605 | 1028 | 1417 | 1619 | 1523 | 1191 | 764 | 386 | 134 | 49 | 759 |
| 64 | 73 | 248 | 582 | 1010 | 1411 | 1621 | 1523 | 1178 | 743 | 363 | 117 | 36 | 745 |
| 65 | 53 | 227 | 560 | 992 | 1404 | 1621 | 1523 | 1164 | 722 | 340 | 100 | 24 | 730 |
| 66 | 41 | 206 | 537 | 980 | 1397 | 1631 | 1522 | 1151 | 700 | 316 | 82 | 11 | 717 |
| 67 | 28 | 184 | 514 | 965 | 1396 | 1643 | 1521 | 1137 | 679 | 293 | 65 | 0 | 705 |
| 68 | 12 | 163 | 491 | 949 | 1393 | 1655 | 1528 | 1122 | 657 | 270 | 46 | 0 | 693 |
| 69 | 6 | 142 | 468 | 932 | 1390 | 1666 | 1540 | 1116 | 640 | 243 | 33 | 0 | 684 |
| 70 | 0 | 117 | 445 | 916 | 1393 | 1676 | 1549 | 1105 | 619 | 222 | 21 | 0 | 675 |
| 71 | 0 | 100 | 421 | 899 | 1395 | 1686 | 1558 | 1093 | 598 | 201 | 8 | 0 | 667 |
| 72 | 0 | 83 | 398 | 882 | 1408 | 1696 | 1567 | 1082 | 576 | 180 | 0 | 0 | 659 |
| 73 | 0 | 65 | 374 | 865 | 1415 | 1705 | 1575 | 1075 | 554 | 159 | 0 | 0 | 652 |
| 74 | 0 | 48 | 350 | 847 | 1422 | 1713 | 1583 | 1068 | 532 | 138 | 0 | 0 | 645 |
| 75 | 0 | 31 | 327 | 838 | 1429 | 1721 | 1591 | 1061 | 510 | 117 | 0 | 0 | 639 |

| Slope | 0 | 1 | 2 | 3 | 4 | 5 | 6 | 7 | 8 | 9 | 10 |
|---|---|---|---|---|---|---|---|---|---|---|---|
| 0 | 100 | 100 | 100 | 100 | 100 | 100 | 100 | 99 | 99 | 99 | 99 |
| 10 | 99 | 99 | 99 | 98 | 98 | 98 | 97 | 97 | 97 | 96 | 96 |
| 20 | 96 | 96 | 95 | 95 | 95 | 94 | 94 | 93 | 93 | 92 | 92 |
| 30 | 92 | 91 | 91 | 90 | 89 | 89 | 88 | 88 | 87 | 87 | 86 |
| 40 | 86 | 85 | 85 | 84 | 83 | 83 | 82 | 81 | 81 | 80 | 79 |
| 50 | 79 | 79 | 78 | 77 | 76 | 76 | 75 | 74 | 73 | 73 | 72 |
| 60 | 72 | 71 | 70 | 70 | 69 | 68 | 67 | 67 | 66 | 65 | 64 |
| 70 | 64 | 64 | 63 | 62 | 61 | 61 | 60 | 59 | 58 | 58 | 57 |
| 80 | 57 | 56 | 55 | 55 | 54 | 53 | 52 | 52 | 51 | 50 | 50 |

Slope multiplication factor (in%) on overcast days
for different slopes

. Fig. 4 - Example of tables for the EEC Atlas on slopes (overcast sky conditions) compoled by S. Aydinly and R. Rattunde.

BELGIQUE                                                          Station :            UCCLE

Latitude : 50 48' N                      Longitude : 04 21' E                          Altitude :    105   m

Global Radiation G, Diffuse Radiation D in (W.h.m-2)         [WRR]        Relative Sunshine Duration S/So in (%)

10 Years Means (1966-1975) of Monthly Means Daily Sums                   Typ of Turbidity :            Urban Area

|  | Jan | Feb | Mar | Apr | May | Jun | Jul | Aug | Sep | Oct | Nov | Dec | Ann |
|---|---|---|---|---|---|---|---|---|---|---|---|---|---|
| S/So | 18 | 26 | 30 | 36 | 39 | 40 | 39 | 43 | 41 | 34 | 23 | 18 | 32 |
| **Horizontal** | | | | | | | | | | | | | |
| Gm | 637 | 1259 | 2202 | 3441 | 4574 | 4925 | 4643 | 4089 | 3021 | 1738 | 833 | 494 | 2662 |
| Dm | 502 | 867 | 1367 | 1999 | 2448 | 2675 | 2621 | 2279 | 1733 | 1145 | 641 | 432 | 1562 |
| Gcl | 1390 | 2498 | 4166 | 6003 | 7435 | 8117 | 7761 | 6553 | 4827 | 3076 | 1685 | 1133 | 4564 |
| Dcl | 492 | 704 | 1014 | 1425 | 1694 | 1842 | 1834 | 1611 | 1255 | 878 | 573 | 437 | 1149 |
| **Vertical North** | | | | | | | | | | | | | |
| Gm | 285 | 499 | 782 | 1147 | 1505 | 1726 | 1632 | 1350 | 993 | 655 | 365 | 245 | 934 |
| Dm | 285 | 499 | 782 | 1143 | 1443 | 1605 | 1541 | 1325 | 993 | 655 | 365 | 245 | 909 |
| Gcl | 401 | 593 | 872 | 1254 | 1823 | 2200 | 2025 | 1456 | 1001 | 697 | 455 | 349 | 1096 |
| Dcl | 401 | 593 | 872 | 1210 | 1513 | 1680 | 1616 | 1333 | 1001 | 697 | 455 | 349 | 979 |
| **Vertical East/West** | | | | | | | | | | | | | |
| Gm | 431 | 839 | 1372 | 2045 | 2537 | 2765 | 2642 | 2461 | 1878 | 1167 | 577 | 365 | 1594 |
| Dm | 313 | 575 | 921 | 1371 | 1691 | 1847 | 1791 | 1621 | 1240 | 795 | 415 | 270 | 1073 |
| Gcl | 1229 | 2031 | 3030 | 3921 | 4546 | 4795 | 4632 | 4108 | 3281 | 2305 | 1388 | 1049 | 3031 |
| Dcl | 510 | 786 | 1167 | 1587 | 1874 | 2010 | 1971 | 1733 | 1360 | 946 | 590 | 445 | 1250 |
| **Vertical East-South** | | | | | | | | | | | | | |
| Gm | 809 | 1400 | 1939 | 2505 | 2780 | 2855 | 2791 | 2891 | 2552 | 1849 | 1044 | 714 | 2013 |
| Dm | 370 | 676 | 1036 | 1478 | 1746 | 1863 | 1823 | 1723 | 1400 | 946 | 502 | 323 | 1160 |
| Gcl | 3020 | 4040 | 4785 | 5015 | 4937 | 4785 | 4789 | 4884 | 4734 | 4178 | 3213 | 2711 | 4257 |
| Dcl | 693 | 993 | 1357 | 1714 | 1905 | 1984 | 1967 | 1825 | 1548 | 1174 | 796 | 619 | 1383 |
| **Vertical South** | | | | | | | | | | | | | |
| Gm | 1021 | 1729 | 2259 | 2656 | 2715 | 2667 | 2659 | 2965 | 2897 | 2247 | 1313 | 907 | 2171 |
| Dm | 403 | 727 | 1090 | 1511 | 1739 | 1799 | 1799 | 1740 | 1472 | 1022 | 550 | 355 | 1188 |
| Gcl | 4067 | 5230 | 5722 | 5224 | 4553 | 4129 | 4260 | 4832 | 5397 | 5276 | 4292 | 3669 | 4716 |
| Dcl | 794 | 1086 | 1431 | 1733 | 1864 | 1909 | 1906 | 1820 | 1619 | 1276 | 905 | 719 | 1423 |
| **30   South** | | | | | | | | | | | | | |
| Gm | 995 | 1822 | 2773 | 3911 | 4658 | 4975 | 4791 | 4646 | 3746 | 2470 | 1302 | 858 | 3085 |
| Dm | 533 | 933 | 1441 | 2064 | 2453 | 2633 | 2594 | 2354 | 1875 | 1265 | 698 | 463 | 1612 |
| Gcl | 3114 | 4588 | 6174 | 7378 | 8046 | 8282 | 8103 | 7565 | 6532 | 5070 | 3461 | 2708 | 5924 |
| Dcl | 700 | 962 | 1299 | 1678 | 1861 | 1952 | 1965 | 1830 | 1550 | 1166 | 804 | 630 | 1368 |
| **60   South** | | | | | | | | | | | | | |
| Gm | 1116 | 1967 | 2773 | 3599 | 3996 | 4119 | 4024 | 4170 | 3671 | 2615 | 1448 | 978 | 2877 |
| Dm | 492 | 875 | 1329 | 1872 | 2176 | 2313 | 2283 | 2147 | 1767 | 1212 | 659 | 431 | 1466 |
| Gcl | 4095 | 5588 | 6745 | 7099 | 6992 | 6825 | 6826 | 6938 | 6740 | 5874 | 4415 | 3637 | 5981 |
| Dcl | 811 | 1102 | 1453 | 1796 | 1928 | 1993 | 2005 | 1914 | 1682 | 1311 | 925 | 733 | 1473 |
| **Latitude South** | | | | | | | | | | | | | |
| Gm | 1102 | 1964 | 2832 | 3771 | 4286 | 4466 | 4340 | 4403 | 3775 | 2627 | 1433 | 961 | 3001 |
| Dm | 511 | 905 | 1380 | 1954 | 2288 | 2435 | 2401 | 2233 | 1825 | 1245 | 680 | 446 | 1528 |
| Gcl | 3893 | 5420 | 6740 | 7367 | 7502 | 7443 | 7389 | 7305 | 6848 | 5775 | 4230 | 3440 | 6114 |
| Dcl | 789 | 1074 | 1423 | 1778 | 1927 | 1994 | 2006 | 1903 | 1663 | 1285 | 902 | 714 | 1456 |

Fig. 5: Example of tables giving the monthly and annual means of irradiation
on different tilted and oriented surfaces for 56 stations of
European Community.

# CLASSIFICATION OF RADIATION SITES IN TERMS
## OF DIFFERENT INDICES OF ATMOSPHERIC TRANSPARENCY

by R. DOGNIAUX and M. LEMOINE

CONTRIBUTION TO ACTION :   *3.2*

CONTRACT NUMBER :   *ESF - 003 - B - (G)*

TITLE :        *Evaluation of radiation on title planes*
               *Action leader for action 3.2*

DURATION :   *3,5     from 1980.01.01 to 1983.06.30*

TOTAL BUDGET :   *1580 kFB = 40 kUC*   CEC CONTRIBUTION :   *20 kUC*

HEAD OF PROJECT :   *R. DOGNIAUX*

CONTRACTOR :     *Commission Administrative du Patrimoine de l'I.R.M.*

ADRESS :         *3, avenue Circulaire,*

                 *B - 1180 BRUSSELS*

## SUMMARY

   Systematic studies were achieved for correlating the global
radiation with the relative sunshine duration and for determining
the variations of their coefficients on a geographical scale as
function of the radiation climatic area.

   The sum A + B of the coefficients of the well - know Angström
regression equation can be regarded as a parameter characterizing
the average monthly or annual conditions of the transparency of the
atmosphere at a given station. This index called "atmospheric
transparency index" (ATI) is a linear function of the latitude and of
the Linke turbidity factor.

   A recapitulative table summarizes the different indices allowing
to characterize a site.  By an appropriate choice of them; it is
possible to estimate with a reasonable accuracy the daily global
irradiation of a site.

# 1.- Introduction

The well known Ångström type regression equation :

$$G/G_\infty = A + B\ S/S_o$$

where :
- G and $G_\infty$ denote respectively, the monthly (or the annual) means of the daily sums of the actual global radiation measured on a horizontal surface at a given station and at the Earth's surface outside the atmosphere,
- S and $S_o$, the actual daily sums of duration of sunshine recorded at the station and the astronomically possible maximum daylength for which the solar elevation angle of the center of the Sun's disk is higher than $0°$,
- A and B, two coefficients estimated by the least squares approximation method,

is a very convenient equation to use for estimating mean global radiation on the basis of measurements of sunshine duration over 10 day or longer periods, in countries where the network of radiation stations does not allow such a direct determination.

The aim of this study is, first, to investigate the possibility of extending the application of these equation to an extended area, by analyzing the variations of A and B as a function of latitude and atmospheric turbidity, and, in addition, to estimate the maximum values of the global radiation at the stations under clear sky conditions, by using an adequate turbidity factor well adapted to the local conditions of the site.

## 2.- Sources of the data

The regression coefficients A and B and the coefficients of correlation used for our investigation, either on a monthly or on an annual basis depending on which data were available, were obtained from the following sources :

- for the countries of the European Community : data communicated by F. Kasten from the Deutscher Wetterdienst – Meteorologisches Observatorium, Hamburg;

- for the countries of the Eastern part of the Mediterranean Basin (Yugoslavia, Greece, Cyprus, Turkey, Middle East) and of the Iberian Peninsula : data derived from the publication by R. Dogniaux and M. Lemoine : "Distribution of solar radiation in the regions bordering on the Eastern and Western Mediterranean Basin", IRM-Misc.Série B, n° 54 (1982);

- for a few countries of Africa (such as Egypt and Sudan) and for some stations from Asia (Kuwait, Iraq) : data published by Ehab M. Abd El Salam (Cairo University, Giza) : COMPLES, 1er semester 1979; pp. 27-32 : "The relation between solar radiation and hours of bright sunshine for the Middle-East region", and by Mohie A. Abbas and M.K. El-Nesr (Physic Department, Al-Mustan-Sirya University, Baghdad) : "Relation between sunshine duration and mean daily solar radiation on a horizontal surface in Iraq";

- for Saudi Arabia : data on G and S collected in the network sponsored by the Department of Water Resources Development of the Ministry of Agriculture and Water, communicated to the authors by the Saudi Arabian National Center for Science and Technology in Ryiadh.

## 3.- Quality control and processing of the data

The values A and B are based on the actual measurements of G and S available from the meteorological stations; they were, for some part, computed and supplied by the workers themselves, as at least as concerns the countries of the European Community. The values for the other countries, with the exception of those of Africa and Iraq, were computed by ourselves. Due to the large variations of both instrumental equipments

and measurement procedures, the available material is far from homogeneous
and the resulting scatter in the calculations reflects the heterogeneity
of the observations.

Measurements of sunshine duration were obtained from Campbell-
Stokes type heliograph while global solar radiation was recorded with
pyranometers of different classes of accuracy. The records from Robitzsch
bimetallic pyranographs equipping quite a number of stations namely in
Italy, Greece, the Middle East, Saudi Arabia and Iraq are of course less
accurate than those obtained with electrical pyranometer devices, which
results in more scatter in the distribution of the points A and B on both
sides of the regression lines.

This fact prompted us to analyze the data according to various
groupings :
[1]  : stations from the European Community Atlas
[2]  : stations from Spain, Yugoslavia, Greece, Cyprus, Lebanon, Syria
        and Israël.
[3]  : [1]  + [2]
[4]  : stations from Egypt, Sudan, Kuwait, Iraq. (only annual sums availa-
        ble).
[5]  : [2]  + [4]
[6]  : [3]  + [4]
[7]  : stations from Saudi Arabia
[8]  : [4]  + [7]
[9]  : [6]  + [7] (all stations together).

## 4.- Analysis of the data with reference to the latitudinal variation

### 4.1  Choice of the regression equations
Two different laws of linear regression were tested for expressing the variations of A and B in terms of the latitude $\phi$ : simply a
function of $\phi$,

$$A = M\phi + N \qquad\qquad B = P\phi + Q$$

and a similar linear function of cos $\phi$ as suggested by different authors.

The later procedure was rejected for the benefit of the first law because it did not entail significantly more accurate results.

The computations were performed for the different months and for the year and the results for group [ 9] are published in [ 3] .

## 4.2  Effect of the selected groupings on the coefficients

The coefficients M, N, P and Q of the regression lines, computed from the annual sums of G and S, are given, for the different groupings, in table 1.

Table 1.- Variations of the coefficients M, N, P and Q according to various groupings of the data (annual sums)

| Groups | M | N | P | Q |
|--------|---|---|---|---|
| [ 1 ] | -0,00145 | 0,29137 | 0,00058 | 0.53150 |
| [ 2 ] | -0,00388 | 0,38799 | 0,00309 | 0.38689 |
| [ 3 ] | -0,00197 | 0,31683 | 0,00293 | 0,41491 |
| [ 4 ] | -0,00196 | 0,32619 | 0,00383 | 0,35641 |
| [ 5 ] | -0,00233 | 0,33348 | 0,00396 | 0,35527 |
| [ 6 ] | -0,00216 | 0,32566 | 0,00378 | 0,37570 |
| [ 7 ] | 0,01120 | 0,05789 | -0.02062 | 0.87497 |
| [ 8 ] | 0,00145 | 0.27611 | -0,00231 | 0.45906 |
| [ 9 ] | -0,00290 | 0.36239 | 0,00491 | 0.31876 |

Due to the differences in the coefficients for the different categories, it was finally decided to consider all the data grouped all together, (case 9), which affords the advantage of covering a large range of latitudes from 60° to 5° N. and of needing only one relationship for expressing the variations of A and B under any conditions of equipment and location.

## 4.3  Effect of the seasonal variation on the coefficients

The influence of the seasonal variation on the coefficients M, N, P and Q can be followed by examining the data presented in table 2.

A comparison of the individual values for the twelve months does not reveal any systematic seasonal effect.

Table 2.- Seasonal variations of the coefficients, M, N, P and Q and of the coefficients of correlation CC compared with their annual values and their means of the monthly values

| Months | I | M | N | CC | P | Q | CC |
|---|---|---|---|---|---|---|---|
| JANUARY | I | -0,00301 | 0,34507 | 0,536 | 0,00495 | 0,34572 | 0,500 |
| FEBRUARY | I | -0,00255 | 0,33459 | 0,383 | 0,00457 | 0,35533 | 0,514 |
| MARCH | I | -0,00303 | 0,36690 | 0,555 | 0,00466 | 0,36377 | 0,529 |
| APRIL | I | -0,00334 | 0,38557 | 0,610 | 0,00456 | 0,35802 | 0,534 |
| MAY | I | -0,00245 | 0,35057 | 0,508 | 0,00485 | 0,33550 | 0,578 |
| JUNE | I | -0,00327 | 0,39890 | 0,555 | 0,00578 | 0,27292 | 0,615 |
| JULY | I | -0,00369 | 0,41234 | 0,603 | 0,00568 | 0,27004 | 0,649 |
| AUGUST | I | -0,00269 | 0,36243 | 0,492 | 0,00412 | 0,33162 | 0,508 |
| SEPTEMBER | I | -0,00338 | 0,39467 | 0,507 | 0,00564 | 0,27125 | 0,618 |
| OCTOBER | I | -0,00317 | 0,36213 | 0,509 | 0,00504 | 0,31790 | 0,534 |
| NOVEMBER | I | -0,00350 | 0,36680 | 0,602 | 0,00523 | 0,31467 | 0,543 |
| DECEMBER | I | -0,00350 | 0,36262 | 0,598 | 0,00559 | 0,30675 | 0,545 |
| YEAR | I | -0,00290 | 0,36239 | 0,530 | 0,00491 | 0,31876 | 0,594 |
| MEANS | I | -0,00313 | 0,37022 | 0,538 | 0,00506 | 0,32029 | 0,555 |

Consequently we have adopted the values corresponding to the means of the monthly values; the actual values A and B can thus be approximated by the values $\overline{A}$ and $\overline{B}$ deduced from the following equations :

$$\overline{A} = 0,37022 - 0,00313\phi \qquad (1)$$
$$\overline{B} = 0,32029 + 0,00506\phi \qquad (2)$$
and
$$\overline{A + B} = 0,69051 + 0,00193\phi \qquad (3)$$

which allow to estimate the global radiation on a horizontal surface by the equation :

$$G/G_\infty = (0,00506 \ S/S_o - 0,00313)\phi + 0,32029 \ S/S_o + 0,37022 \qquad (4)$$

Table 3 and figure 1 have been established according to these regression equations. The symbols used in table 3 have the following meaning :

A : actual values of A, B and (A + B)
C : calculated values of A, B and (A + B) thus $\overline{A}$, $\overline{B}$ and $\overline{(A + B)}$
R : differences between actual and calculated values.

## 5.- Atmospheric turbidity index (ATI)

The quantity A + B is interesting to consider. It corresponds indeed to the ratio between the irradiations of the global radiation on a horizontal surface respectively at the ground for clear sky conditions $(S/S_o = 1)$ and outside the atmosphere.

This number can be regarded as a parameter characterizing the average conditions of the transparency or clearness of the atmosphere at the station and, by extension, over an area surrounding the station where the observations are made.

This quantity is therefore often called *atmospheric transparency index*. Two effects cause spatial variations of this index : an effect of latitude and an effect of local turbidity of the atmosphere.

Therefore we can write :
$$ATI = \overline{A + B} + R \tag{5}$$
where the first term of the right-hand side of the equation expresses the latitudinal variation, and the second term, R, the effect of local atmospheric turbidity.

### 5.1 Latitudinal variations : $\overline{(A + B)}$

This effect finds expression in equation (3) which can be interpreted as representing of the variation of ATI for normalized conditions of atmospheric turbidity which we shall term *clear atmosphere*. Such values follow from a first approximation of ATI by a linear regression equation.

Table 3

# REGRESSION COEFFICIENTS A AND B

## ATMOSPHERIC TRANSPARENCY INDEX A + B

Calculated from the regression equation G/G = A + B S/SO
Means of the monthly values
A = 0,37022 - 0,00313 PHI
B = 0,32029 + 0,00506 PHI
A+B = 0,69051 + 0,00193 PHI

A : Actual values
C : Calculated values
R : Differences between actual and calculated values

| COUNTRY | STATION | LAT | LONG | HEIGHT | A | A C | R | B | B C | R | A | A+B C | R |
|---------|---------|-----|------|--------|-----|-----|------|-----|-----|-------|-----|-------|-------|
| IRL | BIRR | 53 05'N | 07 54'W | 72 | 0,24 | 0,20 | 0,04 | 0,58 | 0,59 | -0,01 | 0,82 | 0,79 | 0,03 |
| IRL | KILKENNY | 52 40'N | 07 16'W | 64 | 0,26 | 0,21 | 0,05 | 0,57 | 0,59 | -0,02 | 0,83 | 0,79 | 0,04 |
| IRL | VALENTIA | 51 56'N | 10 15'W | 20 | 0,23 | 0,21 | 0,02 | 0,62 | 0,58 | 0,04 | 0,85 | 0,79 | 0,06 |
| GB | LERWICK | 60 08'N | 01 11'W | 82 | 0,20 | 0,18 | 0,02 | 0,62 | 0,62 | -0,00 | 0,82 | 0,81 | 0,01 |
| GB | ESKDALEMUIR | 55 19'N | 03 12'W | 242 | 0,19 | 0,20 | -0,01 | 0,60 | 0,63 | -0,00 | 0,79 | 0,80 | -0,01 |
| GB | ALDERGROVE | 54 39'N | 06 13'W | 68 | 0,21 | 0,20 | 0,01 | 0,56 | 0,60 | -0,04 | 0,77 | 0,80 | -0,03 |
| GB | CAWOOD | 53 50'N | 01 08'W | 6 | 0,23 | 0,20 | 0,03 | 0,53 | 0,59 | -0,06 | 0,76 | 0,79 | -0,03 |
| GB | ABERPORTH | 52 08'N | 04 34'W | 133 | 0,22 | 0,21 | 0,01 | 0,54 | 0,58 | -0,04 | 0,76 | 0,79 | -0,03 |
| GB | LONDON | 51 31'N | 00 07'W | 77 | 0,17 | 0,21 | -0,04 | 0,51 | 0,58 | -0,07 | 0,68 | 0,79 | -0,11 |
| GB | KEW | 51 28'N | 00 19'W | 5 | 0,18 | 0,21 | -0,03 | 0,53 | 0,58 | -0,05 | 0,71 | 0,79 | -0,08 |
| GB | BRACKNELL | 51 23'N | 00 47'W | 73 | 0,20 | 0,21 | -0,01 | 0,54 | 0,58 | -0,04 | 0,74 | 0,79 | -0,05 |
| F | TRAPPES | 48 46'N | 02 01'E | 168 | 0,21 | 0,22 | -0,01 | 0,54 | 0,57 | -0,03 | 0,75 | 0,78 | -0,03 |
| F | NANCY | 48 41'N | 06 13'E | 217 | 0,21 | 0,22 | -0,01 | 0,57 | 0,57 | 0,00 | 0,78 | 0,78 | -3,03 |
| F | MACON | 46 18'N | 04 48'E | 217 | 0,20 | 0,23 | -0,03 | 0,57 | 0,55 | 0,02 | 0,77 | 0,78 | -0,01 |
| F | LIMOGES | 45 52'N | 01 11'E | 402 | 0,21 | 0,23 | -0,02 | 0,59 | 0,55 | 0,04 | 0,80 | 0,78 | 0,02 |
| F | MILLAU | 44 07'N | 03 01'E | 720 | 0,21 | 0,23 | -0,02 | 0,62 | 0,54 | 0,08 | 0,83 | 0,78 | 0,05 |
| F | CARPENTRAS | 44 05'N | 05 03'E | 105 | 0,21 | 0,23 | -0,02 | 0,54 | 0,54 | -0,00 | 0,75 | 0,78 | -0,03 |
| F | NICE | 43 39'N | 07 12'E | 10 | 0,21 | 0,23 | -0,02 | 0,55 | 0,54 | 0,01 | 0,76 | 0,77 | -0,01 |
| D | NORDERNEY | 53 43'N | 07 09'E | 13 | 0,22 | 0,20 | 0,02 | 0,58 | 0,59 | -0,01 | 0,80 | 0,79 | 0,01 |
| D | HAMBURG | 53 38'N | 10 00'E | 14 | 0,19 | 0,20 | -0,01 | 0,55 | 0,59 | -0,04 | 0,74 | 0,79 | -0,05 |
| D | BRAUNSCHWEIG | 52 18'N | 10 27'E | 81 | 0,19 | 0,21 | -0,02 | 0,55 | 0,58 | -0,03 | 0,74 | 0,79 | -0,05 |
| D | BRAUNLAGE | 51 43'N | 10 37'E | 601 | 0,19 | 0,21 | -0,02 | 0,59 | 0,58 | 0,01 | 0,78 | 0,79 | -0,01 |
| D | WUERZBURG | 49 48'N | 09 54'E | 259 | 0,23 | 0,21 | 0,02 | 0,56 | 0,57 | -0,01 | 0,79 | 0,79 | 0,00 |
| D | TRIER | 49 45'N | 06 40'E | 265 | 0,20 | 0,21 | -0,01 | 0,59 | 0,57 | 0,02 | 0,79 | 0,79 | 0,00 |
| D | WEIHENSTEPHAN | 48 24'N | 11 44'E | 467 | 0,24 | 0,22 | 0,02 | 0,55 | 0,57 | -0,02 | 0,79 | 0,78 | 0,01 |
| D | HOHENPEISSENBG | 47 48'N | 11 01'E | 975 | 0,22 | 0,22 | -0,00 | 0,61 | 0,56 | 0,05 | 0,83 | 0,78 | 0,05 |
| I | BOLZANO | 46 28'N | 11 20'E | 241 | 0,23 | 0,22 | 0,01 | 0,60 | 0,56 | 0,04 | 0,83 | 0,78 | 0,05 |
| I | TRIESTE | 45 39'N | 13 45'E | 20 | 0,21 | 0,23 | -0,02 | 0,58 | 0,55 | 0,03 | 0,79 | 0,78 | 0,01 |
| I | MILANO | 45 26'N | 09 17'E | 103 | 0,24 | 0,23 | 0,01 | 0,56 | 0,55 | 0,01 | 0,80 | 0,78 | 0,02 |
| I | GENOVA | 44 25'N | 08 51'E | 3 | 0,23 | 0,23 | -0,00 | 0,57 | 0,55 | 0,02 | 0,80 | 0,78 | 0,02 |
| I | ANCONA | 43 37'N | 13 31'E | 104 | 0,25 | 0,23 | 0,02 | 0,59 | 0,54 | 0,05 | 0,84 | 0,77 | 0,07 |
| I | PIANOSA | 42 35'N | 10 06'E | 27 | 0,27 | 0,24 | 0,03 | 0,55 | 0,54 | 0,01 | 0,82 | 0,77 | 0,05 |
| I | ROMA | 41 48'N | 12 35'E | 131 | 0,27 | 0,24 | 0,03 | 0,54 | 0,53 | 0,01 | 0,81 | 0,77 | 0,04 |
| I | NAPOLI | 40 51'N | 14 18'E | 72 | 0,26 | 0,24 | 0,02 | 0,55 | 0,53 | 0,02 | 0,81 | 0,77 | 0,04 |
| I | BRINDISI | 40 39'N | 17 57'E | 10 | 0,28 | 0,24 | 0,04 | 0,54 | 0,53 | 0,01 | 0,82 | 0,77 | 0,05 |
| I | ALGHERO | 40 38'N | 08 17'E | 40 | 0,24 | 0,24 | -0,00 | 0,59 | 0,53 | 0,06 | 0,83 | 0,77 | 0,06 |
| I | MESSINA | 38 12'N | 15 33'E | 50 | 0,25 | 0,25 | -0,00 | 0,58 | 0,51 | 0,07 | 0,83 | 0,76 | 0,07 |
| I | TRAPANI | 37 55'N | 12 30'E | 14 | 0,30 | 0,25 | 0,05 | 0,53 | 0,51 | 0,02 | 0,83 | 0,76 | 0,07 |
| I | PANTELLERIA | 36 49'N | 11 58'E | 170 | 0,28 | 0,25 | 0,03 | 0,55 | 0,51 | 0,04 | 0,83 | 0,76 | 0,07 |
| B | OOSTENDE | 51 12'N | 02 52'E | 4 | 0,22 | 0,21 | 0,01 | 0,52 | 0,58 | -0,06 | 0,74 | 0,79 | -0,05 |
| B | MELLE | 50 59'N | 03 50'E | 15 | 0,20 | 0,21 | -0,01 | 0,48 | 0,58 | -0,10 | 0,68 | 0,79 | -0,11 |
| B | UCCLE-UKKEL | 50 48'N | 04 21'E | 105 | 0,19 | 0,21 | -0,02 | 0,53 | 0,58 | -0,05 | 0,72 | 0,79 | -0,07 |
| B | SAINT HUBERT | 50 02'N | 05 24'E | 563 | 0,19 | 0,21 | -0,02 | 0,59 | 0,57 | 0,02 | 0,78 | 0,79 | -0,01 |
| NL | GRONINGEN-EELD | 53 08'N | 06 55'E | 5 | 0,22 | 0,20 | 0,02 | 0,57 | 0,59 | -0,02 | 0,79 | 0,79 | -0,00 |
| NL | DEN HELDER | 52 55'N | 04 47'E | 2 | 0,23 | 0,20 | 0,03 | 0,58 | 0,59 | -0,01 | 0,81 | 0,79 | 0,02 |
| NL | DE BILT | 52 06'N | 05 11'E | 40 | 0,20 | 0,21 | -0,01 | 0,55 | 0,58 | -0,03 | 0,75 | 0,79 | -0,04 |
| NL | VLISSINGEN | 51 27'N | 03 36'E | 22 | 0,22 | 0,21 | 0,01 | 0,56 | 0,58 | -0,02 | 0,78 | 0,79 | -0,01 |
| NL | MAASTRICHT-BEE | 50 55'N | 05 46'E | 116 | 0,21 | 0,21 | -0,00 | 0,56 | 0,58 | -0,02 | 0,77 | 0,79 | -0,02 |
| DK | HOJBAKKEGARD | 55 40'N | 12 19'E | 30 | 0,20 | 0,20 | -0,00 | 0,57 | 0,60 | -0,03 | 0,77 | 0,80 | -0,03 |
| E | OVIEDO | 43 21'N | 5 52'W | 335 | 0,22 | 0,23 | -0,01 | 0,54 | 0,54 | 0,00 | 0,76 | 0,77 | -0,01 |
| E | TORTOZA | 40 49'N | 0 30'E | 44 | 0,20 | 0,24 | -0,04 | 0,45 | 0,53 | -0,08 | 0,65 | 0,77 | -0,12 |
| E | MADRID | 40 27'N | 3 43'W | 664 | 0,22 | 0,24 | -0,02 | 0,60 | 0,52 | 0,08 | 0,82 | 0,77 | 0,05 |
| E | BADAJOZ | 38 53'N | 6 49'W | 185 | 0,25 | 0,25 | 0,00 | 0,51 | 0,52 | -0,01 | 0,76 | 0,77 | -0,01 |
| E | MURCIA | 37 59'N | 1 07'W | 42 | 0,24 | 0,25 | -0,01 | 0,50 | 0,51 | -0,01 | 0,74 | 0,76 | -0,02 |
| E | MALAGA | 36 40'N | 4 29'W | 7 | 0,23 | 0,26 | -0,03 | 0,55 | 0,51 | 0,04 | 0,78 | 0,76 | 0,02 |
| E | PALMA | 29 34'N | 2 39'W | 32 | 0,05 | 0,28 | -0,23 | 0,67 | 0,47 | 0,20 | 0,72 | 0,75 | -0,03 |
| YU | LJUBLJANA | 46 04'N | 14 31'E | 299 | 0,16 | 0,23 | -0,07 | 0,58 | 0,55 | 0,03 | 0,74 | 0,78 | -0,04 |
| YU | ZAGREB | 45 49'N | 15 59'E | 157 | 0,19 | 0,23 | -0,04 | 0,62 | 0,55 | 0,07 | 0,81 | 0,78 | 0,03 |
| YU | BEOGRAD | 44 47'N | 20 32'E | 243 | 0,25 | 0,23 | 0,02 | 0,51 | 0,55 | -0,04 | 0,76 | 0,78 | -0,02 |
| YU | BANJA LUKA | 44 47'N | 17 13'E | 153 | 0,27 | 0,23 | 0,04 | 0,56 | 0,55 | 0,01 | 0,83 | 0,78 | 0,05 |
| YU | NEGOTIN | 44 14'N | 22 33'E | 42 | 0,27 | 0,23 | 0,04 | 0,56 | 0,54 | 0,02 | 0,83 | 0,78 | 0,05 |

| | Station | Lat | Lon | Elev | | | | | | | | | |
|---|---|---|---|---|---|---|---|---|---|---|---|---|---|
| YU | SARAJEVO | 43 50'N | 18 20'E | 503 | 0,25 | 0,23 | 0,02 | 0,55 | 0,54 | 0,01 | 0,80 | 0,78 | 0,02 |
| YU | ZLATIBOR | 43 44'N | 19 43'E | 1029 | 0,26 | 0,23 | 0,03 | 0,53 | 0,54 | -0,01 | 0,79 | 0,77 | 0,02 |
| YU | SPLIT | 43 31'N | 16 26'E | 122 | 0,28 | 0,23 | 0,05 | 0,52 | 0,54 | -0,02 | 0,80 | 0,77 | 0,03 |
| YU | BAR | 42 06'N | 19 06'E | 6 | 0,23 | 0,24 | -0,01 | 0,57 | 0,53 | 0,04 | 0,80 | 0,77 | 0,03 |
| YU | BITOLA | 41 03'N | 21 22'E | 586 | 0,31 | 0,24 | 0,07 | 0,46 | 0,53 | -0,07 | 0,77 | 0,77 | 0,00 |
| GR | LARISSA | 39 38'N | 22 25'E | 65 | 0,22 | 0,25 | -0,03 | 0,54 | 0,52 | 0,02 | 0,76 | 0,77 | -0,01 |
| GR | AGRINION | 38 37'N | 21 24'E | 60 | 0,25 | 0,25 | 0,00 | 0,48 | 0,52 | -0,04 | 0,73 | 0,77 | -0,04 |
| GR | ALIARTOS | 38 23'N | 23 06'E | 110 | 0,23 | 0,25 | -0,02 | 0,55 | 0,51 | 0,04 | 0,78 | 0,76 | 0,02 |
| GR | PHILADELPHIA | 38 08'N | 23 40'E | 138 | 0,23 | 0,25 | -0,02 | 0,46 | 0,51 | -0,05 | 0,69 | 0,76 | -0,07 |
| GR | ATHENES | 37 58'N | 23 43'E | 107 | 0,21 | 0,25 | -0,04 | 0,49 | 0,51 | -0,02 | 0,70 | 0,76 | -0,06 |
| GR | SPARTE | 37 04'N | 22 26'E | 200 | 0,26 | 0,25 | 0,01 | 0,48 | 0,51 | -0,03 | 0,74 | 0,76 | -0,02 |
| GR | RHODES | 36 22'N | 28 13'E | 9 | 0,24 | 0,26 | -0,02 | 0,54 | 0,50 | 0,04 | 0,78 | 0,76 | 0,02 |
| GR | MITILINI | 36 06'N | 26 33'E | 30 | 0,17 | 0,26 | -0,09 | 0,56 | 0,50 | 0,06 | 0,73 | 0,76 | -0,03 |
| CY | PRODHROMOS | 34 34'N | 32 30'E | 1380 | 0,22 | 0,26 | -0,04 | 0,58 | 0,50 | 0,08 | 0,80 | 0,76 | 0,04 |
| RL | KSARA | 33 49'N | 36 00'E | 920 | 0,25 | 0,26 | -0,01 | 0,63 | 0,49 | 0,14 | 0,88 | 0,76 | 0,12 |
| RL | BEYROUTH | 33 49'N | 35 29'E | 16 | 0,22 | 0,26 | -0,04 | 0,50 | 0,49 | 0,01 | 0,72 | 0,76 | -0,04 |
| SYR | MESSELMIYEH | 36 20'N | 37 13'E | 425 | 0,26 | 0,26 | 0,00 | 0,48 | 0,50 | -0,02 | 0,74 | 0,76 | -0,02 |
| SYR | RAQQA | 35 57'N | 39 00'E | 251 | 0,27 | 0,26 | 0,01 | 0,47 | 0,50 | -0,03 | 0,74 | 0,76 | -0,02 |
| SYR | DOUMA | 33 34'N | 36 24'E | 679 | 0,37 | 0,27 | 0,10 | 0,41 | 0,49 | -0,08 | 0,78 | 0,76 | 0,02 |
| SYR | KHARABO | 33 30'N | 36 28'E | 620 | 0,30 | 0,27 | 0,03 | 0,46 | 0,49 | -0,03 | 0,76 | 0,76 | 0,00 |
| IL | BET DAGAN | 32 00'N | 34 49'E | 35 | 0,19 | 0,27 | -0,08 | 0,56 | 0,48 | 0,08 | 0,75 | 0,75 | -0,00 |
| IRQ | MOSUL | 36 19'N | 43 09'E | 222 | 0,32 | 0,26 | 0,06 | 0,40 | 0,50 | -0,10 | 0,72 | 0,76 | -0,04 |
| IRQ | BAGHDAD | 33 14'N | 44 14'E | 34 | 0,29 | 0,27 | 0,02 | 0,43 | 0,49 | -0,06 | 0,72 | 0,75 | -0,03 |
| IRQ | NASIRIYA | 31 05'N | 46 14'E | 3 | 0,27 | 0,27 | -0,00 | 0,43 | 0,48 | -0,05 | 0,70 | 0,75 | -0,05 |
| IRQ | GIZA | 30 02'N | ' | 21 | 0,25 | 0,28 | -0,03 | 0,47 | 0,47 | -0,00 | 0,72 | 0,75 | -0,03 |
| ET | MERA MATRUH | 31 20'N | 27 18'E | 30 | 0,21 | 0,27 | -0,06 | 0,49 | 0,48 | 0,01 | 0,70 | 0,75 | -0,05 |
| ET | TAHRIR | 30 47'N | 31 00'E | 14 | 0,34 | 0,27 | 0,07 | 0,38 | 0,48 | -0,10 | 0,72 | 0,75 | -0,03 |
| ET | BAHTIM | 30 08'N | 31 34'E | 74 | 0,28 | 0,28 | 0,00 | 0,44 | 0,47 | -0,03 | 0,72 | 0,75 | -0,03 |
| ET | KHARGA | 25 27'N | 30 32'E | 73 | 0,34 | 0,29 | 0,05 | 0,35 | 0,45 | -0,10 | 0,69 | 0,74 | -0,05 |
| SDN | PORT SUDAN | 19 35'N | 37 13'E | 5 | 0,27 | 0,31 | -0,04 | 0,47 | 0,42 | 0,05 | 0,74 | 0,73 | 0,01 |
| SDN | KHARTOUM | 15 36'N | 32 33'E | 380 | 0,36 | 0,32 | 0,04 | 0,37 | 0,40 | -0,03 | 0,73 | 0,72 | 0,01 |
| SDN | MALAKAL | 9 33'N | 31 39'E | 390 | 0,33 | 0,34 | -0,01 | 0,40 | 0,37 | 0,03 | 0,73 | 0,71 | 0,02 |
| SDN | JUBA | 4 52'N | 31 36'E | 460 | 0,26 | 0,35 | -0,09 | 0,39 | 0,34 | 0,05 | 0,65 | 0,70 | -0,05 |
| KWT | KUWAIT | 29 14'N | 47 59'E | 45 | 0,16 | 0,28 | -0,12 | 0,67 | 0,47 | 0,20 | 0,83 | 0,75 | 0,08 |
| SA | SAKAKA | 29 58'N | 40 12'E | 574 | 0,50 | 0,28 | 0,22 | 0,12 | 0,47 | -0,35 | 0,62 | 0,75 | -0,13 |
| SA | HAIL | 27 28'N | 41 38'E | 1010 | 0,30 | 0,28 | 0,02 | 0,35 | 0,46 | -0,11 | 0,65 | 0,74 | -0,09 |
| SA | QATIF | 26 33'N | 50 00'E | 8 | 0,26 | 0,29 | -0,03 | 0,30 | 0,45 | -0,15 | 0,56 | 0,74 | -0,18 |
| SA | ZILFI | 26 18'N | 44 48'E | 605 | 0,39 | 0,29 | 0,10 | 0,28 | 0,45 | -0,17 | 0,67 | 0,74 | -0,07 |
| SA | UNAYZAH | 26 04'N | 43 59'E | 724 | 0,30 | 0,29 | 0,01 | 0,38 | 0,45 | -0,07 | 0,68 | 0,74 | -0,06 |
| SA | UQLATAS-SUQUR | 25 50'N | 42 11'E | 740 | 0,38 | 0,29 | 0,09 | 0,40 | 0,45 | -0,05 | 0,78 | 0,74 | 0,04 |
| SA | HUTATSUDAIR | 25 32'N | 45 37'E | 665 | 0,33 | 0,29 | 0,04 | 0,38 | 0,45 | -0,07 | 0,71 | 0,74 | -0,03 |
| SA | AL-HOFUF | 25 30'N | 49 34'E | 160 | 0,39 | 0,29 | 0,10 | 0,30 | 0,45 | -0,15 | 0,69 | 0,74 | -0,05 |
| SA | SHAQRA | 25 15'N | 45 15'E | 730 | 0,39 | 0,29 | 0,10 | 0,33 | 0,45 | -0,12 | 0,72 | 0,74 | -0,02 |
| SA | RIYADH | 24 34'N | 46 43'E | 564 | 0,29 | 0,29 | -0,00 | 0,41 | 0,44 | -0,03 | 0,70 | 0,74 | -0,04 |
| SA | DAWDAMI | 24 29'N | 44 22'E | 0 | 0,28 | 0,29 | -0,01 | 0,50 | 0,44 | 0,06 | 0,78 | 0,74 | 0,04 |
| SA | DERAB | 24 25'N | 46 34'E | 0 | 0,42 | 0,29 | 0,13 | 0,35 | 0,44 | -0,09 | 0,77 | 0,74 | 0,03 |
| SA | AL-KHARJ | 24 10'N | 47 24'E | 430 | 0,32 | 0,29 | 0,03 | 0,41 | 0,44 | -0,03 | 0,73 | 0,74 | -0,01 |
| SA | TURABAH | 21 24'N | 40 27'E | 1130 | 0,31 | 0,30 | 0,01 | 0,41 | 0,43 | -0,02 | 0,72 | 0,73 | -0,01 |
| SA | TAIF | 21 14'N | 40 21'E | 1530 | 0,21 | 0,30 | -0,09 | 0,45 | 0,43 | 0,02 | 0,66 | 0,73 | -0,07 |
| SA | AL-HEIFA | 19 52'N | 42 32'E | 1090 | 0,32 | 0,31 | 0,01 | 0,53 | 0,42 | 0,11 | 0,85 | 0,73 | 0,12 |
| SA | BIL-JUARSHY | 19 51'N | 41 34'E | 2040 | 0,16 | 0,31 | -0,15 | 0,56 | 0,42 | 0,14 | 0,72 | 0,73 | -0,01 |
| SA | AL-NUMAS | 19 06'N | 42 09'E | 2600 | 0,20 | 0,31 | -0,11 | 0,70 | 0,42 | 0,28 | 0,90 | 0,73 | 0,17 |
| SA | KWASH | 19 00'N | 41 53'E | 350 | 0,25 | 0,31 | -0,06 | 0,27 | 0,42 | -0,15 | 0,52 | 0,73 | -0,21 |
| SA | SIRR-LASAN | 18 15'N | 42 36'E | 2100 | 0,37 | 0,31 | 0,06 | 0,29 | 0,41 | -0,12 | 0,66 | 0,73 | -0,07 |
| SA | NAJRAN | 17 33'N | 44 14'E | 1250 | 0,36 | 0,32 | 0,04 | 0,50 | 0,41 | 0,09 | 0,86 | 0,72 | 0,14 |

The differences R between actual values of ATI and values calculated by the equation $\overline{A + B}$ arise from errors of measurements on the one hand and, on the other hand, from local atmospheric turbidity conditions.

## 5.2 Atmospheric turbidity variation : (R)

Aydinly S. and Rattunde R.[1] proposed a classification of various sites in terms of atmospheric clearness using the following criteria related to the ATI.

| ATI | turbidity type | class of turbidity |
|---|---|---|
| < 0.72 | industrial atmosphere | 4 |
| 0.72 - 0.76 | urban atmosphere | 3 |
| 0.76 - 0.79 | clear atmosphere | 2 |
| > 0.79 | very clear atmosphere | 1 |

As they neglected the latitudinal variations of ATI, these authors met with difficulties for classifying, according the above critera, some stations of certain countries.

We therefore prefered to correlate the correction term R with the well known "Linke turbidity factor T" for which we worked out, as explained in a previous paper [2], a relationship with the water content of the atmosphere, i.e. its precipitable water (w) and the turbidity coefficient ($\beta$).

$$T = [\frac{h + 85}{39,5e^{-w} + 47,4} + 0,1] + (16 + 0,22w)\ \beta \qquad (6)$$

where h denotes the solar altitude.
Adopting for h the value 30°, which corresponds to an optical air mass 2, we defined an *average turbidity factore* $\overline{T}$ the values of xhich are given below for different magnitudes of w and $\beta$.

Average turbidity factor $\overline{T}$

| β | w [cm] | | | |
|---|---|---|---|---|
| | 0.5 | 1 | 2 | 5 |
| 0.05 | 2.5 | 2.8 | 3.1 | 3.4 |
| 0.10 | 3.3 | 3.6 | 3.9 | 4.2 |
| 0.20 | 4.9 | 5.2 | 5.6 | 5.9 |

If specific data on w and β are not available at the stations, the following classification of different types of radiation climate can be adopted.

- polar and desert climates (dry air)     w = 0.5 to 1
- temperates climates                     w = 2 to 4
- tropical climates (humid air)           w = 5
- rural site                              β = 0.05
- urban site                              β = 0.10
- industrial site                         β = 0.20

       The study of the differences R in terms of the estimated values
of $\overline{T}$ for the stations shown in table 3 leads to the following equations
(see fig.2) :

$$R = 0.129 - 0.036 \, \overline{T} \tag{7}$$

and, on account of equations (3) and (5) :

$$ATI = 0.820 + 0.00193 \, \phi - 0.036 \, \overline{T} \tag{8}$$

$$\overline{T} = 22.76 + 0.0536 \, \phi - 27.78 \, ATI \tag{9}$$

Equations (9) permits to determine the average turbidity factor $\overline{T}$ which
is adopted as input parameter in the main methods of predicting the
irradiation on oriented and inclined surfaces, provided that long series
of daily measurements of both global irradiation on a horizontal surface and sunshine duration are available for the calculation of ATI.

       In the absence of any observational data, it is necessary to
proceed with a subjective choice of indexes of turbidity. The following
section presents some information allowing to determine this choice.

# 6.- Classification of sites in terms of different indices of transparency

By combining the various site classifications that have been attempted, it is possible to present a synoptic table of the indices of transparency in use. Table 4 summarize the different possibilities for characterizing the atmospheric clearness at a site based on either an *objective method* assuming the availability of global irradiations and sunshine duration or on a *subjective method* based on the estimation of some atmospheric parameters such as w and β.

# 7.- References

[1] - ADYNLY S. and RATTUNDE R., final report on calculation methods for the radiation data on inclined surfaces and calculated atlas data. EEC Solar Energy Programme - Project F - Action 3.2 - Sept. 1982.

[2] - DOGNIAUX R. and LEMOINE M., Programme de calcul des éclairements solaires énergétiques et lumineux des surfaces orientées et inclinées. I.R.M. Misc. Série C, n° 14; 1976.

[3] - DOGNIAUX R. and LEMOINE M., Classification des sites radiatifs en fonction d'indices de transparence atmosphérique : I.R.M. - Public. Série B; 1982 - à paraître.

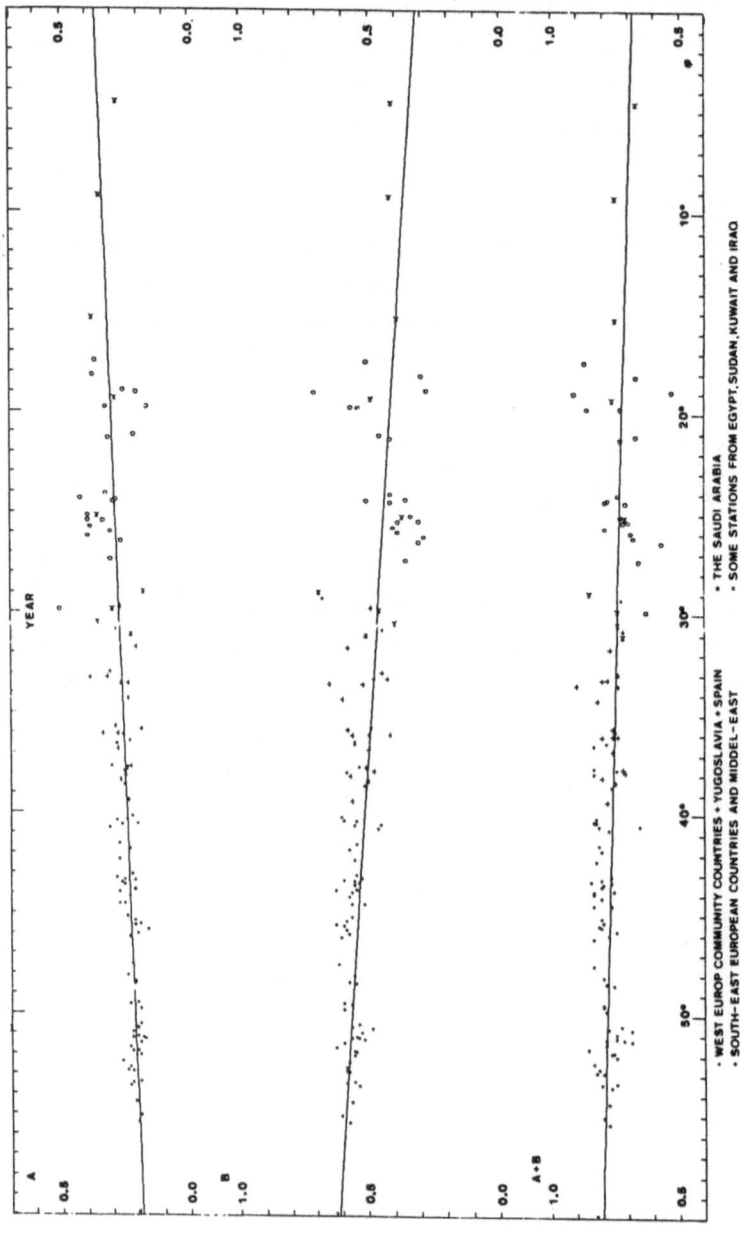

Fig. I - Variations in term of latitude of the coefficients of the Angstrom's equation $G/G_\infty = A+B \; S/S_0$

- 106 -

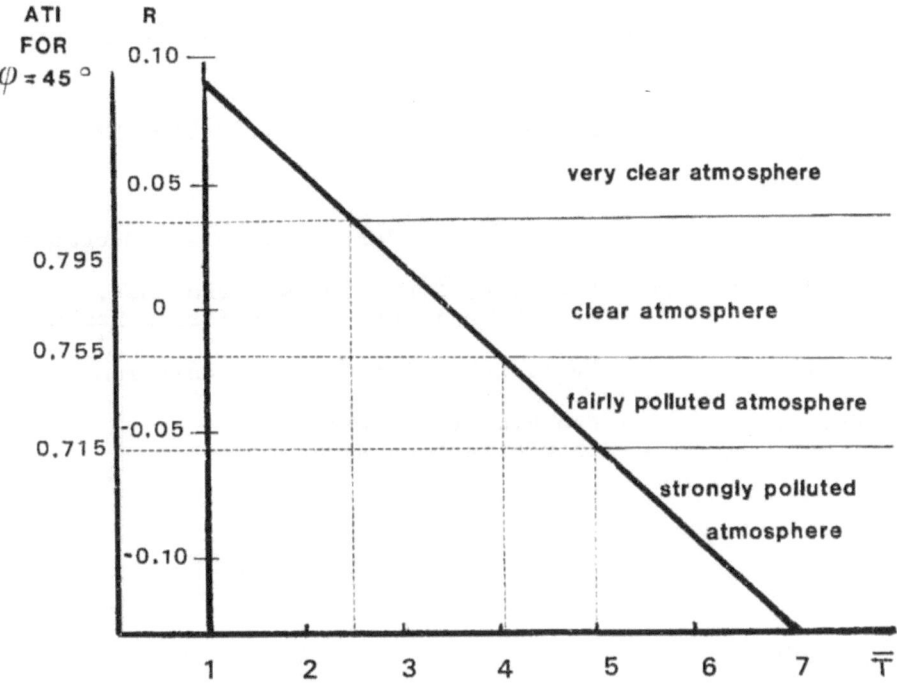

$$ATI = \overline{A+B} + R$$

$$\overline{A+B} = 0.00193\,\varphi + 0.69051$$

$$R = 0.129 - 0.036\,\overline{T}$$

$$\overline{T} = 3.58 - 27.78\,R$$

Fig. 2 - Correlation between correction index (R) and average linke turbidity factor $(\overline{T})$

# PARAMETERIZATION OF RADIATION FLUXES AS FUNCTION OF SOLAR ELEVATION, CLOUDINESS AND TURBIDITY

Authors            : F.KASTEN, H.J.GOLCHERT, M.STOLLEY

Contract number : ESF-004-80 D  (B)

Duration           : 3 years     1 January 1980 - 3 December 1982

Head of project : F.Kasten, Deutscher Wetterdienst,
                   Meteorologisches Observatorium Hamburg

Contractor         : Deutscher Wetterdienst, Zentralamt

Address            : Frankfurter Strasse 135
                     6050 Offenbach am Main

## Summary

The parameterization formula for global radiation as function
of total cloud amount found earlier for Hamburg is confirmed
to be valid for 7 other stations, the constants being but
slightly different. A simple parameterization formula is
found for global radiation from cloudless sky as function of
Linke turbidity factor. Both formulae combined describe global radiation as function of solar elevation, cloud amount
and turbidity factor. - Eight different parameterization formulae for atmospheric radiation from cloudless sky as function of temperature and water vapor pressure are shown to
fairly agree with 12 year records of measurements. Further
analysis including comparison to the results of rigorous radiative transfer calculations is required.

# 1. Introduction

Besides solar elevation, cloudiness is the factor which most influences solar as well as terrestrial radiation. An investigation of the dependence of solar and terrestrial radiation flux densities on cloud amount and cloud type has been published in the past (1).

The aim of the present study is a) to incorporate turbidity as a kind of second order effect; b) to refine the dependence of the terrestrial radiation flux densities on cloudiness by taking temperature and water vapor content as additional parameters; c) to apply the methods developed for the pilot station, Hamburg, to other stations, and to eventually modify these methods in order to take account of different climatic conditions.

In reference (1), 10 year continuous records at Hamburg of hourly sums of solar and terrestrial radiation flux densities had been evaluated with regard to simultaneous hourly cloud observations. The following results obtained are pertinent to the present study:

a) The ratio of global radiation at total cloud amount N okta, $G(N)$, to global radiation from cloudless sky, $G(0)$, at the same solar elevation $\gamma$ turned out to be independent of $\gamma$ and could be parameterized by

$$G(N)/G(0) = 1 - a \cdot (N/8)^b. \qquad (1)$$

b) For the global radiation from cloudless sky, $G(0)$, the following linear parameterization could be concluded:

$$G(0) = A \cdot \sin\gamma - B \qquad (2)$$

where $\gamma$ = solar elevation. Of course, this equation represents an average over all occuring atmospheric turbidities.

c) The influence of cloud type was demonstrated by the ratio of irradiance from skies overcast by a specific cloud type, $G(8)$, to the irradiance from cloudless sky, $G(0)$. These ratios $G(8)/G(0)$ which may be interpreted as the transmittances of the specific cloud types for global radiation turned out to be independent of solar elevation and had the following mean values: Cirrus 0.61, Altus 0.27, Cumulus 0.25, Stratus 0.18, Nimbostratus 0.16.

d) Atmospheric radiation A increased with total cloud amount N; the increase was more pronounced at low solar elevations $\gamma$ i.e. at times of day and year with relatively low temperatures.

# 2. Methods
## 2.1 Application to other stations of the parameterization methods found for Hamburg

For the 12 year period 1964-1975, complete data of hourly sums of global radiation were available on magnetic tape from 8 German stations: Norderney, Hamburg, Braunschweig, Braunlage, Würzburg, Trier, Weihenstephan, Hohenpeissenberg. For the same time period and from all 8 stations, hourly selected synoptic weather data were manually taken from the meteorological observation diaries of the stations and transferred onto magnetic tape. The cloud data on tape were subjected to quality control procedures based on meteorological reasoning.

On the basis of these data, the coefficients a, b in Eq. (1) and A, B in Eq(2) were determined for each of the 8 stations mentioned.

## 2.2 Consideration of turbidity

From the records at Hamburg of the 12 year period 1964 -1975, 3799 cloudless hours were selected on the following conditions: a) relative sunshine duration $S/S_0 = 1$; b) total cloud amount $N \leqslant 1$ okta at both the beginning and the end of the respective hour. The hourly sums of global radiation G and diffuse solar radiation D in these cloudless hours were used to compute the hourly sums or mean hourly irradiances, respectively, of normal direct solar radiation $I(\gamma)$ by means of

$$I(\gamma) = (G - D)/\sin\gamma \tag{3}$$

where $\gamma$ = solar elevation at the middle of the respective hour. From these hourly data of $I(\gamma)$, hourly values of the Linke turbidity factor $T_L$ were computed with the help of the quite accurate parameterization derived in reference (2):

$$T_L = (0.9 + 9.4 \cdot \sin\gamma) \cdot \ln \left[I_0/I(\gamma)\right] \tag{4}$$

where $I_0$ = extraterrestrial solar irradiance at the time of observation.

Since a well defined value of the turbidity factor $T_L$ is now assigned to each hourly value of global radiation from cloudless sky, $G(0)$, the plot of the latter versus solar elevation $\gamma$ can be discriminated with respect to $T_L$, cf. Fig.1.

## 2.3 Dependence of atmospheric radiation on temperature and water vapor content

Several parameterization formulae for the atmospheric radiation from cloudless sky have been proposed in the literature; a selection is compiled on Table I. These formulae were checked on the basis of the hourly records at Hamburg of the 12 year period 1964-1975, screened for cloudless hours which amounted to 8197 values, day and night as well.

From the hourly data of temperature T and water vapor pressure e, the mean hourly irradiances of atmospheric radiation, $A_c$, were computed according to each of the parameterization formulae listed on Table I. These computed values $A_c$ were plotted versus the corresponding measured values $A_m$, separately for daylight hours and night hours. An example is shown in Fig.2. The correlation coefficients for the correlations between $A_c$ and $A_m$ were also determined separately for day and night; they are listed as $R_{day}$ and $R_{night}$ on Table I.

## 3. Results
### 3.1 Global radiation dependent on total cloud amount, and global radiation from cloudless sky

The parameterizations Eq.(1) and Eq.(2) are valid for all 8 stations considered. The constants for each station have the following numerical values; a and b are dimensionless, A and B have the dimension $kW \cdot m^{-2}$:

| Station | a | b | A | B |
|---|---|---|---|---|
| Norderney | 0.75 | 3.4 | 1.02 | - 0.05 |
| Hamburg | 0.75 | 3.4 | 0.97 | - 0.04 |
| Braunschweig | 0.70 | 3.4 | 0.95 | - 0.04 |
| Braunlage | 0.75 | 2.6 | 0.98 | - 0.02 |
| Würzburg | 0.70 | 3.0 | 0.97 | - 0.03 |
| Trier | 0.70 | 3.0 | 0.97 | - 0.03 |
| Weihenstephan | 0.70 | 3.4 | 0.97 | - 0.02 |
| Hohenpeissenberg | 0.70 | 3.4 | 1.05 | - 0.02 |

## 3.2 Global radiation from cloudless sky as function of turbidity

Since global radiation from cloudless sky, $G(0)$, is considered to be proportional to the extraterrestrial radiation on horizontal plane, $I_0 \cdot \sin\gamma$, the ratio $G(0)/I_0 \cdot \sin\gamma$ was analyzed as function of turbidity factor $T_L$. The values of ln $[G(0)/I_0 \cdot \sin\gamma]$ plotted versus $T_L \cdot m$ with $m$ = relative air mass $\approx 1/\sin\gamma$ turned out to be arranged along a straight line. From that, the following equation was derived:

$$G(0)/kW \cdot m^{-2} = 0.83 \cdot \overline{I}_0 \cdot \sin\gamma \cdot \exp(-0.026 \cdot T_L/\sin\gamma) \qquad (5)$$

where $\overline{I}_0 = 1.367$ kW·m$^{-2}$ = the solar constant. When the exponential function in Eq.(5) is developed into a series, Eq.(5) reduces to a linear function of $\sin\gamma$ i.e. to the simpler parameterization Eq.(2).

The quality of the parameterization may be judged from Fig.1 where the three solid curves represent the function in Eq.(5) for $T_L$ = 2, 3, and 4. Two other diagrams drawn for $T_L$ = 5, 6, 7 and $T_L$ = 8, 9, 10, respectively, also show good agreement between the measured values and the parameterized curves.

## 4. Conclusions

The combined effect of total cloud amount and turbidity factor on global irradiance can be described by Eq.(1) combined with Eq.(5). Whereas the constants a and b in Eq.(1) describing the dependence on cloud amount are slightly different for different stations, Eq.(5) which parameterizes the dependence on turbidity is supposed to be of a more general nature. Validation of Eq.(5) by applying it to measured data of other stations will be performed in the future.

Many known parameterization formulae for atmospheric radiation fairly agree with the measured data. It is planned to compare the measured and the parameterized atmospheric radiation data to values which are derived from radio soundings of the vertical temperature and humidity profiles by rigorous radiative transfer calculations.

## References
(1) F.Kasten, G.Czeplak: Solar and terrestrial radiation dependent on the amount and type of cloud. Solar Energy 24, 177-189 (1980).
(2) F.Kasten: A simple parameterization of the pyrheliometric formula for determining the Linke turbidity factor. Meteorologische Rundschau 33, 124-127 (1980).

| Author | | Formula | R_day | R_night |
|---|---|---|---|---|
| Brunt | (B) | $A/\sigma T^4 = 0.60 + 0.056 \cdot (e/mbar)^{0.5}$ | 0.79 | 0.93 |
| Ångström | (A) | $A/\sigma T^4 = 0.79 - 0.174 \cdot 10^{-0.04125 \cdot (e/mbar)}$ | 0.79 | 0.93 |
| Feussner | (F) | $A/\sigma T^4 = 1 - 10^{-0.5(e/mbar)^{0.2}}$ | 0.79 | 0.93 |
| Idso + Jackson | (IJ) | $A/\sigma T^4 = 1 - 0.261 \cdot \exp(-7.77 \cdot 10^{-4} \cdot (t/^\circ C)^2)$ | 0.79 | 0.91 |
| Staley + Jurica | (SJ) | $A/\sigma T^4 = 0.718 \cdot (e/mbar)^{0.0875}$ | 0.79 | 0.93 |
| Swinbank | (S) | $A/Wm^{-2} = 5.31 \cdot 10^{-13} \cdot (T/K)^6$ | 0.79 | 0.91 |
| Cole | (C) | $A/Wm^{-2} = 222 + 4.94 \cdot t/^\circ C$ | 0.78 | 0.91 |
| Unsworth + Monteith | (UM) | $A/Wm^{-2} = -119 + 1.06 \cdot \sigma T^4/Wm^{-2}$ | 0.79 | 0.91 |

Table I: Parameterization formulae for atmospheric radiation according to:

(B) Quart.J.Roy.Met.Soc. 58, 389 (1932); cited after R.Dogniaux, prelim.note (1981).

(A) Astrophys.J. 37, 305 (1918). - Met.Z. 33, 529 (1916). - Gerl.Beitr.Geophys. 21, 145 (1929).

(F) cited after F.Möller, Einführung in die Meteorologie II, 53, Mannheim (1973).

(IJ) J.Geophys.Res. 74, 5397 (1969).

(SJ) J.Appl.Met. 11, 349 (1972).

(S) Quart.J.Roy.Met.Soc. 89, 339 (1963).

(C) Solar Energy 22, 459 (1979).

(UM) Quart.J.Roy.Met.Soc. 101, 13 (1975).

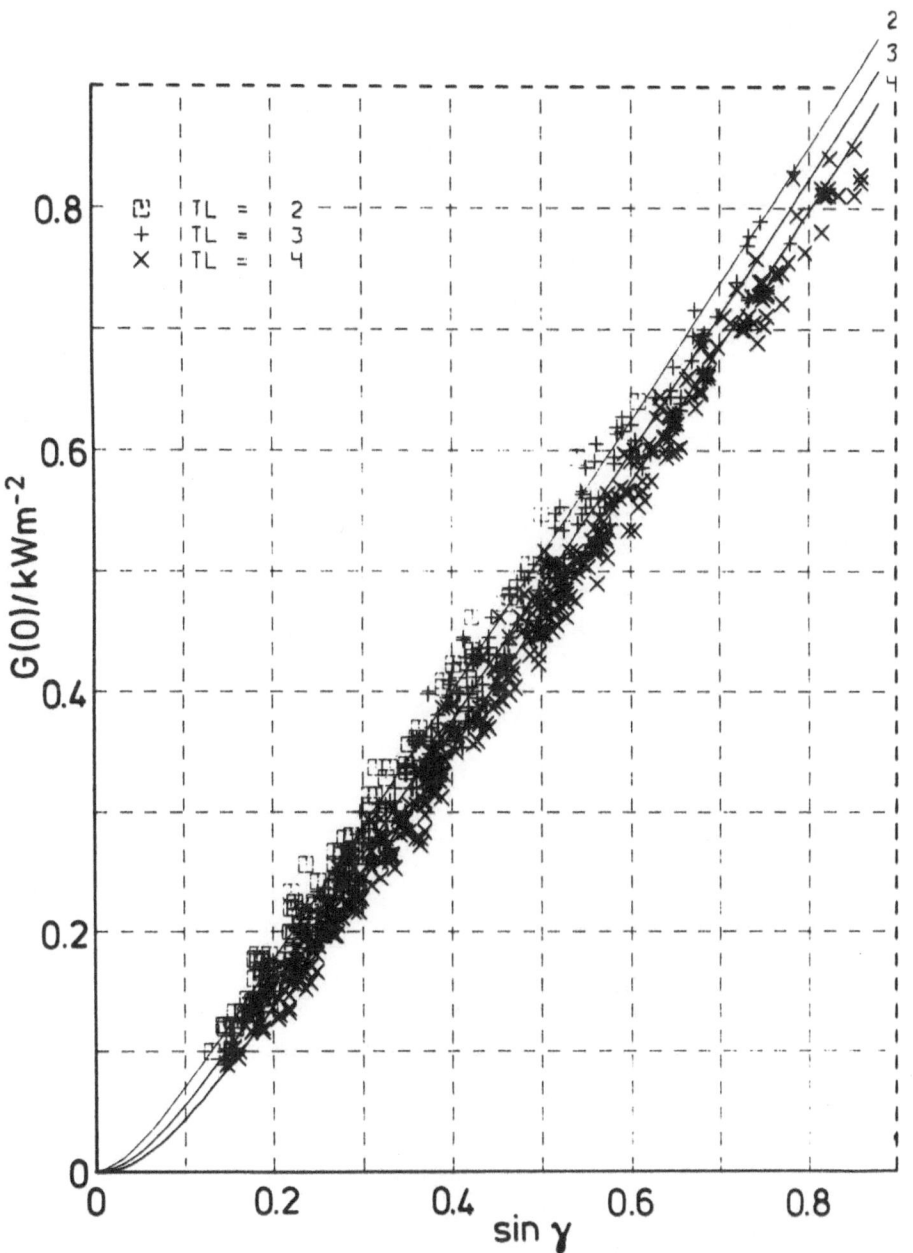

Fig.1: Global radiation from cloudless sky, G(0), versus the sine of solar elevation $\gamma$, separated according to Linke turbidity factor TL. Hourly means, Hamburg 1964-1975. Solid lines: Parameterization function

$$G(0)/kW \cdot m^{-2} = 0.83 \cdot I_0 \cdot \sin\gamma \cdot \exp(-0.026 \cdot T_L/\sin\gamma).$$

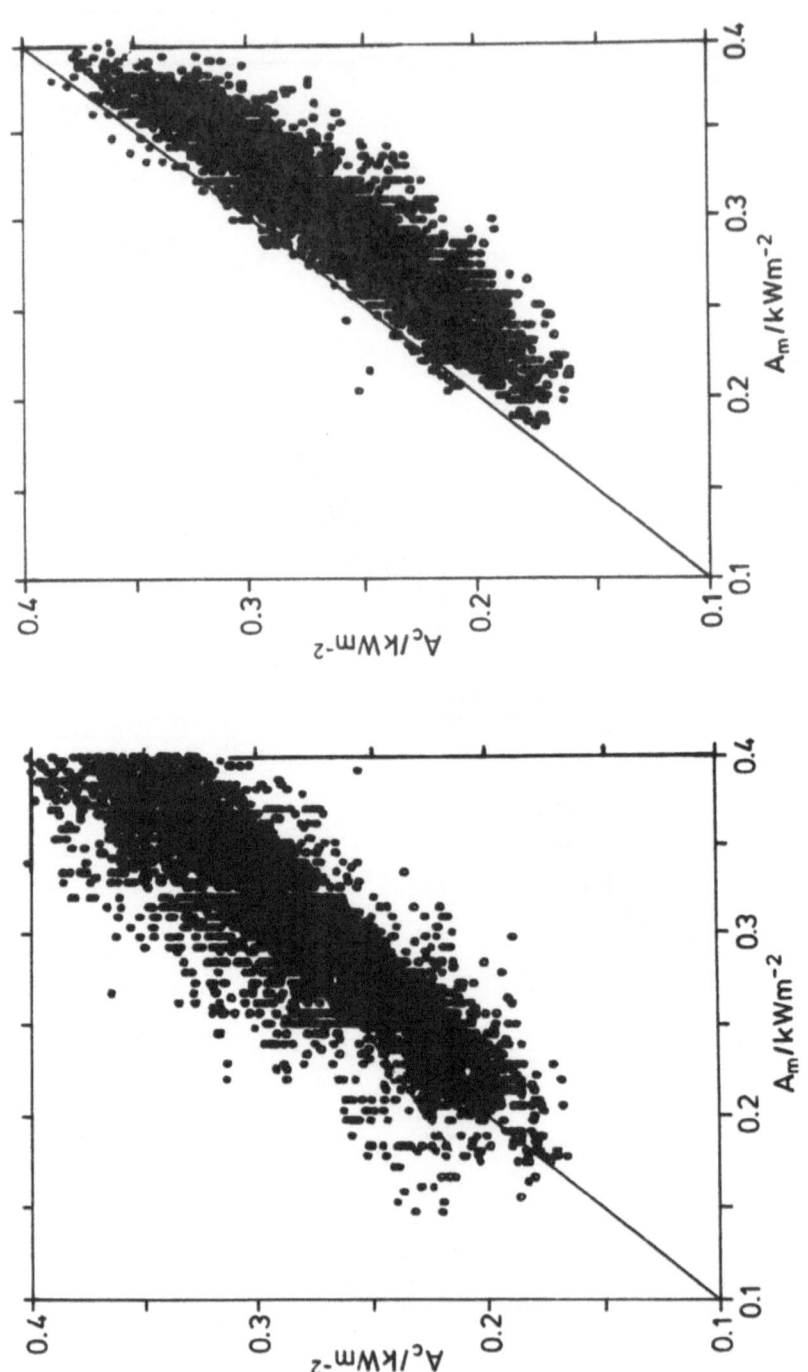

Fig.2: Atmospheric radiation from cloudless sky, computed values $A_c$ versus measured values $A_m$. Hourly means, Hamburg 1964-1975. $A_c$ computed from Brunt's formula, see Table I Left: daytime values, right: nighttime values.

# DIFFUSE SKY RADIATION ON TILTED SURFACES

Contract number : ESF - 005 - F
Authors         : J.A. BEDEL, J. JAN, S. JANICOT
Total budget    : 662 098 FF - CEC contribution 50 %
Duration        : 1.01.1980 to 30.06.1983

## I)- AIM OF THE WORK.

Climatological study of the diffuse sky radiation on tilted surface.
The aim is to improve the models of assessment of the global radiation received on tilted surfaces. For that purpose, the diffuse sky radiation is resolved into 3 components, a direct component (DIR), an isotropic component (DISO) and a zenithal component (DZ). These 3 components enable to reconstitute the diffuse sky radiation on slopes. For example, we have, in case of horizontal surface :
DH = DISO + DZ + DIR. sin(h), h is the solar elevation.

The first part of this study is to find a method to assess these components.

Secondly, the validity of the model is assessed with real hourly data at Trappes and Carpentras, by computing cumulative frequency curves. We also compare the model with one based on isotropic diffuse sky radiation. Finally, statistical tables of the daily global radiation on slopes are edited. These tables are obtained, in case of validation of the model, from global (on horizontal surface), and direct (normal incidence) radiations measured at Trappes and Carpentras during the period 1971-1980, and by using the proposed method for calculating diffuse sky radiation.

## II)-TIME SCHEDULE

1980-1981: Constitution of file for the study of diffuse sky radiation.
Study of decomposition of diffuse sky radiation.

1982      : Estimation of components of diffuse sky radiation. Test of the method.

1983(first half-year): Presentation of results and draft of the final report.

## III- MEANS AND METHODS

III.1) 6-minutes data of direct and global radiation on
tilted surfaces are available for Trappes (from 3 april
1979 to 31 january 1982), and Carpentras (from 4 january
1979 to 31 january 1982). We study the hourly irradiation
on the following surfaces :

- horizontal surface,
- vertical surface directed eastward, westward, southward
  andnorthward,
- tilted surface at 45° southward.

III.2) We have decomposed the diffuse sky radiation into
3 components :
- "direct" component which comes directly from the circumsolar area;
- isotropic component;
- zenithal component, the latter can be either positive or
  negative.
It is negative, in particular for high values of the atmospheric turbidity (for clear skies conditions).

III.3 We try to set formulae for estimating these 3 components by using 2 parameters :

- the solar elevation above the horizon (taken at half-
  hour),
- the "extended" turbidity factor T' : T' is a generaliza-
  tion of the Linke's factor for cloudy skies. It can be
  defined from Linke's turbidity factor T by the relation:
  $T' = T + \frac{\delta_c}{\delta_r}$ where $\delta_c$ is the mean value of the scattering
  optical thickness of the clouds and $\delta_r$ the optical
  thickness of a clean and dry atmosphere (Rayleigh
  scattering).

The value of T' is obtained by using the hourly values of
direct solar radiation in the formulae used to calculate
the turbidity factor T, without taking account of the
sunshine duration. It is calculated only when the direct
solar radiation is over 1J/cm2.

Note that, a priori, calculation of these 3 components is
available only at 12 o'clock.

III.4 Finally, we compare, for each month and each of the
6 surfaces, the cumulative frequency curves of global radiation obtained from :

- the hourly measurements of global radiation at Trappes
  and Carpentras,
- the model based on isotropic diffuse sky and ground
  radiation to which we add the direct radiation,

- the model computing the diffuse sky radiation by the
  proposed method (extended turbidity) to which we add
  the direct radiation and the diffuse radiation from the
  ground (isotropic one).

IV - <u>RESULTS</u>

IV.1 - The 3 components, DISO, DIR, DZ are represented (see figures) by classes of sun elevation, according to extended turbidity T', in order to parametrize these 3 components.

IV.2 - We have edited the cumulative frequency curves of hourly of global radiation computed with the model based on isotropic diffuse radiation.

IV.3 - The edition of statistical tables of the daily global radiation on slopes is in progress (see example).

V - <u>FUTURE WORK</u>

V.1 - Choose for the assessment of the different components of diffuse sky radiation between a parameterization according to the extended turbidity and solar elevation, and more simply, mean values (contingency tables).

V.2 - Assess the validity of the proposed model for hours different from 12 o'clock.

V.3 - Compare the hourly sums with the 2 elaborated models by cumulative frequency method.

V.4 - Draft of the final report.

VI - <u>OPERATING DIFFICULTIES</u>

We have some problems about tests used for the validity of the hourly sums. May be, they are too strict : for the computation of cumulative frequencies, we have only 24 days at Trappes (for a 2 -years- period) and 210 days at Carpentras.

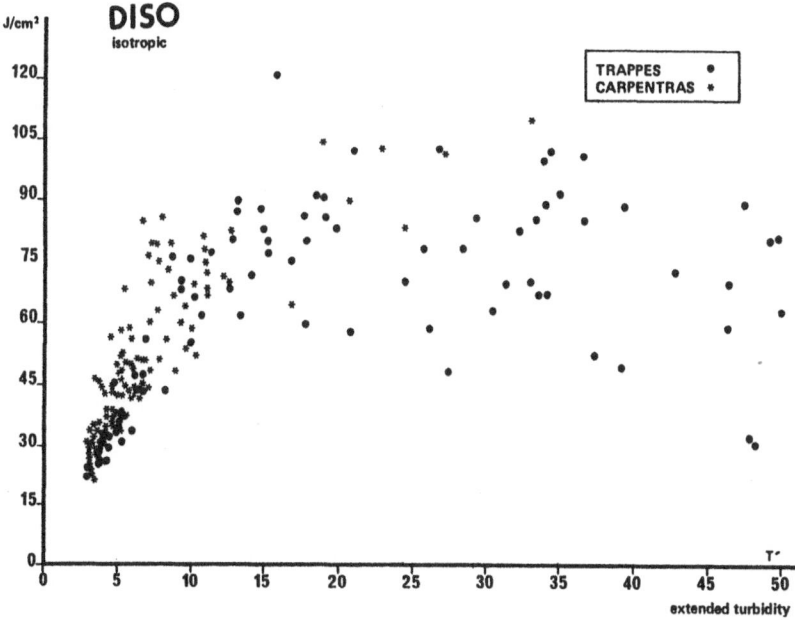

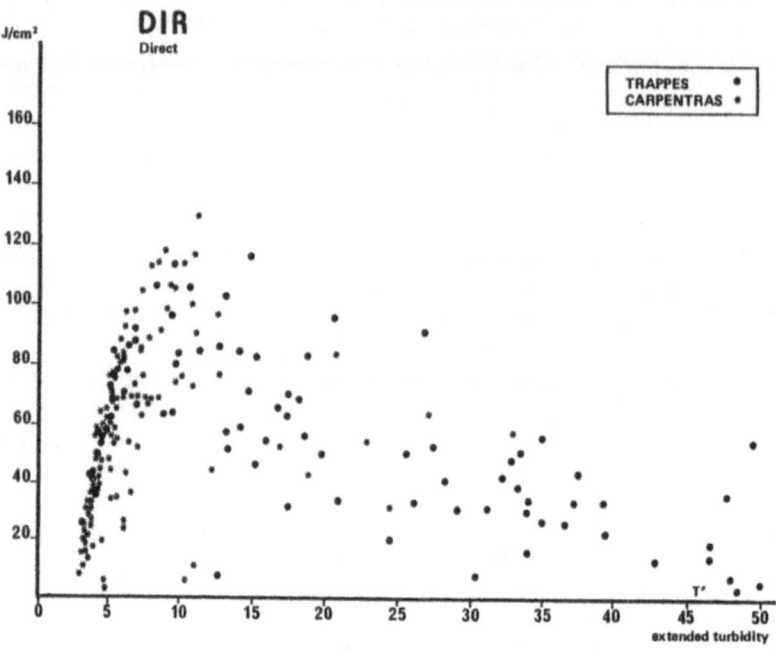

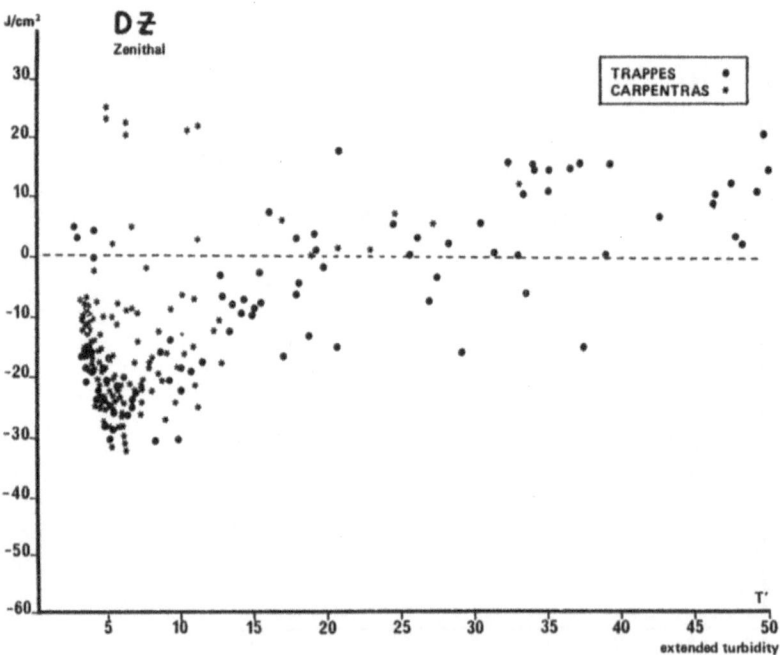

# STATISTICAL TABLES OF IRRADIATION

## example of presentation

METEOROLOGIE ( FRANCE ) - SMM/CLIM/DEV

PARICKAMF

DOCUMENT ETABLI D APRES L ETAT DU FICHIER AU 82/04/21.

ALTITUDE : 20 METRES
LATITUDE : 43:27 NORD (DEG.MIN)
LONGITUDE : 5.13 EST (DEG.MIN)

RAYONNEMENT SOLAIRE GLOBAL RECU SUR UNE SURFACE HORIZONTALE

DISTRIBUTION STATISTIQUE AU NIVEAU MENSUEL
PERIODE DU 1/ 3/1979 AU 31/12/1980

1. DISTRIBUTION STATISTIQUE DES IRRADIATIONS QUOTIDIENNES - UNITE : JOULE/CM2/JOUR

| | JANV | FEVR | MARS | AVRI | MAI | JUIN | JUIL | AOUT | SEPT | OCTO | NOVE | DECE | ANNEE |
|---|---|---|---|---|---|---|---|---|---|---|---|---|---|
| NOMBRE D OBS. | 31 | 28 | 29 | 59 | 56 | 58 | 57 | 59 | 58 | 59 | 60 | 62 | 616 |

2. FREQUENCES POUR MILLE DES CAS OU L IRRADIATION QUOTIDIENNE A DEPASSE LES SEUILS INDIQUES EN JOULE/CM2/JOUR

3. DUREE MOYENNE PAR JOUR (EN HEURES X 10) DURANT LAQUELLE L ECLAIREMENT MOYEN HORAIRE A DEPASSE LES SEUILS INDIQUES (EN WATTS/M2)

4. RAPPEL : L IRRADIATION EXTRATERRESTRE MOYENNE RECUE SUR UNE SURFACE HORIZONTALE - UNITE : JOULE/CM2/JOUR

| | JANV | FEVR | MARS | AVRI | MAI | JUIN | JUIL | AOUT | SEPT | OCTO | NOVE | DECE | ANNEE |
|---|---|---|---|---|---|---|---|---|---|---|---|---|---|

- 119 -

# CLIMATOLOGICAL VALUES OF SOLAR IRRADIATION ON THE HORIZONTAL AND SEVERAL INCLINED SURFACES AT DE BILT

Authors            : W.H. Slob

Contract number    : ESF-006-80NLB

Duration           : 1-10-79  -  1-7-1982

Total budget       : ƒ 422.000,--          CEC contribution: ƒ 153.000,--

Head of project    : Drs. A.P. van Ulden

Contractor         : The Royal Netherlands Meteorological Institute

Subcontractor      : Institute of Applied Physics TNO-TH

Address            : Wilhelminalaan 10
                     3730 AE De Bilt

## Summary

Climatological values of the direct and the diffuse solar irradiation on the horizontal surface at De Bilt are calculated. A data set of about $2\frac{1}{2}$ years on 11 different orientations in Cabauw is used to relate the diffuse solar irradiation on the orientations to the diffuse solar irradiation on the horizontal surface. These relations were used together with the climatological diffuse solar irradiation on the horizontal surface and sunshine climatology to calculate monthly climatological diffuse irradiations on these orientations. Monthly climatological values of the direct irradiation on each orientation were calculated assuming an even distribution of the sunshine over the day and a constant Linke turbidity factor for each month during sunshine. The groundreflected irradiation on the orientations was calculated for an albedo of 0.2. Halfmonthly climatological values for the global, the direct, the diffuse and the groundreflected daily irradiation are calculated on the following orientations at De Bilt: horizontal, east $90°$, south $90°$, west $90°$, north $90°$, east $45°$, south-east $45°$, south $45°$, south-west $45°$, west $45°$, south $22,5°$ and south $67,5°$.

# Climatological values of solar irradiation on the horizontal and several inclined surfaces at De Bilt

## 1. Introduction

The aim of this study was to produce daily climatological solar irradiation data on several inclined surfaces using the relative sunshine duration as input parameter. At De Bilt the following measurements were available:
a) Sunshine duration (about 80 years with Cambell Stokes).
b) Global solar irradiation on the horizontal (about 16 years with pyrano-meter).
c) Direct solar irradiation on the horizontal (about 10 years with pyrhe-liometer).
d) Measurements at several inclined surfaces at Cabauw (about $2\frac{1}{2}$ years).
The long set of sunshine duration data was used to get:
- a good climatological monthly mean of daily relative sunshine duration
- the occurrence probability as a function of relative sunshine duration for each month.
The measurements of global and direct solar irradiation were used to cal-culate a quardratic regression as a function of the relative sunshine duration.

## 2. Calculation for the horizontal surface

The daily data of global, direct and diffuse were assigned to:
- the 12 months of the year
- 11 classes of daily relative sunshine duration $S/S_o$
  $(0, 0-0.1, 0.1-0.2, ..., 0.9-1.0)$.
Then we calculated for the direct, the diffuse and the global daily irra-diations the best quardratic regression equation with $S/S_o$ as parameter for each month.
From the long term sunshine data we calculated the occurrence probability in each class of $S/S_o$ for each month. The regression equations were com-bined with the occurrence probabilities to calculate the climatological average values of the direct, the diffuse and the global daily irradiation on the horizontal surface. Fig. 1 shows the result of the calculation for De Bilt.

## 3. Calculation on the inclined surface

The global solar irradiation on an inclined surface is the sum of 3 components namely:
a) the direct irradiation.
b) the diffuse sky irradiation.
c) the ground reflected irradiation.
Each of these components is calculated separately.

### 3.1. The direct solar irradiation

When we assume that the Linke turbidity factor is constant during sunshine (for the calculations we used an average monthly value) and that the sun-shine duration is evenly distributed over the day, we can calculate the ratio of the direct irradiation on the inclined and the direct irradiation on the horizontal surface. This ratio only depends on the geographical position, the date, the Linke turbidity factor and the orientation and inclination angle of the surface. Half monthly values of this ratio were calculated for several orientations and inclination angles for De Bilt. A table of these results is available and makes it possible to calculate the average direct irradiation on any inclined surface when the direct irradiation on the horizontal surface is known.

## 3.2. The diffuse sky irradiation

From the measurements in Cabauw on the inclined surfaces we got:
- the diffuse daily irradiation on the horizontal surface $D_H$.
- the diffuse daily irradiation on the inclined surface $D_S$.
Now we defined a diffuse multiplication factor $f_{DS}$, which was calculated from

$$D_S = f_{DS} \cdot D_H \cdot (\cos \frac{\alpha}{2})^2 . \tag{1}$$

$\alpha$ is the inclination angle of the inclined surface S.
The factor $f_{DS}$ accounts for the anisotrophy of the sky irradiance, because

$D_H \cdot (\cos \frac{\alpha}{2})^2$ is the expected diffuse sky irradiation on the surface S when the sky radiance was istropic.
The factor $f_{DS}$ depends on
- the orientation and inclination angle of S.
- the date.
- the relative daily sunshine duration $S/S_o$.
For each month and each orientation $f_{DS}$ was plotted as a function of $S/S_o$. (From the measurments in Cabauw). A linear regression equation for each orientation and month fitted the data very well.

$$f_{DS} = a_{MS} + b_{MS} \cdot S/S_o \tag{2}$$

The values of $a_{MS}$ and $b_{MS}$ were calculated for each month and each of the 11 different orientations. The values of $a_{MS}$ are nearly constant during the year and shows the factor for an overcast day.
The value of $(a_{MS} + b_{MS})$ shows the factor for a totally clear day.
In Fig. 2 $a_{MS}$ and $(a_{MS} + b_{MS})$ are plotted for the south $45^o$ orientation. In this case the isotropic approximation seems rather good for overcast skys $(S/S_o = 0)$. For the clear days however $(S/S_o = 1)$ the isotropic approximation only gives reasonable results in the summer months. In winter $(a_{MS} + b_{MS})$ shows values of about 2.5. The equations (1) and (2) make it possible to estimate $D_S$ when $D_H$ and $S/S_o$ are known. $D_H$ can be calculated as a function of $S/S_o$ with the quardratic equations mentioned in II. Combining this with the occurrence probablities of $S/S_o$ in the 11 classes of sunshine duration the climatological diffuse sky irradiation for the surface S can be calculated.

## 3.3. The ground reflected irradiation

The ground reflected irradiation on a surface with an inclination angle $\alpha$ and an albedo $\rho$ can be written as:

$$R = \rho \ G_H \ (\sin \frac{\alpha}{2})^2 \tag{3}$$

$G_H$ is the horizontal global solar irradiation, which can be calculated with the regression equations mentioned in II. For the albedo we took a value of 0.2.

## 3.4. The global irradiation

The global irradiation is the sum of the 3 components mentioned before. Fig. 3 and Fig. 4 show the results of the climatological calculation for the orientations east $90^{\circ}$, south $90^{\circ}$, north $90^{\circ}$ and south $45^{\circ}$, but the calculations were also carried out for the orientations east $45^{\circ}$, south-east $45^{\circ}$, south $67,5^{\circ}$, south $22,5^{\circ}$, south-west $45^{\circ}$, west $45^{\circ}$ and west $90^{\circ}$.

## 4. Conclusion

A long set of sunshine data in combination with a relative short set of measurements and the regression equations resolved from these measurements can help to get a better long term average of the measured values.

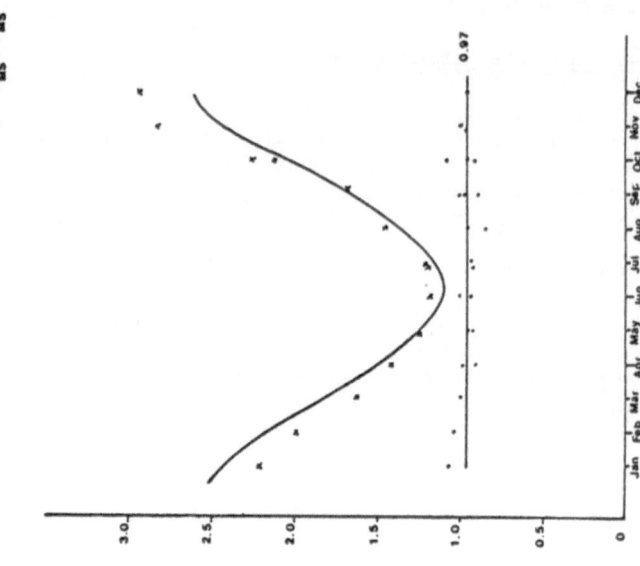

Fig. 2 – Values of the regression coefficients for South 45°

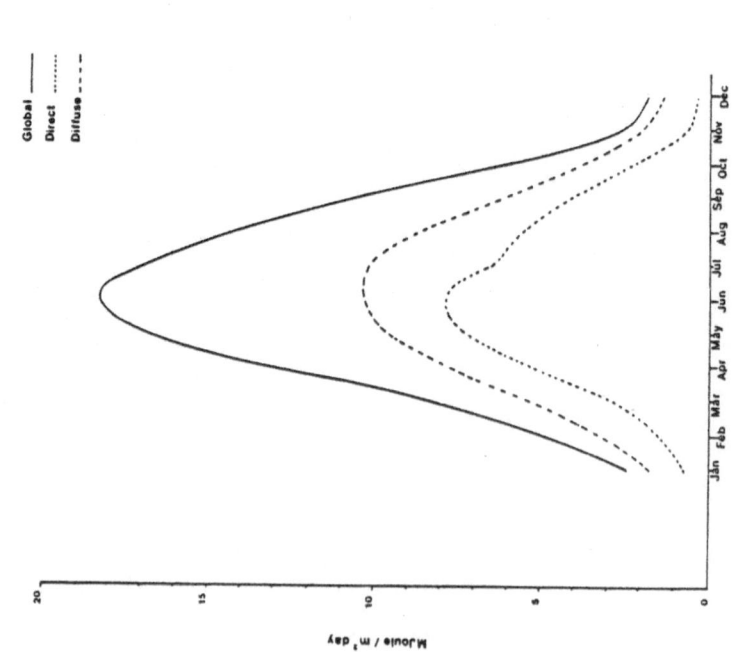

Fig. 1 – Climatological irradiation at De Bilt for the horizontal surface

- 124 -

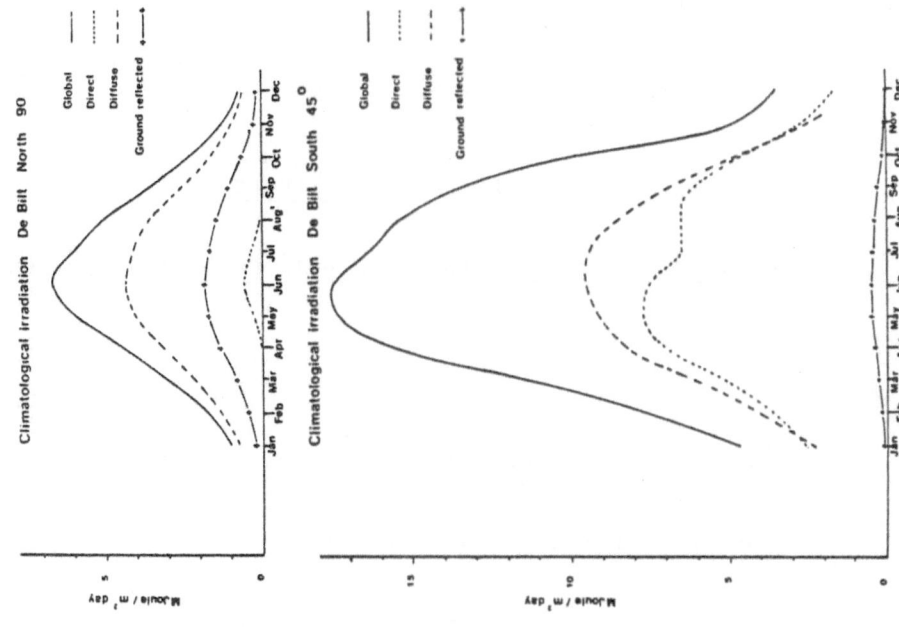

Fig. 3

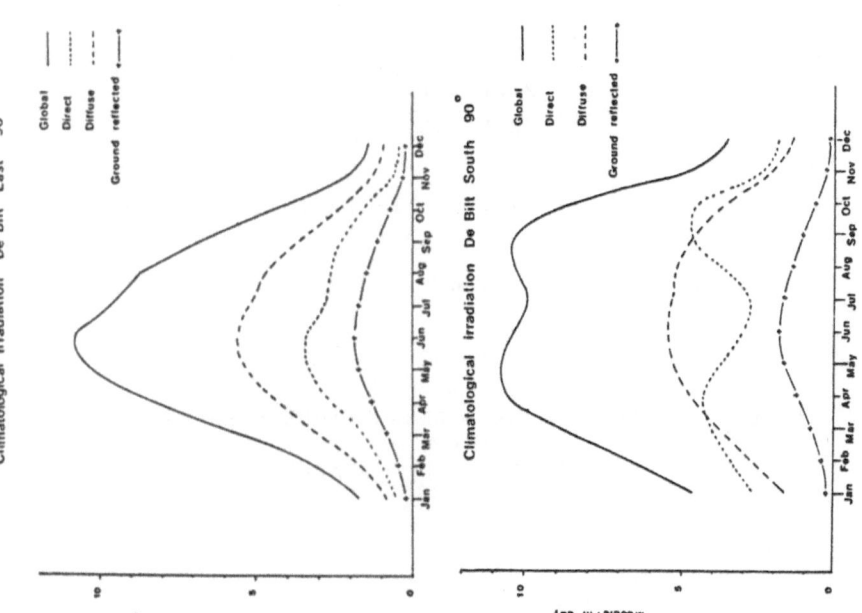

Fig. 4

- 125 -

DEVELOPMENT METHODS

## ANALYSIS OF SOLAR RADIATION DATA OF A HIGH ALTITUDE
## STATION (1 508 m)


Contribution to : Action 3 - 2


Contract number : ESF - 009 - F


Duration : 2 years        from July 1981        to June 1982


Total budget : 40 KFF                CEC contribution : 20 KFF
(in national currency)


Head of project : Monsieur A.VIALARON
                (ask J.F.TRICAUD)


Authors : J.F.TRICAUD, Mme LE PHAT VINH, Mme PARRA


Contracting organisation : C.N.R.S.



Address : LABORATOIRE D'ENERGETIQUE SOLAIRE
          BP N° 5 ODEILLO
          66120 FONT ROMEU - FRANCE

## GOAL OF THE WORK

1) Study of the distribution of diffuse radiation for different conditions of turbidity and environment.

2) Checking and analysis of hourly and daily values of global radiation for vertical surfaces facing south, east and west and southward oriented surface with slope angle equal to the latitude of ODEILLO.

3) Analysis of relationships between direct, diffuse and global radiation in ODEILLO with regard to duration of sunshine, slope of the surface, orientation and reflected radiation.

## STATUS AND FIRTS RESULTS

The equipments required for the data logging are operative since October 1981. Two disks files are available, one in tenth-hour the other in hourly values. A model of monthy presentation is given TABLE 1.

The daily values of the diffuse and global radiation for horizontal surface, vertical and inclined south-facing surfaces has been analysed for three months of 1982.

A lot of curves (linear regression) is given. It represent the monthly variation of diffuse radiation for horizontal surface, vertical and inclined south facing surfaces in regard with respectively : relative sunshine duration "$\sigma$" and the expression representing the gap between latitude of the sation and declension affected of the relative sunshine duration "$(\Phi - \delta).\sigma$".

Terminology :

GH : global radiation for horizontal surface

DH : Diffuse radiation for horizontal surface

Di$\Phi$ : Diffuse radiation for inclined surface south-facing with slope equal to the latitude.

DVS : Diffuse radiation for vertical south-facing surface.

gRVS : ground reflected radiation for vertical south-facing surface.

$r_2$ : correlation coefficient

For each month are given the extrem values of the ratio gRVS/GH. This value represent the effective effect of ground radiation for the vertical face of a building.

For ODEILLO we have to note that the measurement is made at fifty meters of the ground. As soon as possible we should make a comparison with ground reflected radiation with a sensor at five meters of the ground.

In regard with "$\sigma$" and "$(\Phi - \delta).\sigma$ we have the next linear regression equation :

I) In January 1982, 31 set of daily values
- For horizontal surface :

$$DH/GH = 0,990 - 0,826\sigma \qquad \text{(Fig : 1)}$$
$$r_2 = 0,93$$

- For inclined surface :

$$Di\Phi/DH = 0,751 + 0,008 \ (\Phi - \delta).\sigma \qquad \text{(Fig.: 2)}$$

- For vertical surface :

$$DVS/DH = 0,466 + 0,016 \ (\Phi - \delta) \ \sigma \qquad \text{(Fig : 3)}$$
$$r_2 = 0,77$$

Variation of ground reflected radiation effect :
$$0,09 < gRVS/GH < 0,28$$

II) In MARCH 1982, 28 set of daily values
- For horizontal surface :

$$DH/GH = 0,909 - 0,808 \ \sigma \qquad \text{(Fig : 4)}$$
$$r_2 = 0,90$$

- For inclined surface :

$$Di\Phi/DH = 0,788 + 0,008 \ (\Phi - \delta).\sigma \qquad \text{(Fig : 5)}$$
$$r_2 = 0,71$$

- For vertical surface :

$$DVS/DH = 0,468 + 0,019 \ (\Phi - \delta).\sigma \qquad \text{(Fig : 6)}$$

$$r_2 = 0,82$$

Variation of ground reflected radiation effect :

$$0,08 < gRVS/GH < 0,23$$

    III) In   JUNE 1982, 30 set of daily values

- For horizontal surface :

$$DH/GH = 0,799 - 0,757 \, o \qquad\qquad (Fig : 7)$$

$$r_2 = 0,80$$

- For inclined surface :

$$Di\Phi/DH = 0,830 + 0,006 \, (\Phi - \delta).\sigma \qquad (Fig : 8)$$

$$r_2 = 0,39$$

- For vertical surface :

$$DVS/DH = 0,457 + 0,030 \, (\Phi - \delta).\sigma \qquad (Fig : 9)$$

$$r_2 = 0,60$$

Variation of ground reflected radiation effect

$$0,08 < gRVS/GH < 0,09$$

    In our conditions of observations wa can observe that the ratio gVRS/GH pass, by a maximum (0,23 to 0,29) when there is snow on ground and for overeast conditions ($0 < \sigma < 0,6$) and by a minimum (0,08 to 0,09) for elear sky conditions ($\sigma > 0,8$).


FUTURE WORK UNTIL THE END OF THE CONTRACT

    Fot the final report the data of all month available would be checked (for same parameters for the period 1979 - 1982) and analysed also as function of $T_L$ (the Linke Turbidity factor). The dependence of the diffuse radiation with solar altitude will be given.

# TABLE 1

## ODEILLO

**RELEVE MENSUEL DE RAYONNEMENT GLOBAL HORIZONTAL ET D'INSOLATION**
**INSOLATION**

| | |
|---|---|
| PYRANOMETRE KIPP&ZONEN   NO:5730 | HELIOGRAPHE: PYRHELIOMETRE |
| ENREGISTREUR MECI | AVEC SEUIL: 110W/M2 |
| INTEGRATEUR HF85 | UNITE EMPLOYEE: 1/10 HEURE |
| UNITE EMPLOYEE: J/CM2   RRMBO | |

ANNEE: 1982  
MOIS: JANVIER

LATITUDE : 42 29' N  
LONGITUDE: 02 07' E  
ALTITUDE : 1580 M

INTERVALLES HORAIRES EN TEMPS SOLAIRE VRAI — INTERVALLES HORAIRES EN TEMPS SOLAIRE VRAI — INSOLATION

*Note: morning intervals 0-1 through 6-7 are all 0; afternoon intervals -18 through -24 are all 0.*

| J/U/R | 7-8 | 8-9 | 9-10 | 10-11 | 11-12 | 12' MAT | -13 | -14 | -15 | -16 | -17 | 24' SOIR | TOT | M | S | T |
|---|---|---|---|---|---|---|---|---|---|---|---|---|---|---|---|---|
| 1 | 2 | 42 | 97 | 130 | 159 | 430 | 162 | 147 | 110 | 39 | 6 | 464 | 894 | 33 | 35 | 68 |
| 2 | 4 | 55 | 96 | 133 | 155 | 443 | 150 | 132 | 100 | 55 | 5 | 442 | 885 | 39 | 40 | 79 |
| 3 | 5 | 35 | 96 | 143 | 160 | 439 | 133 | 97 | 43 | 16 | 4 | 293 | 732 | 30 | 17 | 47 |
| 4 | 5 | 40 | 69 | 131 | 133 | 377 | 157 | 134 | 76 | 40 | 6 | 411 | 788 | 22 | 32 | 54 |
| 5 | 4 | 45 | 98 | 137 | 157 | 441 | 159 | 131 | 97 | 43 | 6 | 436 | 877 | 39 | 43 | 82 |
| 6 | 4 | 46 | 84 | 126 | 136 | 397 | 121 | 131 | 89 | 36 | | 377 | 774 | 39 | 37 | 76 |
| 7 | 1 | 7 | 11 | 22 | 30 | 71 | 33 | 75 | 87 | 51 | 7 | 252 | 323 | 0 | 24 | 24 |
| 8 | 3 | 37 | 66 | 139 | 168 | 413 | 133 | 90 | 68 | 39 | 8 | 336 | 749 | 36 | 15 | 51 |
| 9 | 3 | 45 | 101 | 147 | 157 | 453 | 129 | 91 | 114 | 53 | 8 | 395 | 848 | 37 | 36 | 73 |
| 10 | 4 | 37 | 74 | 109 | 121 | 345 | 138 | 135 | 93 | 53 | 5 | 424 | 769 | 21 | 36 | 57 |
| **TD** | 35 | 389 | 792 | 1217 | 1376 | 3809 | 1315 | 1151 | 877 | 432 | 55 | 3830 | 7639 | 295 | 314 | 609 |
| 11 | 3 | 30 | 121 | 136 | 160 | 450 | 96 | 110 | 83 | 60 | 7 | 356 | 806 | 22 | 20 | 42 |
| 12 | 3 | 51 | 106 | 144 | 166 | 470 | 167 | 148 | 109 | 55 | 8 | 487 | 957 | 38 | 39 | 77 |
| 13 | 2 | 56 | 105 | 122 | 166 | 451 | 165 | 146 | 83 | 38 | 8 | 440 | 891 | 38 | 37 | 75 |
| 14 | 3 | 16 | 33 | 47 | 33 | 132 | 37 | 52 | 48 | 19 | 6 | 152 | 284 | 0 | 0 | 0 |
| 15 | 3 | 23 | 36 | 28 | 6 | 96 | 7 | 5 | 3 | 5 | 2 | 22 | 118 | 0 | 0 | 0 |
| 16 | 3 | 25 | 21 | 51 | 61 | 161 | 47 | 60 | 27 | 26 | 3 | 166 | 327 | 0 | 0 | 0 |
| 17 | 3 | 22 | 41 | 59 | 99 | 226 | 92 | 67 | 32 | 17 | 3 | 211 | 437 | 1 | 0 | 1 |
| 18 | 7 | 20 | 0 | 0 | 0 | 27 | 0 | 0 | 0 | 25 | 1 | 26 | 53 | 0 | 0 | 0 |
| 19 | 7 | 61 | 115 | 129 | 110 | 422 | 108 | 165 | 111 | 53 | 6 | 443 | 865 | 27 | 30 | 57 |
| 20 | 8 | 62 | 114 | 154 | 175 | 513 | 176 | 156 | 123 | 78 | 11 | 544 | 1057 | 41 | 44 | 85 |
| **TD** | 44 | 366 | 692 | 870 | 976 | 2948 | 895 | 909 | 611 | 376 | 58 | 2847 | 5795 | 166 | 170 | 336 |
| 21 | 5 | 21 | 44 | 80 | 104 | 254 | 100 | 81 | 74 | 51 | 9 | 315 | 569 | 0 | 12 | 12 |
| 22 | 6 | 33 | 36 | 40 | 62 | 177 | 59 | 76 | 71 | 29 | 5 | 240 | 417 | 0 | 1 | 1 |
| 23 | 5 | 60 | 131 | 171 | 149 | 536 | 173 | 147 | 124 | 73 | 17 | 534 | 1070 | 29 | 41 | 70 |
| 24 | 5 | 31 | 54 | 79 | 118 | 287 | 106 | 147 | 118 | 70 | 16 | 412 | 699 | 0 | 27 | 27 |
| 25 | 4 | 11 | 19 | 33 | 42 | 109 | 51 | 53 | 54 | 31 | 7 | 196 | 305 | 0 | 0 | 0 |
| 26 | 5 | 55 | 101 | 97 | 134 | 393 | 163 | 167 | 119 | 62 | 11 | 522 | 915 | 25 | 22 | 47 |
| 27 | 5 | 21 | 55 | 67 | 65 | 213 | 52 | 44 | 40 | 20 | 4 | 160 | 373 | 0 | 0 | 0 |
| 28 | 6 | 64 | 135 | 178 | 199 | 582 | 199 | 202 | 135 | 78 | 20 | 608 | 1190 | 38 | 47 | 85 |
| 29 | 12 | 70 | 113 | 181 | 214 | 590 | 221 | 202 | 166 | 89 | 9 | 687 | 1277 | 42 | 39 | 81 |
| 30 | 11 | 68 | 126 | 170 | 191 | 566 | 192 | 171 | 131 | 76 | 20 | 590 | 1156 | 42 | 48 | 90 |
| 31 | 12 | 85 | 137 | 179 | 200 | 613 | 200 | 178 | 136 | 80 | 22 | 616 | 1229 | 42 | 48 | 90 |
| **TD** | 77 | 519 | 951 | 1275 | 1498 | 4320 | 1516 | 1468 | 1168 | 659 | 140 | 4880 | 9200 | 218 | 284 | 502 |
| **TM** | 156 | 1274 | 2435 | 3362 | 3850 | 11077 | 3726 | 3449 | 2656 | 1472 | 254 | 11557 | 22634 | 679 | 769 | 1448 |
| **MM** | 5 | 41 | 79 | 108 | 124 | 357 | 120 | 111 | 86 | 47 | 8 | 373 | 730 | 22 | 25 | 47 |

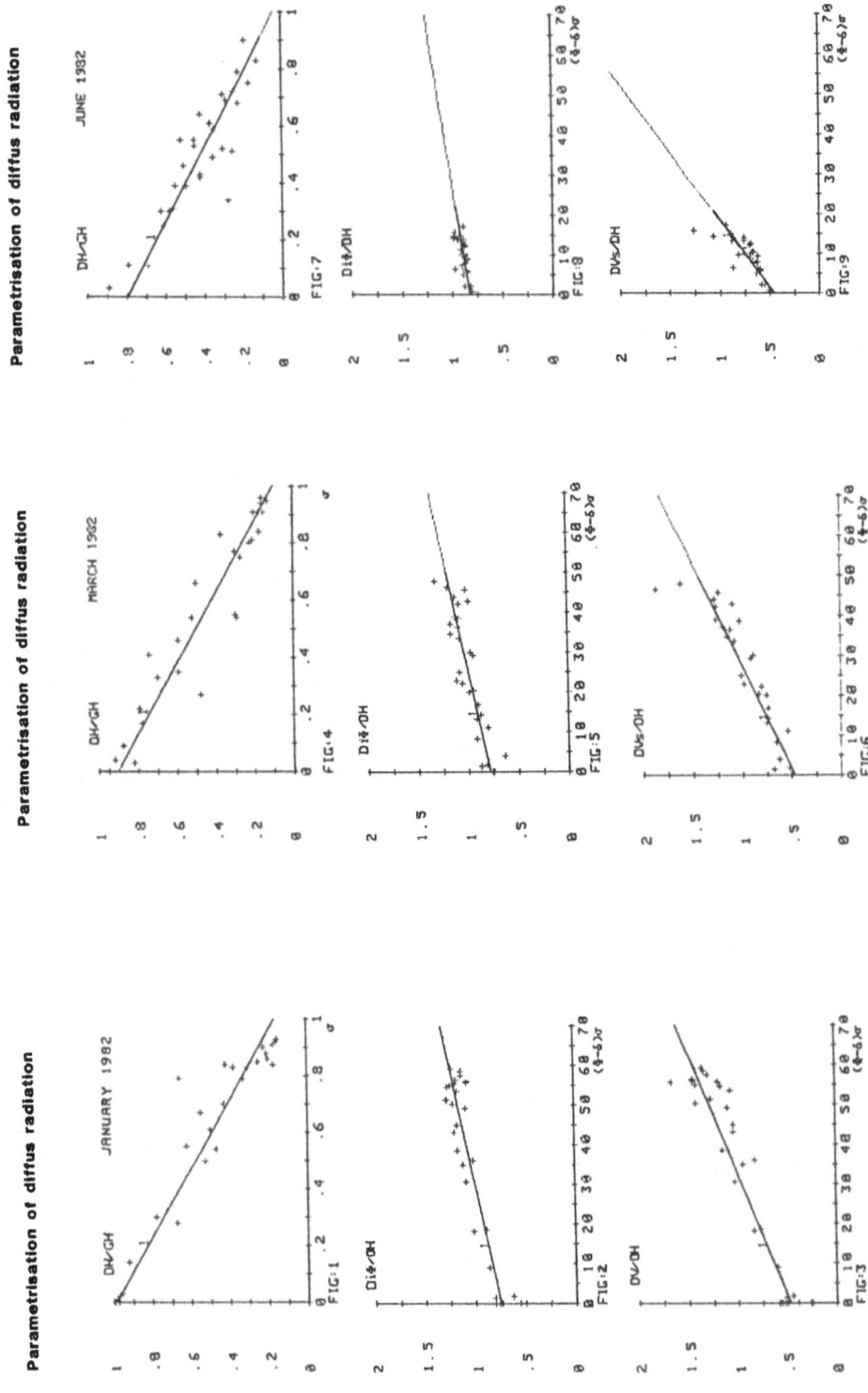

Parametrisation of diffus radiation

- 131 -

# A STATISTICAL ANALYSIS OF THE COMPONENTS OF SOLAR RADIATION ON A VERTICAL SOUTH-FACING SURFACE

Author               : K.G. Commins

Contract number      : ESF-018-EIR(G)

Duration             : 36 months     1 January 1980 - 31 December 1982

Total budget         : IR£7,452     CEC contribution   : IR£3,726

Head of project      : E.J. Murphy

Contractor           : Irish Meteorological Service

Address              : Glasnevin Hill
                       Dublin 9
                       IRELAND

## Summary

Measured hourly values of the diffuse and ground reflected radiation components on a vertical south-facing surface at Valentia Observatory are compared with the theoretically expected values and are analysed mainly with respect to solar elevation (E) and D/G.

Data are available since May 1981 but this preliminary survey was confined to June and December 1981.  An asymmetry between morning and afternoon values was found for both components.  The diffuse component asymmetry may be accounted for by the screening effect of a range of hills south-east of the Observatory while the ground-reflected component asymmetry may possibly be explained by the fact that the ground to the south-east of the pyranometers slopes upwards but is level to the south-west.

The following equations were found to fit the data for June:

$$F_{R morning} = -0.908 + 0.0430E + (1.630 - 0.04E)D/G$$

$$F_{R afternoon} = 3.344 - 0.0198E + (-2.72 + 0.0236E)D/G$$

$$K_{morning} = 4.612 - 0.0586E + (-3.770 + 0.062E)D/G$$

$$K_{afternoon} = -2.015 + 0.0476E + ( 3.579 - 0.056E)D/G$$

where $F_R$ and K are the ratios of the measured to expected diffuse and ground-reflected components respectively.  It was not possible to derive reliable equations for December as there were insufficient ranges for elevation available for analysis.

# 1. Project report

## 1.1 Introduction

Measurements of the direct, diffuse and ground-reflected components of solar radiation on a vertical south-facing surface commenced at Valentia on 1st May 1981. For the purpose of this report, the analysis will be confined to June and December 1981. It is intended to utilise data for the 18 month period 1st May 1981 to 31st October 1982 for the complete analysis and final report.

## 1.2 Measuring equipment

A Kipp and Zonen CM6 pyranometer and a BD8 Flatbed recorder have been installed to monitor the ground-reflected radiation.

Radiation on a vertical south-facing surface is monitored by two Kipp and Zonen CM6 pyranometers, one of which is screened by a 2 foot square honeycombed mat black metal surface to eliminate the ground-reflected radiation. The 'screened' radiation is recorded on a Kipp and Zonen BD8 Flatbed recorder and an Eppley/Digitec electronic integrator/printer while the unshielded radiation is recorded by means of a Kipp and Zonen CC2 integrator.

Direct sun radiation at normal incidence is recorded using an Eppley pyrheliometer with a solar tracker and an Eppley/Digitec electronic integrator/printer. The component of direct radiation incident of the south-facing vertical surface is given by: $ICosECosA$

where I is the measured direct radiation at normal incidence
E is the solar elevation and A is the solar azimuth.

The component of diffuse radiation incident on the south-facing vertical surface is obtained by subtracting $ICosECosA$ from the total radiation recorded by the screened pyranometer - (S).

The component of ground reflected radiation incident on the south-facing vertical surface is given by the difference between the radiation recorded by the screened and unscreened pyranometers.

## 1.3 Method of analysis

Theoretically, the radiation on a vertical south-facing surface is given by: $G_v = \text{Direct}_v + \text{Diffuse}_v + \text{Reflected}_v$
$$= ICosECosA + \tfrac{1}{2}D + \tfrac{1}{2}\rho G$$
where: $G_v$ = Global radiation on a south-facing vertical surface
$I$ = Direct radiation at normal incidence,
$E$ = Solar elevation, $A$ = Solar azimuth, $\rho$ = Ground albedo
$D$ = Diffuse sky radiation on a horizontal surface
$G$ = Global radiation on a horizontal surface.

In the above formula the assumption is made that the radiation is isotropically distributed for both the diffuse component ($\tfrac{1}{2}D$) and the ground-reflected component ($\tfrac{1}{2}\rho G$).

Global and diffuse radiation on a horizontal surface are recorded at Valentia. The albedo $\rho$ is given by GR/G where GR is the measured ground-reflected radiation. Hence it is possible to compare the 'theoretical' values $\tfrac{1}{2}D$ and $\tfrac{1}{2}\rho G$ with the measured values of the diffuse and ground reflected components. This analysis and comparison is done

- 133 -

mainly with respect to solar elevation (E) and the relative duration of bright sunshine. However it is considered that the hourly ratio D/G is more meaningful and representative of the duration of bright sunshine than the value measured from a Campbell-Stokes sunshine card and so this ratio is used in the analysis.

The ratio of the measured values of the diffuse component on the vertical surface (S - ICosECosA) to the theoretically expected value ($\frac{1}{2}$D) has been computed for each hour of June and December 1981 and is designated $F_R$. For truly isotropic conditions this ratio should be 1.00.

A similar ratio has been found for the ground reflected component and has been designated K i.e. the ratio of $G_v$ - S to $\frac{1}{2}\rho$ G. Again this should be 1.00 for truly isotropic conditions.

$F_R$ and K have been analysed with respect to elevation and D/G.

## 1.4 Results

An asymmetry between morning and afternoon values of $F_R$ was immediately apparent. This asymmetry is possibly due to the range of hills which rise to approximately 360 metres at a distance of 1.5 Km south-east of the pyranometers. Graphs of $F_R$ against D/G were plotted for $5^o$ Ranges of elevation and using linear regression based on the method of least squares, a line was fitted to the data.

Fig. 1. shows the situation for elevation angles of $40^o$ and $60^o$ in June 1981. As expected $F_R$ increases as D/G decreases but only in the afternoon. In the morning $F_R$ is almost constant - probably due to mountain screening. For all elevation angles between $30^o$ and $60^o$ a good fit with the data was obtained particularly for the afternoon data. For elevation angles less than $30^o$ there was no direct component incident on the south-facing pyranometer and these angles will be examined in the final report.

For each range of elevations there is a linear relationship between the slope of these curves ($B_F$) and the elevation (Fig. 2) and between $F_R$ at D/G = 0 ($F_{RO}$) and the elevation (Fig. 3). An expression for $F_R$ in terms of D/G and E has been calculated.

For December there were two ranges of elevation angle $10^o$ and $15^o$. The variation of $F_B$ with D/G is much greater than in June reflecting the lower elevation angles. It is not possible to get a reliable fit for the variation of the regression coefficients with elevation. Figs. 2 and 3 show that the December coefficients do not lie on the fit for the June data.

For the ground reflected ratio K morning and afternoon data were again analysed separately and a similar procedure followed as for $F_R$. Morning-afternoon asymmetry is quite apparent. Expressions for K were calculated in terms of E and D/G for morning and afternoon.

## 2. Conclusions

The asymmetry between morning and afternoon data for both $F_R$ and K is probably due to local factors. A seasonal variation is indicated by the fact that the December data do not lie on the fit for the June data. This variation has yet to be determined. More concrete conclusions may be drawn from an analysis of the full years data. Other regression analyses of K and $F_R$ against D/G such as log K, $K_e$ etc will be examined. Values of the albedo obtained from G and GR lie mainly in the range 20 to 25 and variations in $\rho$ will be examined in conjunction with other meteorological parameters.

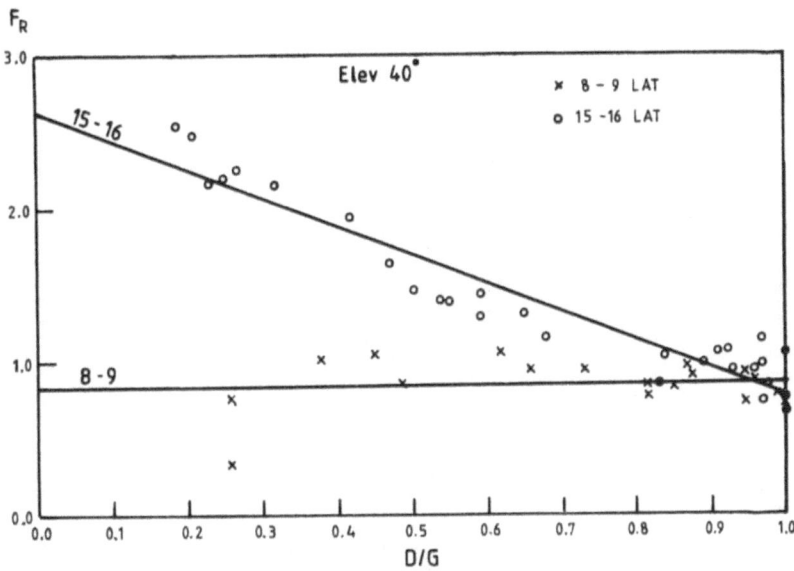

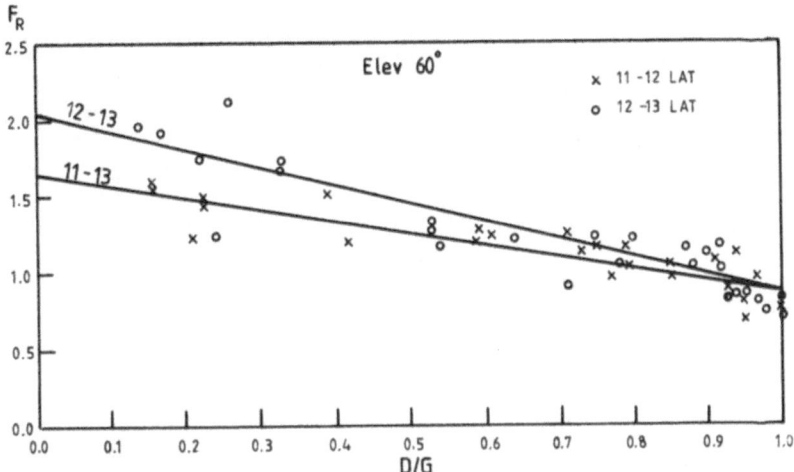

Fig. 1 Ratio of measured diffuse components on a vertical south - facing
surface to the theoretically expected values $(F_R)$ with corresponding
values of D/G for elevation angles 40° and 60° - June.

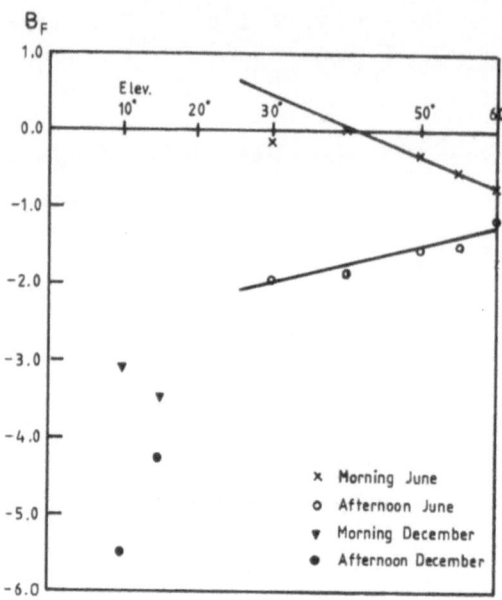

Fig. 2 Slopes of regression curves of $F_R$ against D/G for range of elevation angles – June and December.

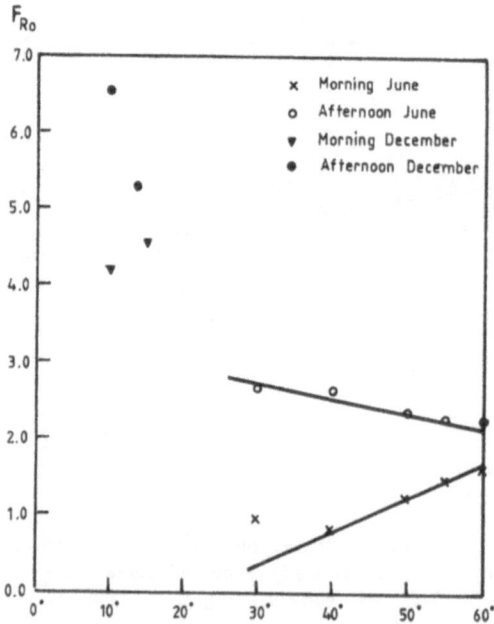

Fig. 3 Intercepts of regression curves of $F_R$ against D/G for range of elevation angles - June and December.

# DEVELOPMENT OF A SYSTEMATIC METHOD FOR THE PREDICTION OF SOLAR RADIATION ON INCLINED PLANES

Authors            : J.K. PAGE AND R.J. FLYNN

Contract number : ESF-021-80/UK/H

Duration:          : Originally 15 months    Dates: 1st April, 1980 –
                     Extension under                30th June, 1981
                     discussion
Total budget     : £7,457                  CEC contribution : £7,457

Head of project : Professor J.K. Page, Department of Building Science

Contractor       : University of Sheffield, Sheffield, S10 2TN.

Address          : Department of Building Science,
                     University of Sheffield,
                     Sheffield.    S10 2TN.
                     UK.

## Summary

A comprehensive report has been prepared on the prediction of monthly mean hourly and daily values of the irradiation on inclined planes using as the key inputs the observed monthly mean duration of bright sunshine, and a monthly value of the Linke Turbidity Factor selected to match the site characteristics. Five different types of site have been identified, industrial/large city, urban, clear, very clear maritime and very clear mountainous. The model derives the hourly values of the monthly mean irradiance on the basis of sunshine duration using suitable algorithms from the values produced by the clear day model and the overcast day model, so the one model produces the three key outputs needed for solar design. All algorithms used will be described in the final full report. A practical methodology for desk top calculator predictions has been developed with associated tables. Microcomputer programs have been developed. Tables for about 100 stations in the Community area for a new inclined plane atlas are being produced.

# 1. Introduction

This report provides a brief summary of the work carried out to produce a systematic methodology for the prediction of the irradiance on inclined planes. The work has been carried out jointly with the Lichtmesstechik Berlin. A full report on the methods developed for the prediction of inclined surface radiation will be produced as a later volume in this series. The methodology developed is being used to produce standard irradiation tables for inclined planes for 100 stations in Europe. These tables form the basis of new maps which will appear in the Inclined Surface Radiation Atlas. The tables will be included in the Atlas. The areas covered not only include member countries, but also adjacent territories in Europe.

# 2. Basic methodology

The methodology developed has three main components:

a)   Estimation of clear day hourly slope irradiance and daily slope irradiation

b)   Estimation of overcast day hourly slope irradiance and daily slope irradiation

c)   Estimation of monthly mean hourly slope irradiance and daily slope irradiation.

The direct beam and the sky diffuse components are estimated separately. The ground contribution is assessed assuming the ground reflects isotropically with an albedo of 0.20. The global irradiance and global irradiation are assessed by adding the three components. The modelling technique thus not only provides data about average and extreme conditions, but also provides data on the relative magnitude of the components of radiation, which is critical information in certain applications.

# 3. Modelling typical clear day direct beam irradiance

A suitable clear day direct beam irradiance model that would provide hourly estimates of the clear day direct beam irradiance to be obtained from a description of basic atmospheric conditions at the site was evolved. Five atmospheric conditions were identified:

a)   Very clear mountainous

b)   Very clear maritime

c)   Clear

d)   Urban

e)   Industrial/large city centre.

The values of the clear day irradiance were generated using the Linke Turbidity Factor. An allowance for the effect of solar altitude on the Linke Turbidity Factor was incorporated. The air mass 2 Linke Turbidity Factors used are given in Table I. These values were evolved from a study of monthly average global and diffuse radiation data for 20 sites in the Community area considered in conjunction with the somewhat earlier data of Steinhauser,    published in 1935. The direct beam irradiation was estimated by subtracting the diffuse radiation from the global.

Table I    Monthly values of typical clear day Linke Turbidity Factors
           adopted for the production of standard tables of irradiation
           on inclined surfaces for Europe

| Site type | J | F | M | A | M | J | J | A | S | O | N | D |
|---|---|---|---|---|---|---|---|---|---|---|---|---|
| Very clear mountainous | 1.9 | 2.0 | 2.1 | 2.3 | 2.5 | 2.6 | 3.0 | 2.8 | 2.4 | 2.2 | 2.0 | 1.9 |
| Very clear maritime | 2.3 | 2.3 | 2.4 | 2.5 | 2.5 | 2.6 | 3.0 | 2.8 | 2.6 | 2.5 | 2.4 | 2.3 |
| Clear | 2.3 | 2.3 | 2.6 | 2.9 | 3.6 | 3.6 | 3.9 | 3.5 | 2.9 | 2.5 | 2.4 | 2.3 |
| Urban | 3.0 | 3.0 | 3.2 | 3.7 | 4.1 | 4.3 | 5.0 | 4.3 | 3.4 | 3.0 | 3.0 | 3.0 |
| Industrial/ large city | 4.3 | 4.3 | 4.3 | 4.5 | 5.1 | 5.2 | 6.0 | 5.5 | 4.5 | 4.0 | 4.0 | 4.0 |

## 4.  Modelling clear day diffuse sky irradiance

The second stage was the evolution of a suitable clear sky diffuse
prediction model to estimate the clear sky diffuse radiation on inclined
planes.  This modelling is carried out in two parts:

a)   the determination of the horizontal diffuse sky irradiance

b)   the conversion of the horizontal value to the corresponding
     slope value for a plane of the required tilt and orientation
     with respect to the sun's direction and altitude.

## 5.  Clear sky radiance model

A non isotropic sky radiation model was used which was evolved from
the work of Liebelt     and of Steven and Unsworth.     The radiance model
was used to produce numerical estimates of the diffuse sky irradiance
relative to the horizontal diffuse irradiance.  The horizontal diffuse
irradiance model is sensitive to the Linke Turbidity Factor and the clear
sky diffuse horizontal irradiance increases as the atmosphere becomes more
turbid.

The relative sky radiance model used was as follows:

$$\frac{L_{cl}(\theta,\alpha_s)}{L_{clz}} = \frac{(1 - e^{-0.088\,T_L/\sin\theta})(X_1 + X_2(\exp(-3Q.\pi/180) + X_3\cos^2 Q)}{(1 - e^{-0.088\,T_L})(X_1 + X_2(\exp(-3z.\pi/180) + X_3\cos^2 z)}$$

where $L_{cl}(\theta,\alpha_s)$ = radiance at specific point of sky at angle $\theta$ from
                    horizontal plane at an azimuth angle $\alpha_s$ from sun's
                    meridian

$L_{clz}$        = zenith radiance

$Q$            = angle in degrees between sun's position and selected
                 point of sky

$T_L$           = Linke turbidity

$z$ = zenith distance in degrees (angle between centre of solar disc and zenith)

$\alpha_s$ = azimuth angle of surface from the sun's meridian

$\theta$ = angle between the element of the sky at point p and the horizontal plane

and where $X_1 = 0.8995 - 0.0053 \, Y$, $X_2 = 0.6155 + 1.9687 \, T_L$, $X_3 = 0.409 - 0.0096 \, Y$

It was subsequently shown that this model produced results that were close to but not identical with the results of Valko. Valko's observations are now being used to re-evaluate the radiance model, but it is not anticipated large changes will be needed.

## 6. Overcast day model

The work on this aspect of the project was carried out by Lichtmesstechnik in Berlin. It was shown that the overcast sky horizontal irradiance could be predicted from the formula:

$$D_{oc} = (2.6 + 182.6 \sin Y) \, K_d \quad Wm^{-2}$$

where $K_d$ is the correction to mean solar distance.

It was also shown that the conversion to slopes could be achieved using the Moon and Spencer radiance formulation:

$$\frac{L_{oc}(\theta)}{L_{ocz}} = \frac{1 + b \sin \theta}{1 + b}$$

where $L_{oc}(\theta)$ is the radiance from the sky at an angle $\theta$ from the horizontal plane

$L_{ocz}$ is the zenith radiance

$b$ is a constant with a value of 2.

## 7. Estimation of monthly mean direct irradiance

The estimation of monthly mean direct irradiance is based on the clear day value in conjunction with observed monthly mean sunshine duration. The monthly mean sunshine duration is used to calculate the relative sunshine duration on a card burning basis. A sunshine recorder threshold of 200 $Wm^{-2}$ was adopted and used together with the Linke Turbidity Factors in Table I to estimate the card burning daylength. This varies with turbidity, latitude and time of year. An uncorrected value of the direct beam irradiance is obtained by multiplying the clear day direct beam irradiance by $\sigma_{cb}$ where $\sigma_{cb}$ is the relative daily duration of card burning bright sunshine. A correction function $R_b$ is then applied to allow for the over-estimate of bright sunshine by sunshine recorders at higher solar elevations and for the underestimate at lower solar altitudes. The correction function developed from the original Berlin studies for altitudes above $10^0$ is:

$$R_b = (((-3.34 \, \sigma_{cb} + 6.92)\sigma_{cb} - 4.066)\sigma_{cb} + 1.48)/(0.84 + 0.0025(Y - 10))$$

where $\sigma_{cb}$ is the relative daily duration of card burning bright sunshine and $Y$ is the solar altitude. When $Y < 10$, the value is set for $10^o$.

## 8. Predicting monthly mean diffuse irradiation

The monthly mean diffuse irradiation at any hour cannot be found by simple interpolation between the clear day and overcast day diffuse irradiation because the effect of clouds is to raise the amount of diffuse irradiation considerably above the values that would be expected with simple linear interpolation. A methodology of interpolation had to be developed. The original Berlin interpolation method was found to give poor results at an hourly level. Furthermore, it underestimated diffuse irradiation from very clear and clear sites. A new diffuse interpolation model has been developed which uses relative daily sunshine duration on an astronomical basis as the basis for interpolation. The new diffuse radiation correction function was developed for the observed diffuse radiation data at 20 EEC sites. It was found that the interpolation correction function varied according to site clarity. The diffuse sky slope irradiation is estimated by splitting, according to relative daily sunshine duration, the horizontal diffuse into two parts. The conversion of the overcast horizontal component to slope values uses the overcast model and the clear sky part the clear day model.

## 9. Accuracy of prediction

It was found that the new model predicted hourly values of both direct and diffuse irradiation on the horizontal plane to a typical accuracy of about ±5%. The prediction errors tend to be greatest in winter when the sun is low. The checks on slopes gave a similar order of accuracy. The new model is considerably more accurate at the hourly level than the original Berlin model.

## 10. Standard tables

Using data in the EEC Horizontal Surface Atlas as inputs plus additional data for other sites in Europe, 100 standard tables of monthly values of slope irradiation have been produced. The tables are being used for mapping of solar radiation on inclined surfaces and will be published in the Inclined Surface Atlas together with the maps.

## 11. Conclusion

A validated model for clear days, overcast days and for monthly mean hourly values of the hourly irradiance in the Community area has been successfully developed. The methodology of the modelling process with all algorithms will be published as a later book in this series.

Mainframe and microcomputer programs have been developed to carry out the necessary calculations. These programs will be made available in due course. At present the microcomputer program runs in Basic on an Apple II computer. The mainframe programs are written in Fortran.

# CALCULATION METHOD FOR THE RADIATION DATA ON INCLINED SURFACES AND CALCULATED ATLAS DATA

Author            : S. AYDINLI

Contract number   : ESF-020-80 D(B)

Duration          : 18 months           1 January 1981 - 30 June 1982

Total budget      : DM 36.500,--         CEE contribution : 100 %

Head of project   : Prof.Dr.-Ing. J. Krochmann, Institut für Lichttechnik
                    der Technischen Universität Berlin

Contractor        : LMT Lichtmeßtechnik Berlin

Address           : LMT Lichtmeßtechnik Berlin
                    Helmholtzstraße 9
                    D-1000 Berlin 10

## Summary

In cooperation with the University of Sheffield, Prof. Page, a method for
the calculation of radiation data on inclined surfaces is developed, which
can be used for any station within the European Community, if the relevant
input data are known:
- the mean values of monthly sunshine duration of the location
- the type of the atmospheric turbidity of the location
The method agrees with measured data in general with deviations of $\pm$ 10 %.
More accurate results can be obtained, if monthly mean hourly values of
sunshine duration are known.
Radiation data are calculated for overcast sky, clear sky, and average sky
conditions for 56 stations of the European Community for a horizontal and
7 inclined surfaces. These data are the basis of an EEC-Atlas on slopes.

# 1. Introduction

For the practical use of solar energy the long time average radiation data on horizontal and inclined surfaces must be known. At many stations within the European Community the radiation data on a horizontal surface and the mean values of the monthly sunshine duration are measured. For only few stations the monthly mean hourly values of sunshine duration are known.

In general the radiation data on horizontal and inclined surfaces depend on:

- the geographical location
- the atmospheric turbidity
- the sunshine duration
- the ground reflectance
- the obstraction
- the type, thickness, the local distribution and daily and yearly variation of cloudiness.

The accuracy of a prediction method depends on these data, of which some cannot be measured accurately. Therefore a calculation method must be developed, which uses as input data these only, which are known. It must be expectrd that calculation results of such a method cannot agree better than $\pm$ 10 %, with measured radiation data, especially if the measuring accuracy for radiation and sunshine duration are taken into account.

# 2. Bases of the method

The bases of the method are the radiation data of the two extreme sky conditions, the completely overcast and the completely clear sky. The radiometric conditions for these two conditions of the sky are defined. The radiation data for the average sky conditions are calculated by a non-linear superposition of the two extreme conditions according to the sunshine duration. The geographical position, the ground reflectance, the obstruction, and the turbidity are taken into account.

## 2.1 Overcast sky

The irradiances on inclined surfaces by the overcast sky depend on the slope of surface and on the solar altitude (1).
On the basis of a study on the observed values of the irradiances on overcast days for different European stations, the particular values show a large scattering for the horizontal and the slope irradiances. The mean values of the horizontal and the slope irradiances present a good agreement with the calculation method, which is shown in figures.

## 2.2 Clear sky

The irradiances on inclined surfaces on clear sky are specified by:

- the atmospheric turbidity
- the solar altitude
- the slope of surface
- the azimuth angle between sun and surface

For the horizontal irradiances by clear sky a formula (2) is used, which is corrected on the basis of a study on the irradiances on clear days in the University of Sheffield (3).
For the calculation of the slope radiation data by clear sky, the relative radiance distribution of the clear sky according to Page (3) is taken into account.
For the atmospheric turbidity the monthly mean values of Linke Turbidity Factor of the 4 classifications of the atmosphere are considered.

## 2.3 Average sky conditions

The radiation data for the average sky conditions are calculated by the extreme sky conditions in consideration of the monthly mean hourly values of the sunshine duration (2), which can be derived from the monthly mean values according to (4).

## 3. Atlas data

Radiation data on inclined surfaces as monthly and yearly mean radiant exposure are calculated for the 56 stations within the European Community for overcast, clear, and average sky conditions.

The values are given for

- horizontal surface
- vertical surfaces faced N, E/W, SE/SW, S
- 30°, 60°, latitude inclined South.

Each table contains the values of the radiant exposure for the mentioned surfaces for:

- global radiation for average sky conditions
- diffuse radiation for average sky conditions
- global radiation for clear sky conditions
- diffuse radiation for clear sky conditions

References
(1) J. Krochmann    Bestrahlungsstärken auf geneigten Flächen
                    1. Deutsches Sonnenforum. Deutsche Gesellschaft für
                    Sonnenenergie e.V. (DGS) Bd. II, S. 47-58 (1977)

(2) S. Aydinli      Über die Berechnung der zur Verfügung stehenden Solar-
                    energie und des Tageslichts
                    Dissertation 1981 TU Berlin. Fortschrittsberichte der
                    VDI-Zeitschriften. Reihe 6 Nr. 79 (1981)

(3) J.K. Page       Final report on "EEC Solar Energy Programme, Project F -
    R.J. Flynn      Solar Radiation Data, Action 3.2. Radiation data on in-
                    clined surfaces". Sheffield, August 1982

(4) S. Aydinli      3. Report on "Final-Berlin Model on the availability of
                    solar radiation and daylight". EEC Solar Energy Pro-
                    gramme. Project F-Solar Radiation data Action 3.2, Ra-
                    diation data on inclined surfaces. Berlin-West October
                    1981

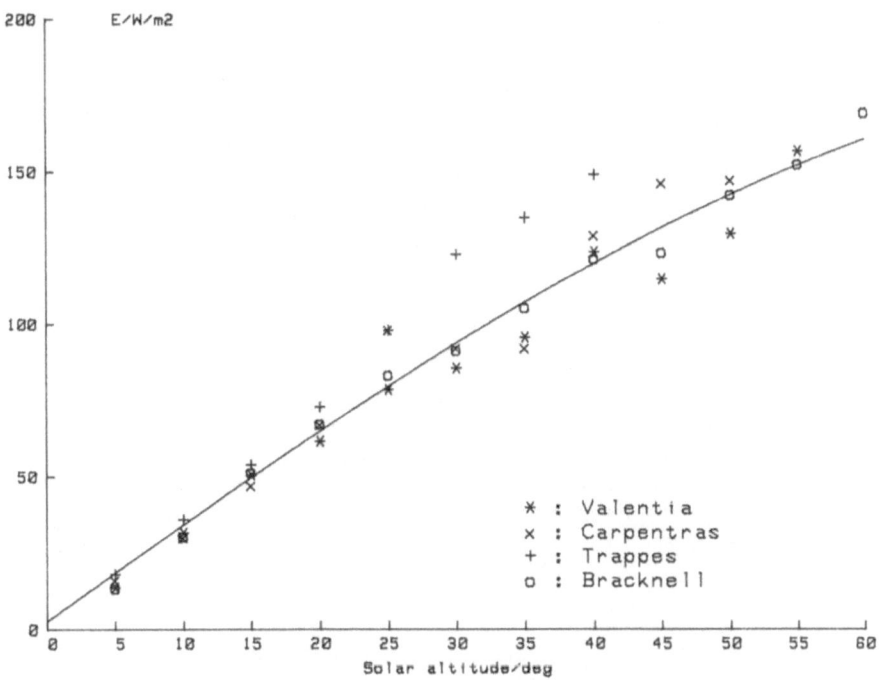

Comparison of the mean values of the horizontal irradiances
with the calculation method on overcast sky

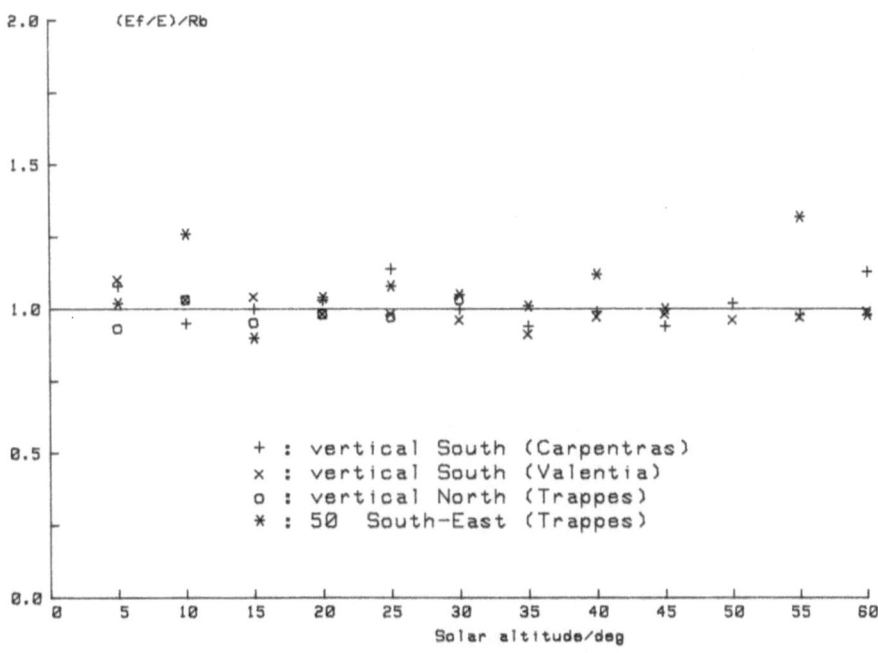

Comparison of the mean values of the relative irradiances on slopes
with the calculation method on overcast sky

# EMPIRICAL STUDY OF THE ANGULAR DISTRIBUTION OF SKY RADIANCE AND OF GROUND REFLECTED RADIATION FLUXES

Author        :    P. Valko

Contract number:    ESF-020-80-D

Duration      :    2 years   1.7.1980 - 30.6.1982

Address       :    Swiss Meteorological Institute
                    Krähbühlstrasse 58
                    CH-8044 Zurich

## Summary

Measuring campaigns at several places under different atmospheric conditions have been made using the Swiss Mobile Solar Radiation Research System. Extensive data on global and diffuse irradiance on surfaces of various tilts and aspects as well as of the angular distribution of both sky radiance and ground reflectance were collected. Information on type, amount and spatial distribution of clouds were made available by digitalizing fish-eye sky photographs.

Ratios of diffuse irradiance on tilted surfaces to that on the horizontal surface and similarly ratios of radiance at a point in the sky to that in the zenith have been analyzed. These ratios have been related to solar height and solar azimuth, the angles of surface tilt and orientation respectively the angular coordinates of the point in the sky, as well as to atmospheric turbidity. For clear skies site independent solutions of the multivariate relationships could be found. A close functional link between zenith radiance and diffuse irradiance on the horizontal surface could also be detected. These empirical functions permit computation of the sky component of real time irradiance on a surface of arbitrary tilt and orientation by using only commonly accessible input parameters.

## Introduction

The objectives of the study (1) may be summarized as follows:
- to measure the instantaneous angular components of both radiance and irradiance over the whole $4\pi$ steradian-sphere;
- to parametrize and find empirical relationships between radiance and irradiance on the one hand and angular, atmospheric and site specific quantities on the other;
- to integrate the established relationships over hours and days in order
  to compute irradiation on surfaces of any tilt and orientation.

## Means and Methods

Using an absolute radiometer and other sun-tracking devices, a set of pyranometers and silicon-diode radiometers of $5^0$ full view angle, the following
data were gained (2, 3, 4):
- Diffuse irradiance for the tilt angles $0^0$ (horizontal surface), $18^0$, $39^0$,
  $60^0$, $75^0$ and $90^0$ (vertical surface) in a number of azimuth steps giving
  alltogether 77 surface expositions;
- Diffuse irradiance on a corresponding number of overhangs - tilts between vertical and downward-facing horizontal - measured from 10 m height
  above ground;
- Radiance was scanned at 121 points in the sky at the elevations $0^0$ (horizontal circle), $18^0$, $36^0$, $54^0$, $72^0$ and $90^0$ (zenith) at every $15^0$ azimuth;
- Ground reflectance - also from a 10 m height - was scanned at the elevations $0^0$, $-18^0$, $-36^0$, $-54^0$, $-72^0$ and $-90^0$ (nadir) at different azimuths,
  alltogether at 65 points.

In all 2544 series of irradiance measurements have been made on 187 days.
Similarly sky radiance have been scanned 1578-times. 1141 series have been
compiled on ground reflectance and on irradiance on overhangs as well. All
these measurements have been taken at several sites in Switzerland up to
2575 m a.s.l., at Carpentras (F) and Hamburg (FRG).

## Results

Polar diagrams displaying the directional distribution of the different
quantities as well as graphs of the diurnal change of irradiance for constant surface positions and such of radiance at fixed points were the standard presentation forms of a first analysis. Multivariate relationships could
be found by plotting both absolute and normalized irradiance and radiance
respectively against angular and atmospheric parameters as explained in the
Summary above. Until now regression studies of this kind have been made only
for clear skies.

The main results are (2, 5):
- Diffuse irradiance on any surface normalized to that on the horrizontal
  surface decreases for all solar heights and all slopes with increasing
  surface-solar azimuth and increasing turbidity (Fig. 1). For clean air
  steep surfaces show a higher ratio for all aspects than slight tilts;
  with increasing turbidity the hierarchy of tilt angles tends to turn into
  an inverse sequence.
- Radiance at a point in the sky normalized to that at the zenith depends
  on solar height, the elevation angle of the point, the azimuthal distance
  between that point and the sun as well as on atmospheric turbidity. The

dependence on turbidity is negligible near the sun and is strong at great angular distance away from the sun (Fig. 2, 3). Turbidity can be considered as a function of solar height and the ratio of diffuse to global irradiance on the horizontal surface.

- Zenith radiance and diffuse irradiance on the horizontal surface are closely linked with each other; the relationship is dependent on solar height but does not depend on turbidity (Fig. 4).

## Conclusions

The significance of these results are:
- The angular distribution of clear sky radiance can, both absolutely and relatively, be determined in real time at any place where global and diffuse irradiance on the horizontal surface is known.
- The sky component of the diffuse irradiance on a surface of arbitrary tilt and orientation can be determined in real time - by numerical integration of sky radiance over the sky hemisphere respective to that surface - at any place where global and diffuse irradiance on the horizontal surface is known.

## References

(1) Proposal for Contribution of Switzerland to Study No 4 of Action 3.2 within Project F of the EEC Solar Energy Programme. Zurich, September 1979.

(2) Report, prepared for the Project F, Action 3.2 Meeting, Reading, February 1982

(3) Valko, P. 1980: Some empirical properties of solar radiation and related parameters. Chapter 8, pp. 8-1 - 8-46 in the Task-IV/1-Handbook "An Introduction to Meteorological Measurements and Data Handling for Solar Energy Applications" of the Int. Energy Agency Solar R & D Programm, Washington.

(4) Report, prepared for the Project F Coordination Meeting, Brussels, April 1981.

(5) Report, prepared for the Project F, Action 3.2 Meeting, Odeillo, September 1982.

(6) Schüepp, W. 1949: Die Bestimmung der Komponenten der atmosphärischen Trübung aus Aktionometermessungen. Archiv. Met. Geoph. Biokl. Vol. B1. pp 257 - 346.

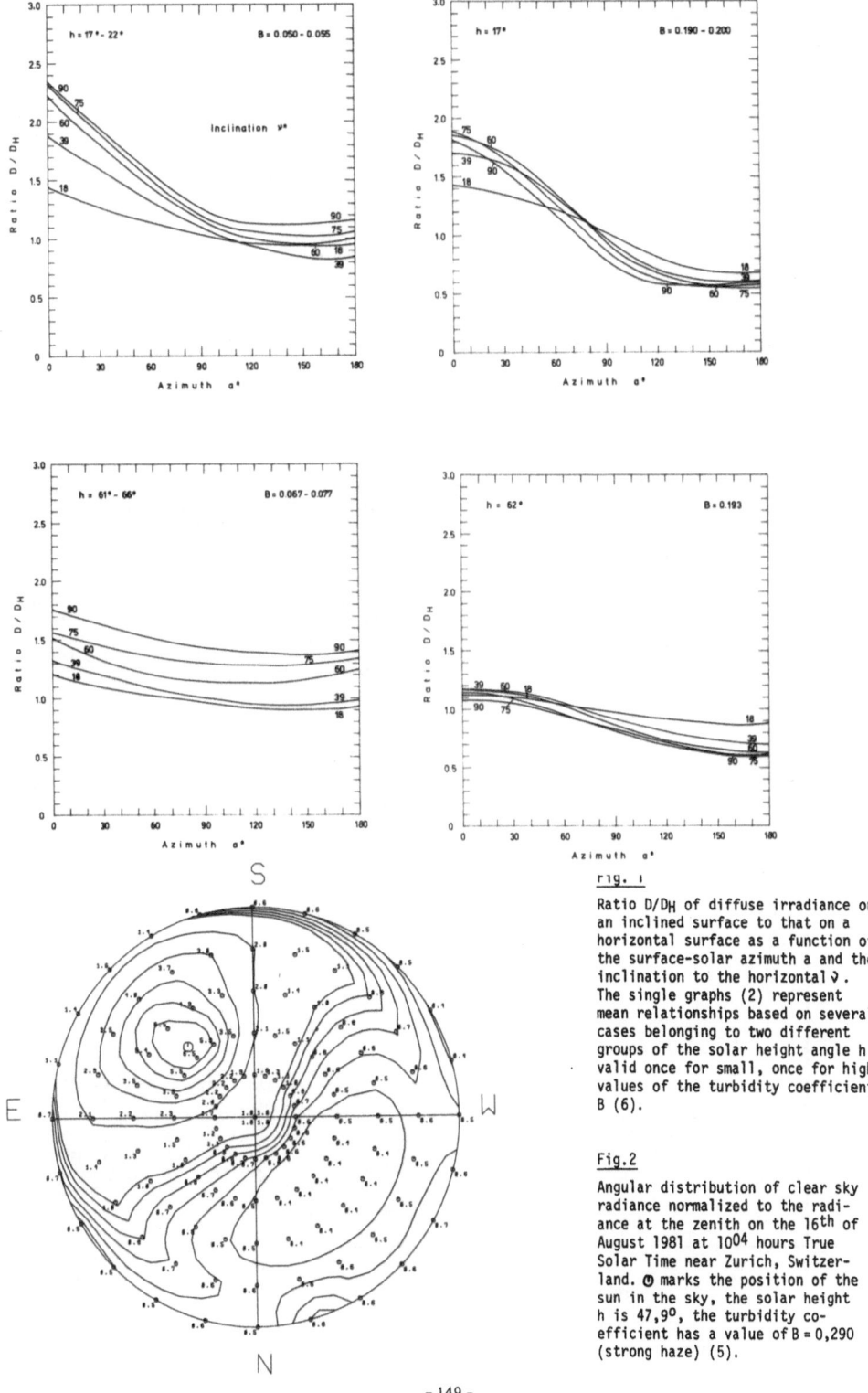

fig. 1

Ratio D/D$_H$ of diffuse irradiance on an inclined surface to that on a horizontal surface as a function of the surface-solar azimuth a and the inclination to the horizontal ϑ. The single graphs (2) represent mean relationships based on several cases belonging to two different groups of the solar height angle h valid once for small, once for high values of the turbidity coefficient B (6).

Fig.2

Angular distribution of clear sky radiance normalized to the radiance at the zenith on the 16th of August 1981 at 10[04] hours True Solar Time near Zurich, Switzerland. ☉ marks the position of the sun in the sky, the solar height h is 47,9[o], the turbidity coefficient has a value of B = 0,290 (strong haze) (5).

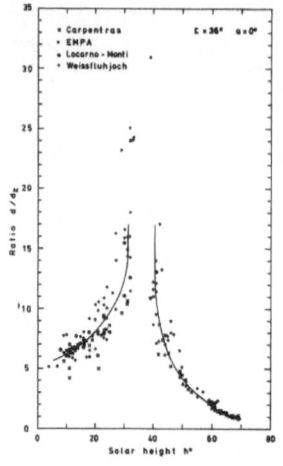

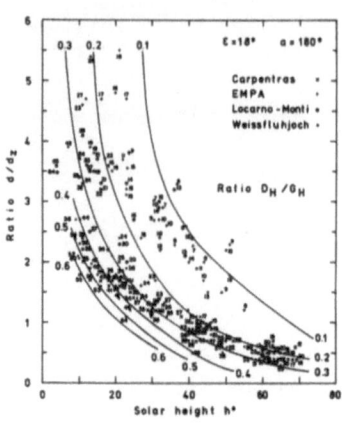

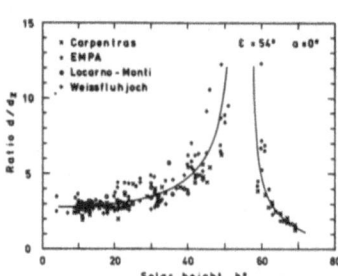

Fig. 3

Normalized sky radiance $d/d_Z$ as a function of the solar height for two points in the sun's vertical $(a = 0^0)$ having the elevation angles above the horizon $\mathcal{E} = 36^0$ (top left) and $\mathcal{E} = 54^0$ (bottom left) as well as for the point opposite the sun $(a = 180^0)$ of the elevation $\mathcal{E} = 18^0$ (graph on the right). This latter shows strong dependence also on turbidity expressed as ratio of diffuse to global irradiance $D_H/G_H$ on the horizontal surface. The 182 scans involved in the analysis are from four different sites (see legend) (5).

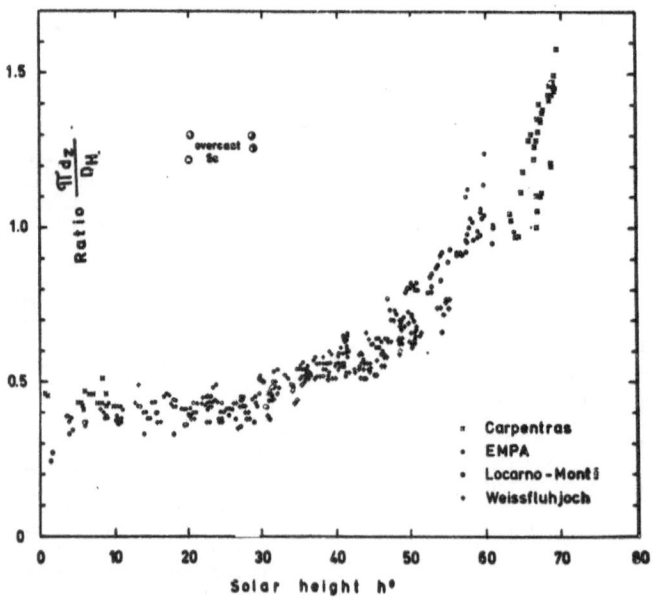

Fig. 4

Ratio of zenith radiance $d_Z$, related to the full view angle "seen" by the horizontal surface, to the diffuse irradiance $D_H$ on the horizontal surface as a function of the solar height h. The relationship is fairly independent of site and turbidity (5).

# ESTIMATION OF HOURLY SOLAR IRRADIATION OVER THE UK

Author            : F.RAWLINS

Contract no.      : ESF - 024 - 80 - UK(H)

Duration          : 42 months          1 January 1980 - 30 June 1983

Total budget : £ 55 255      CEC contribution  : £ 24 000

Head of project : G.J.Jenkins, Meteorological Office UK.

Contractor        : Ministry of Defence (Meteorological Office)

Address           : Sec Met O

                    Meteorological Office

                    London Road

                    Bracknell, Berks., RG12 2SZ,  UK.

## Summary

The aim is to provide hourly estimates of the components of solar
irradiation -- global, diffuse and direct -- on horizontal surfaces
averaged over a month (or longer period) from records of bright sunshine
duration. A method was devised to determine global irradiation on an
hourly basis using coefficients from the (daily) Angstrom equation and
gave improved results compared to a method which ignored hourly sunshine
statistics, although long term averages were similar. Little progress was
achieved in the estimation of hourly diffuse irradiation for different
geographical sites from sunshine records alone. Linke turbidities at Kew
and Bracknell were compared with those of actual measurements and were
found to give similar seasonal variations. The ratios of average to 'clear'
direct irradiations (calculated from the derived turbidities) were
presented as a function of the bright sunshine fraction, which allows the
average hourly direct irradiation to be determined from sunshine statistics
providing that a representative turbidity is known. Maps of global
irradiation could be prepared for regions where there are numerous sunshine
recording stations but maps of direct, and hence diffuse, irradiation
require information concerning the mean turbidity. The extension of schemes
of this type, involving simple regression from sunshine fractions, is
deemed to be an unpromising approach to the investigation of irradiation on
vertical and inclined surfaces.

## 1.1 Introduction

One approach to the problem of determining the average irradiation received on an inclined slope at a particular hour of the day is to estimate the mean hourly components on a horizontal surface (global,diffuse and direct) and apply simple models, such as (1). Radiometric sites within the EEC region are sparsely distributed, hence estimation from the more numerous stations which record the duration of bright sunshine was explored, initially for the UK.

A variant of the Angstrom equation (2), calculating hourly global irradiation by regression from sunshine statisics, was tested for several stations. The application of turbidity values, derived from hourly data, to the determination of mean hourly direct irradiation was investigated.

## 1.2 Measurements

Records of daily global irradiation (Kipp and Zonen pyranometers) and bright sunshine (Campbell-Stokes recorder) exceeding 10 years have been obtained from 25 stations in the UK network, of which 11 reported hourly. Hourly measurements of direct irradiation have been made at Bracknell (1974-1982) and Kew (1947-1980) with Angstrom pyrheliometers. Occasional observations of broadband turbidity have been performed at both sites (with some bias towards clear atmospheric conditions); these have been supplemented by values obtained from the inspection of records of direct irradiation at minute intervals.

## 1.3 Results

Average daily global irradiation on a horizontal surface, G, was estimated throughout the UK in (3) using the relations :

$$G/G_o = A + B\, N/N_o \qquad \text{for } N/N_o > 0 \qquad \text{Eq. 1}$$

$$G/G_o = A' \qquad \text{for } N/N_o = 0 \qquad \text{Eq. 2}$$

where $G_o$ is the extraterrestial irradiation and $N/N_o$ is the daily fraction of possible sunshine; the coefficients $A, A', B$ were derived by regression from long term statistics and mapped for the UK region. It was found that the daily coefficients did not give the best estimate of mean hourly global irradiation, g, from the hourly sunshine fraction, n, as indicated in Figure 1 where the daily regression line is not a good fit to the long term averages of hourly values for a particular month.

A method was devised to obtain global irradiation for a particular hour (the $j^{th}$) averaged over a month from hourly bright sunshine records using the following expression (excluding hours near dawn or dusk) :

$$g_j/g_{oj} = a_j + b_j\, n_j \qquad \text{Eq. 3}$$

where the hourly coefficients $a_j, b_j$ are determined from the daily coefficients $A, A', B$ such that
a) the number of overcast days (with N=0) in the month were considered
b) the long term averages (over many years) were consistent with the hourly to daily ratios, g/G, found by (4)
c) the sum of hourly estimates were normalised to agree with the daily estimate given by Eq.s 1 and 2.

The hourly root mean square errors obtained after applying the method, labelled HGSUN, to 14 years of Bracknell data are shown in Figure 2 for June, where they are compared with the errors found when daily estimates were given by Eq.s 1 and 2 to provide hourly values from the ratios tabulated by (4), labelled LJ. Smaller differences were found in the comparison for winter months, when there is less information content from the hourly sunshine. Similar results were shown for Eskdalemuir and Aberporth data, with r.m.s. errors of approximately 10%.

Little progress was made in deducing hourly diffuse irradiation from sunshine statistics alone and attention was addressed towards the estimation of direct irradiation, from which the diffuse could be

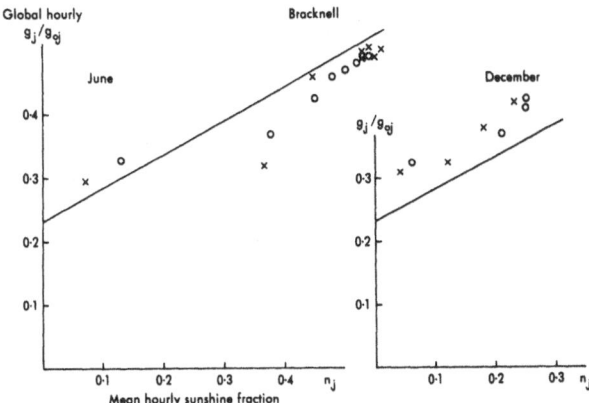

Fig. 1
The ratio of measured to
extraterrestial global
radiation for June and
December averaged for each
hour from Bracknelldata
1965-1979 is shown as a
function of the average
sunshine fraction with
morning and afternoon
values indicated by x and o
respectively. The lines
shown apply to the re-
gression from <u>daily</u> totals
only.

inferred, given the global component.

The behaviour of Linke turbidity values derived from the frequency
distributions of hourly direct irradiation, similarly to the method of (5),
was examined for Kew and Bracknell. It was found that a suitable definition
of derived turbidity was given by $T_{95} = -\ln(i/i_o)_{95}/\tau_s m$      Eq. 4
(for monthly averages), where $(i/i_o)_{95}$
is the 95 percentile level of the hourly ratio of measured to extraterrestial direct irradiation and a convenient expression for $1/\tau_s m$ is
quoted in (6). Figure 3 compares the turbidity derived in this fashion
from 10 years of Kew data with the averages of actual measurements for the
same period and indicates similar seasonal variations in both, subject to
the errors expected from both sets of observations. However, the differences
between the derived turbidities at noon and at other hours were found to be
largerthan those predicted by the turbidity corrections (for solar altitude)
given by (7). Afternoon $T_{95}$ exceeded morning values by 0.2 on average.

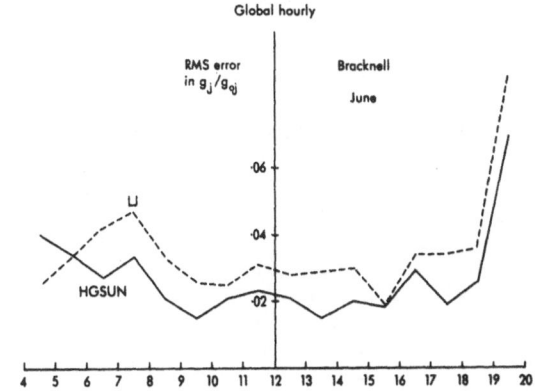

Fig. 2
The root mean square error
in monthly average irra-
diation(as a fraction of
the extraterrestial) is
shown for June from Bracknell data 1965-1979 for
each hour of Local Appar-
ent Time. The proposed
method (HGSUN) is compared
to that of Liu and Jordan
(LJ).

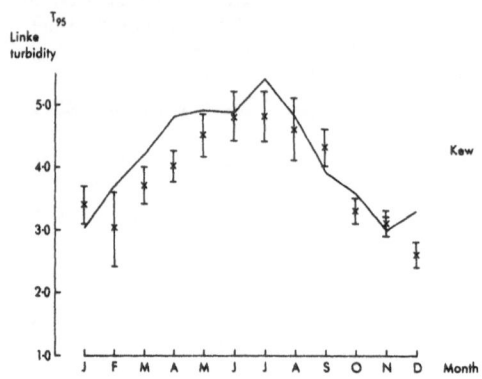

Fig. 3
The continuous line rep-
resents the linke turbidity
derived from the 95 percentile level of the ratio of
measured to extraterrestial
hourly direct irradiation
from a sample comprising of
6 hours (0900-1500 LAT)
daily for 10 years (1970-79)
for each month at Kew. The
vertical bars mark the
monthly averages of actual
turbidity, together with
their standard errors,
measured over a similar
period for the same site.

The long term averages of derived turbidity were inverted to obtain
'clear' direct irradiation $i_T$ for each hour in every month from :
$$i_T = i_0 \exp( -T_{95}\, \gamma_s\, m) \qquad\qquad Eq.\ 5$$
The ratio $i/(i_T\, n)$ is shown as
a function of the hourly sunshine fraction n in Figure 4 for the combined
data of Bracknell and Kew, where i is the average of hourly direct
irradiation over several years. A convenient fit to these data points for
$n < 0.6$ is given by $\qquad i/(i_T\, n) = 0.645 + 5(0.6 - n)^3 \qquad\qquad Eq.\ 6$
and is also shown.
Similar results emerged from the two sites in all the analyses discussed.

1.4 Comments    The use of regression coefficients to determine global
irradiation is an alternative to the modelling approach and circumvents
deficiencies in understanding of the physical processes involved. The
calculation of coefficients for each hour, month and station would be a
laborious undertaking and not justified by the accuracy of the scheme. the
method described requires only the daily coefficients to be found but
makes use of hourly sunshine statistics and could be employed for any EEC
area for which daily coefficients of the Angstrom equation have been
compiled and hourly sunshine data are available. The results show an
improvementbin the accuracy for individual months of hourly global
irradiation compared to the method of (4), though the long term averages
are, by definition, very similar.
    The Linke turbidity $T_{95}$, provides a measure of the variation of
turbidity and is not expected to be identical to observed values;
alternative definitions may be preferred. However $T_{95}$ allows monthly mean
conditions to be quantified for each hour provided that a sufficiently
large sample of hourly data exists, with a proportion of hours being
completely clear. A sample of more than 5 years, with an hourly sunshine
fraction exceeding an average of 0.15,is usually adequate.
    The average hourly direct irradiation can be estimated from the
sunshine fraction using Eq.6 with an assumed turbidity taken together with
an hourly correction. The form of the curve (in Figure 4) arises from the
presence of high cloud and from the deficiencies of the Campbell-Stokes
recording of bright sunshine.

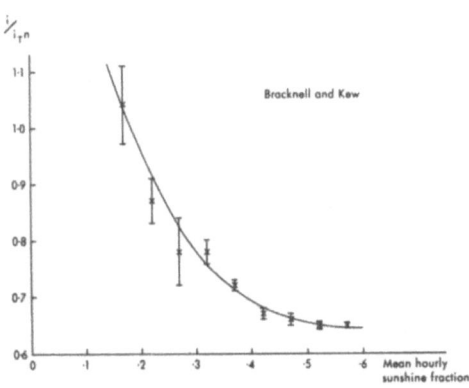

Fig. 4
The ratios of $i/(i_T n)$ are shown separated into groups of n, with vertical bars denoting the standard errors, where $i$, $i_T$ are the average and 'clear' direct irradiation and n is the sunshine fraction, all refering to long term averages in monthly and hourly categories. The figure combines measurements from Bracknell and Kew and includes a curve representing a suitable fit to the data.

## 2. Conclusions

Hourly coefficients can be obtained from daily regression coefficients of the Angstrom equation, which are more readily available, to provide estimates of hourly global irradiation on a horizontal surface from bright sunshine records.The hourly estimates for different years of a particular month showed an improvement over the method of (4), which does not employ hourly sunshine data, for summer months when there is sufficient sunshine to furnish additional information; the r.m.s. error was approx. 10%, long term averages were very similar to those of (4).

Hourly diffuse irradiation is not simply related to hourly sunshine. The Angstrom equation is also not suitable for large scale estimates of direct irradiation due to the scarcity of measurement sites. However the Linke turbidity derived from hourly direct irradiation distributions exhibited a seasonal variation consistent with that of actual measurements and allow the 'clear' direct values to be tabulated both by hour and by month. The average direct irradiation can be related to the 'clear' using a function of the sunshine duration so that estimation is possible from an assumed turbidity, though more work is required to examine the universality of the function. The average hourly sunshine fraction must exceed about 0.15 for the derived turbidity concept to be applicable.

The development of similar methods, employing simple regression from the fraction of sunshine duration, is considered to offer little promise for calculating the irradiation on vertical and inclined surfaces due to the complicating factors of turbidity and the cloud dependent anisotropy of the sky radiance distribution.

## 3. References

1. Hay,J.E.                      1979  Canadian Climate Centre Report 79-12.
2. Angstrom,A.                   1924  Q.J.Royal Met.Soc.,50.
3. Cowley,J.P.                   1978  Met.Mag.,107.
4. Liu,B.Y.H. and Jordan,R.      1960  Solar Energy,4.
5. Page,J.K.                     1978  BS46,University of Sheffield.
6. Kasten,F.                     1980  Met.Rund.,33.
7. Dogniaux,R. and Doyen,P.      1968  Inst.Royal Met.Belgium,A65.

ACTION 3.3


STATISTICAL ANALYSIS OF METEOROLOGICAL DATA


Action leader's progress report
  J.A. BEDEL, Direction de la Météorologie, Paris, France


Reports of action participants

- Analysis of short time data of irradiation on inclined
  planes

- Solar energy recovered by a flat plate collector

- Statistical distribution of solar radiation (hourly sums) :
  cumulative frequency curves

- Statistical analysis of solar radiation data of a high
  altitude station

- Studies of relationships between inclined surface irradiation, temperature and wind speed

## ACTION 3.3 : <u>STATISTICAL ANALYSIS OF METEOROLOGICAL DATA</u>

<u>ACTION BUDGET</u> (CEC CONTRIBUTION) :

<u>DURATION OF PROJECT</u> : until 1983-06-30

<u>ACTION LEADER</u> : J.A. BEDEL
Direction de la Météorologie
2, Avenue Rapp
F-75340 PARIS Cedex 07

<u>PARTICIPANTS</u> :

- Institut Royal Météorologique de Belgique
  Contract Nr. ESF-003
  R. DOGNIAUX

- Deutscher Wetterdienst
  Contract Nr. ESF-004
  F. KASTEN

- Direction de la Météorologie
  Contract Nr. ESF-005
  J.A. BEDEL

- Royal Dutch Meteorological Institute
  Contract Nr. ESF-006
  W.H. SLOB

- Ecole des Mines de Paris
  Contract Nr. ESF-008
  B. BOURGES

- Centre National de la Recherche Scientifique
  Contract Nr. ESF-009
  J.F. TRICAUD

- University of Sheffield
  Contract Nr. ESF-033
  J.K. PAGE

<u>TASKS</u> :

- Cumulative frequency distribution.
- Time sequence of irradiation.
- Correlation between irradiance and other meteorological parameters
  e.g. temperature.

1. <u>SUMMARY</u>
   This intermediate report of Action 3.3 concerns :
- The general task of Action 3.3
- The short review of the available results
- The special reports on Action 3.3

## 2. GENERAL TASK OF ACTION

Action 3.3 of project F concerns the statistical analysis of Solar Energy Data. Within the second programme of project F, the contracts concerning Action 3.3 are grouped under the following items :
- Cumulative frequency curvers (CPC);
- Time sequence of solar irradiation;
- Correlation between irradiance and other meteorological parameters.

### 2.1. Cumulative frequency curves

The results of the previous programme of project F have shown the usefulness of CFC for solar energy applications. And, in his final report, the "Ecole des Mines de Paris" proposes a method for :
- the parameterization of the cumulative frequency curves;
- The calculation of the cumulative frequency curves with the monthly means of sunshine duration.
But the results are available only for the global radiation on horizontal plane.

The work in progress about CFC concerns the 3 following items :
- Computation of CFC with data set of hourly measurements of global radiation on tilted planes;
- Reconstitution of CFC on tilted planes with usual network measurements;
- Sensitivity of CFC at the integrated time scale of the data.
This point concerns 5 contracts.

### 2.2. Time sequence of irradiation

Statistical analysis of time sequences of irradiation are very useful, especially for the computation of the storage system.

The fluctuations of solar energy depend on :
- The astronomical conditions (season variations, daily variations);
- The weather conditions.

The time scales of the astronomical variations are well known : 365 days and 24 hours.

But the time scales of the weather events which influence the solar radiation are depending on the meteorological situation.

For the solar energy applications, statistical analyses of the persistence of the solar radiation are established separately for the
- Short time intervals (1 minute - 24 hours);
- Long time intervals (2 days - 20 days ...).

A statistical analysis of the mean maximum numbers of consecutive days on which the daily irradiation persists above given thresholds has been processed by F. Kasten. The results of 49 stations of E.C are available in a special report of project F, entitled : "Frequency distributions of global radiation on horizontal surface for the region of the EC". This report is available but not published.

This point concerns 3 contracts during the second programme.

## 2.3. Correlation between irradiance and other meteorological parameters

The correlations between solar radiation and temperature or other meteorological parameters (wind speed, humidity, ...) are important for some solar energy applications, especially for those concerning the passive systems. This point concerns 4 contracts.

## 3. WORK PERFORMED AND RESULTS

The available intermediate reports of contractors are (period 1982) :
- Contract ESF-004 - F. KASTEN       May 1982
- Contract ESF-004 - F. KASTEN       September 1982
- Contract ESF-005 - J.A. BEDEL      May 1982
- Contract ESF-008 - B. BOURGES      May 1982
- Contract ESF-009 - J.F. TRICAUD    May 1982
- Contract ESF-009 - J.F. TRICAUD    September 1982
- Contract ESF-033 - J.K. PAGE       September 1982

The available final report is :
- Contract ESF-006 - W.H. SLOB       January 1982

## 3.1. Cumulative frequency curves

### 3.1.1. Computation of CFC on tilted planes with data set of data of global radiation on tilted planes

Experimental stations of solar radiation measurements on inclined surfaces have been developed in a few countries of the Community. Three contracts intend to analyse those new data sets in terms of frequency distributions. Those new analyses complete CFC's, calculated by B. Bourges in the first research programme.

This item affects 4 contracts :
- ESF-009 F  - J.F. TRICAUD
- ESF-006 NL - W.H. SLOB
- ESF-005 F  - J.A. BEDEL
- ESF-008 F  - B. BOURGES

The work needs no further development. The method of computation of CFC is the method developed by B. Bourges in the first research programme.

The following results are available :
- ODEILLO (France) : 3 cumulative frequency curves of daily sums of direct radiation; January, February, March 1982. (Report contract 009 - September 1982).
- CABAUW (Netherlands) : 14 tables with the frequencies per thousand of occurrencies of 6 minutes values of global solar radiation on 7 selected planes, at and above specified limits; January and July 1981.

The 7 selected planes are :

22.5 , South
90.  , South
67.5 , South
45.  , South
45.  , South-east
45.  , South-west
(Final report 006 - January 1982).

- CARPENTRAS (France) : 8 cumulative frequency curves of hourly sums of hour-
  ly sums of global radiation on the 4 following planes :
  Horizontal,
  East vertical,
  West vertical,
  South vertical
  (Report contract 008 - May 1982).

### 3.1.2. Reconstitution of CFC with usual network measurements :

The item concerns 1 contract : ESF-008 B. Bourges

The available result is :
- A method for the calculation of the cumulative frequency curves of the
  global radiation on the horizontal plan. (Report contract 008 - May 1982).

### 3.1.3. Sensitivity at the integrated time scale of the data :

The aim is to compare the differences between the cumulative frequency
curves calculate with
- Hourly data and
- short time interval data (6 minutes or less).

The item concerns 2 contracts :
- ESF-004 F. Kasten
- ESF-005 J.A. Bedel

The following results are available :
- Comparison of the CFC calculated with actual irradiance and with the mean
  hourly irradiance, using Hamburg's data. The comparison is made only for
  selected thresholds : 600 $W/m^2$, 400 $W/m^2$, 200 $W/m^2$? 1PP $W/m^2$.
  (Report contract 004, April 1981, October 1981).

- Comparison of the CFC calculated with 6 minutes data and with hourly sums,
  using trappes and Carpentras data. (Report contrat 005, May 1982).

### 3.2. Time sequence of irradiance

### 3.2.1. Persistence of solar irradiance

This item concerns 1 contract :
- ESF-004 F. Kasten

- The results :
  Development of methods for the presentation of the persistence of the solar
  irradiance on the following selected time intervals :

    0 -  5 minutes
    5 - 15 minutes
  15 - 30 minutes
  30 - 60 minutes
   1 -  2 hours
   2 -  4 hours
     4    hours

For each month and each tilted planes, the data are arranged on 8 different tables. Most of the data are also displayed on diagrams. The list and the
contents of the tables and diagrams are given in F. Kasten's report.

Complete data series have been obtained and processed of 24 months of the years 1980-82 so far. Some examples of diagrams are given in the following reports of Action 3.3 :
- ESF-004 April 1981
- ESF-004 November 1981
- ESF-004 May 1982
- ESF-004 October 1982

3.2.2. Persistence of daily sums os solar radiation :

Statistical analysis of time sequence of daily sums of solar radiation are very useful, especially for the computation of the storage system.

The item concern 2 contracts :
- ESF-005 J.A. Bedel
- ESF-009 J.F. Tricaud

The following results are available :
- Analysis, for 50 French stations, of the sequences of consecutive days with a sunshine duration :
- Equal To 0
- Less than 0.5 hour
- Less than 1    hour

And maps of the number of consecutive days with a statistical return period of :
- Once every 5 years
- 3 times every years
For the 3 preceding sunshine duration conditions (3.3 Action leader report November 1981).

- ODEILLO : 3 monthly diagrams of the number of consecutive days with daily sums of direct radiation above given thresholds. (Report - contract 009 - October 1982).

3.3. Correlation between irradiance and other meteorological parameters

The point concerns 3 contracts :
- ESF-003 R. Dogniaux
- ESF-005 J.A. Bedel
- ESF-033 J.K. Page.

The following results are available :
- Relation between the monthly means of temperature and the monthly sums of global radiation on horizontal plane, from Hamburg. (Report - contract 004 - April 1981).
- Development of a method of parameterization of the daily variations of temperature and the global radiation (Report - contract 003 - April 1981).
- Development of a mapping presentation of the correlations between solar radiation and air temperature, at the hourly level, and calculation of tables of the mean number of hours by day during which
- The air temperature is between 2 thresholds and
- The global solar radiation is between 2 thresholds.
  (Report - contract 005 - April 1981 - May 1982).

## 4. SPECIAL REPORTS

The "strategy paper" defines 4 reports or "products" for the Action 3.3 during the 2nd R & D programme of project F.

- Cumulative frequency distribution, daily values, horizontal planes for 49 stations; time sequence of irradiation, daily values, horizontal planes, for 49 stations.
- Time sequence of irradiation, daily values, inclined planes for 49 stations.
- Report on cumulative frequencies, hourly data, horizontal and inclined surfaces.
- Report on correlation between irradiance and other meteorological parameters.

During his last meeting, the group notices, concerning these 4 reports :

4.1. - Cumulative frequency distribution; daily values, horizontal planes, for 49 stations; and time sequence of irradiation, daily values, horizontal planes; for 49 stations.

The report has been prepared by F. Kasten. It is available since November 1980. The group recommends its publication by E.E.C.

4.2. - Time sequence of irradiation, daily values, inclined planes for ·49 stations.

Action 3.2 will not produce daily sums (day by day) of global solar radiation on tilted surfaces. Consequently, the report will not be able to be produce during the 2nd programme.

Further the group thinks that this report will be very big and not very useful to the applications. The group proposed to develop a method to estimate the time sequences on tilted surfaces. The input data of the method would be the statistical data on horizontal plane which will be available in the report of F. Kasten (see 6.1).

J.K. Page will study the possibility to develop a such method before June 1983, and the development of the method would be proposed for the next research programme.

4.3. - Report on cumulative frequencies; hourly data; horizontal and inclined surfaces.

The following contents is approved by the group of Action 3.3.
- A user's guide with definition and utilization of CFC's for computing solar systems.
- An atlas of CFC's of global radiation on horizontal surface which will include :
  . Formulas
  . Charts of reduced CFC's and utilizability curves.
  . Tables for 29 European stations.
  . CFC's of 29 stations.
- An atlas of CFC's of global radiation on tilted planes with :
  . Tables and plots of CFC's for location where measurements on tilted surfaces are available (Bracknell, Locarno ...).
  . Tables and plots of CFC's for inclined surface (30°, 60°, 90° and latitude) facing due south, computed from measurements of global and diffuse radiation. Time period and locations are to be selected presently.
  . If possible a reconstitution method will be given.
  . Temperature data for selected stations will be given in the report (monthly means of minimum and maximum temperature, period 1931-1960).
  . Part 1 and 2 could be ready for publication at the begining at 1983. First version of part 3 (to be completed after) could take place in June 1983.

4.4. - Report on correlation between irradiance and other meteorological
        parameters.
      The following contents are approved :
- Introduction with :
  . Effects of other meteorological parameters for some solar applications.
  . Time scale of the study.
- Short summary of some methods with :
  . Author's name; references; summary of the method; (input data - time
    scale - presentation of the results with some examples if possible);
    applications.
- The methods or presentations developed in Action 3.3 (R. Dogiaux, F. Kasten,
  J.A. Bedel) and the available results.

# ANALYSIS OF SHORT TIME DATA OF IRRADIATION

## ON INCLINED PLANES

Authors          : F. KASTEN, K. DEHNE, W. BRETTSCHNEIDER

Contract number : ESF-004-80 D (B)

Duration         : 3 years, 1 January 1980 - 31 December 1982

Head of project : F. Kasten, Deutscher Wetterdienst,
                   Meteorologisches Observatorium Hamburg

Contractor       : Deutscher Wetterdienst, Zentralamt

Address          : Frankfurter Strasse 135
                   6050 Offenbach am Main

## Summary

Global radiation on 4 south-facing pyranometers tilted by different inclination angles and equipped with zero-albedo screens has been measured and one-line analysed in terms of irradiance and persistence intervals. 8 types of frequency distributions were generated for each month and each inclined plane. The frequency distribution of persistence of given irradiance exhibits maxima and minima the contributions of which to the total irradiation received per day appreciably varies with season and inclination angle.

# 1. Introduction

In former times, almost all meteorological radiation records were evaluated and stored in terms of daily sums. About twenty years ago, a few observatories began to determine hourly sums on a routine-basis. Apart from noticing the scattered measuring points on a strip-chart recorder, very little was known on the short-time behaviour on time scales below 1 h of global radiation much less of solar irradiation on inclined planes.

The aim of the present study is to measure global radiation on inclined planes with high time resolution and to analyse these data with respect to threshold and persistence.

# 2. Methods
## 2.1 Hardware

A measuring system had been devi ed which is capable of determining global solar radiation on 4 south-oriented planes of different inclinations withsregard to
- hourly and daily irradiation, in kWh $m^{-2}$,
- duration of given irradiance, in min,
- fraction contributed by given irradiance of given duration to hourly or daily irradiation, in percent,
- frequency of given irradiance of given duration, in hours per day.

The 4 pyranometers were tilted by the following inclination angles ß:

ß = $0^{o}$: Horizontal plane as reference.

ß = $30^{o}$: Corresponding to the inclination of the roofs of most houses; relatively high gain of solar energy in summer and under clouded sky.

ß = $53.6^{o}$: Equal to the geographical latitude $\varphi$ of the location of Meteorological Observatory Hamburg; high gain of solar energy in spring and fall.

ß = $90^{o}$: Vertical plane, representing the walls and windows of buildings. High gain of solar energy in winter under cloudless sky, especially when the ground is snow-covered.

Since global radiation on an inclined plane comprises solar radiation reflected from the foreground of the plane, it depends on the spatial distribution of the reflectance of the ground surface including buildings, trees etc. as seen from the receiving plane. In order to eliminate this dependence on the very specific surroundings, so-called zero-albedo screens which artificially reduce the albedo of the whole foreground to approximately zero, were developed and attached to each inclined pyranometer.

## 2.2 Software

The output of the 4 pyranometers were on-line discriminated into 4 intervals of irradiance and 7 intervals of duration. The intervals of irradiance were selected with emphasis on the medium range of irradiance prevailing at Hamburg:

100-200 W $m^{-2}$: Range of irradiance which is probably too low for solar energy use.

200-400 W $m^{-2}$: Range of low solar energy gain.

400-600 W $m^{-2}$: Medium range with sufficient gain; the upper limit corresponds to the mean irradiance of global radiation on horizontal plane at noon in June.

$\geq$600 W m$^{-2}$: High gain; maximum values may reach 1100 W m$^{-2}$ on rare occasions.

The intervals of duration were chosen with regard to both the meteorological variability caused by changing cloudiness, and the assumed inertness of typical solar energy receivers:
 0- 5 min: Passage of single clouds; fast receivers like solar cells.
 5-15 min: Passage of fields of clouds; receivers of low inertness like aluminum collectors.
15-30 min: Passage of cloud layers; medium inert receivers like iron collectors.
30-60 min: Passage of large cloud layers, or the times around sunrise and sunset; receivers of fairly large inertness.
Durations of 1-2 h, 2-4 h and >4 h were derived from the usual hourly records.


3. Results
Complete data series have been obtained and processed of 24 months of the years 1980-1982 so far. For each month and each tilt angle, the following types of data have been generated and arranged on 8 tables each:
- Numbers of cases per day when the global radiation persisted within the irradiance ranges and within the time intervals given in section 2.2.
- Corresponding relative numbers of cases per day; they were obtained by dividing the absolute numbers by the possible maximum numbers of the respective time intervals between sunrise and sunset.
- Corresponding total irradiances per day, in Wh m$^{-2}$, contributed by each irradiance range within each time interval mentioned above.
- Numbers of cases per day when the global radiation persisted above the threshold irradiances 100, 200, 400, or 600 W m$^{-2}$, respectively, and within one of the given time intervals.
- Corresponding relative numbers per day.
- Corresponding total irradiations per day.
- Cumulative frequencies, h/d, of global irradiances above the thresholds 100, 200, 400, or 600 W m$^{-2}$, irrespectively of time persistence.
- Cumulative frequencies as just mentioned but based on mean hourly irradiances derived from the hourly irradiation sums.
Most of the data have also been displayed on diagrams. As examples, Figs. 1 and 2 show the total irradiations per day, in kWh m$^{-2}$, contributed by the different persistence intervals, for given irradiance thresholds. On the left, the results of January 1981 are shown, on the right of July 1981. The data are from the horizontal pyranometer, tilt angle 0$^{o}$. Figs. 3 and 4 show the corresponding data of the vertical pyranometer.


4. Conclusions
The distributions of persistences of given irradiance levels are quite different in the different seasons and also on differently inclined planes. On horizontal plane, short

time persistences much prevail in summer as a consequence of
scattered cloudiness. A second maximum is observed at about 40
min persistence which may be explained by the passage of cloud
layers. In winter, the elevation of the sun is so low that
direct solar radiation very often does not penetrate the
cloud holes so that its contribution to the total irradiation
per day is small.

The corresponding distributions of persistence on vertical plane are distinctly different from those of the horizontal plane. In summer, the second maximum at about 40 min is
not pronounced because most of the radiation received on the
vertical plane in summer is diffuse anyway. To the contrary,
the winter curves exhibit a second maximum at about 100 min
which considerably contributes to the total irradiation received on the vertical plane during one day.

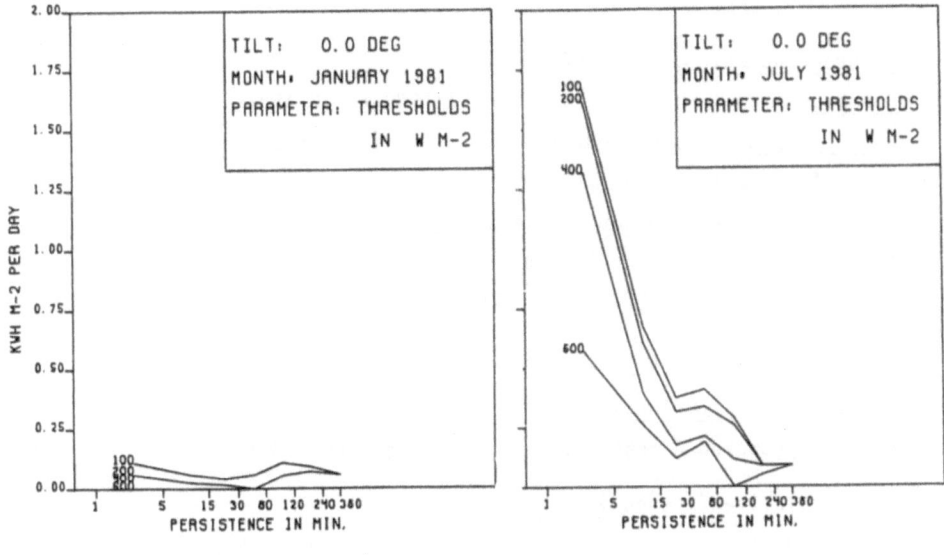

Fig. 1                    Fig. 2

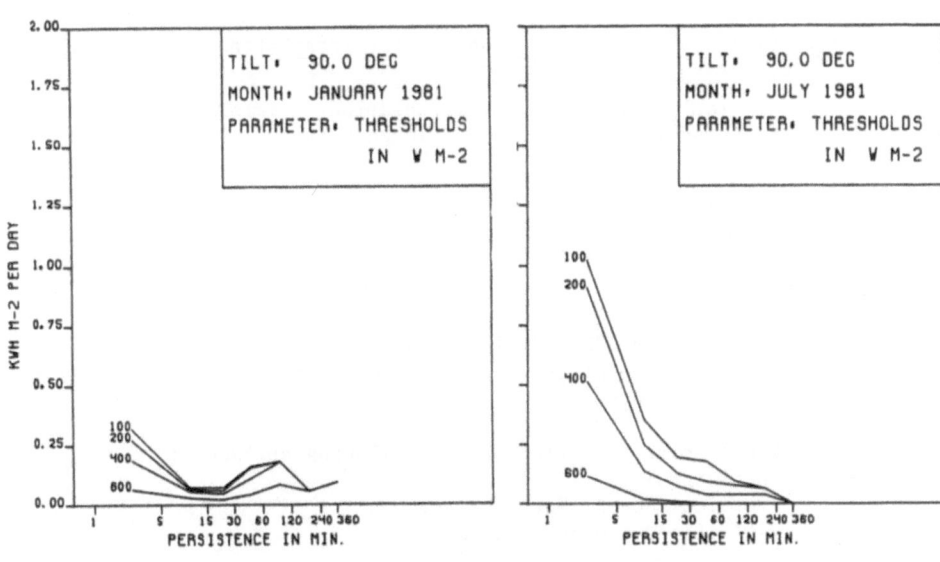

Fig. 3                    Fig. 4

# SOLAR ENERGY RECOVERED BY A FLAT PLATE COLLECTOR

Contract number : ESF - 005 - F

Duration       : 3 years    from 01/1980 to 12/1982

Total budget   : 662 098 FF        CEC contribution : 50 %

Head of project : J.A. BEDEL

Contracting organisation : Direction de la Météorologie

Address : 73-77, rue de Sèvres
          92 106 BOULOGNE France


## 1. AIMS OF THE CONTRACT

We have proposed 2 contributions to the action 3.3 :

### 1.1 Cumulative frequencies

Computation of the cumulated frequencies curves applied to global
solar irradiation on the following surfaces :
- vertical surfaces oriented towards the South, West, North and East;
- 45° tilted surfaces oriented towards the South.

Comparison between the cumulative frequencies obtained from the 6
minute data and the hourly sums.

### 1.2 Correlation between the global solar irradiation and the air - temperature

Statistical analysis of the irradiation - temperature distribution :
calculation of the monthly contingency tables with the hourly data.

Research of a mathematical or empirical adjustment of distribution
surfaces.

## 2. CUMULATIVE FREQUENCIES

### 2.1 Data

Since July 1979, "Météorologie Nationale" has been performing, at
Trappes and Carpentras, the measurements of the following parameters :
- sunshine duration
- direct solar radiation
- global solar radiation on the following surfaces :
  . horizontal
  . vertical South, North, East, West
  . 45° tilted towards the South.

The time-step of measurements is 6 minutes (values integrated over
6 minutes).

2.2 We have analyzed the hourly sums of the global radiation measurements (horizontal; vertical S-N-E-W; 45°-S).
The cumulative frequencies have been calculated with the same method used during the first research programme of EEC. The results will be available on magnetic tapes with the same formats : mean number of hours by day with solar global radiation above the following thresholds : from 0 to 1 200 $W/m^2$ by 25 $W/m^2$ step.

2.3 We have plotted the cumulative frequency curves. Some examples are presented in figures 4 to 7.
The plotting method is a simple linear interpolation between the calculated values and not the method developed by Mr BOURGES (Polynomial interpolation).

## 3. SENSIBILITY OF THE CFC AT THE INTEGRATED TIME SCALE OF THE MEASUREMENTS

3.1 In order to compare the CFC obtained with the hourly sums and with shorter time interval integrated values, we have calculated the CFC of the global solar radiation with the 6 minutes data and the hourly sums.

This work completes the study of Mr KASTEN with data from Hamburg.

The 6 minutes cumulative frequencies and the hourly cumulative frequencies have been calculated with the same data set on the same period : the 6 minutes data file from Carpentras (see paragraph 2.1).

2 forms of presentation are available to show the results :

3.2 Figures (fig. 1 to fig. 3)

The figures present the CF obtained with the hourly sums and the CF obtained with the 6 minutes data for the following thresholds : from 0 to 1 200 $W/m^2$ by step of 25 $W/m^2$.

The hourly CF are plotted using a linear interpolation : the 6 minutes data CF are indicated by points ( ).

3.3 Tables (tab. 1 to tab. 3)

The tables give :
- the mean number of hours by day with global solar radiation above selected thresholds;
- the mean useful energy above selected thresholds.

The tables show the results obtained with the 6 minutes data and the corresponding hourly sums.

3.4 The figures and the tables are examples of those 2 presentations. They show the results for Carpentras and for the following surfaces.
- horizontal
- vertical South
- 45° inclined South.

## 4. CORRELATION BETWEEN THE GLOBAL SOLAR IRRADIATION AND THE AIR - TEMPERATURE

4.1 The main part of this study is an analyse fo the relations between the solar radiation data and the air-temperature at the hourly level. The results have been presented in the preceding meetings (Paris, December 1980; Brussels, April 1981).
We recall the steps of the method.

## 4.2 The data

The available data are :
- the global solar radiation at the hourly (solar time) level;
- the air temperature at 00, 03, 06, 09, 12, 15, 18 and 21 GMT.

In order to have a file of the couple : global solar radiation, air temperature, at the hourly level (solar time), we have interpolated the 3 hours temperature data by a simple linear interpolation method. (Other methods have been tested).

## 4.3 Monthly contingency tables

The tables give the mean number of hours by day during which :
- the air-temperature is between two thresholds and
  the global solar radiation is between two thresholds.

Examples are given in tables 4 and 5.

## 4.4 Monthly diagrams

We have proposed a presentation of the function of the density of probability of the couple global solar radiation - air temperature (see report Paris, December 1980, Brussels, April 1981).

TABLE 1

| | Puissance (Watts/m²) | Energie (Joules/cm²/Jour) | | a - b | $\frac{a-b}{b}$ (%) | Différence de Durée (Heures x 10) |
|---|---|---|---|---|---|---|
| | | Données 6 minutes ⓐ | Données horaires ⓑ | | | |
| JUILLET (11 jours) | >800 | 159 | 152 | 7 | 5% | - 0,7 |
| | >600 | 581 | 573 | 8 | 1% | - 3,9 |
| | >400 | 1177 | 1172 | 5 | 0,5% | - 1,2 |
| | >200 | 1934 | 1930 | 4 | 0,2% | - 0,1 |
| JANVIER (28 jours) | >400 | 16 | 13 | 3 | 2% | 0,1 |
| | >300 | 101 | 92 | 9 | 10% | 1,4 |
| | >200 | 251 | 246 | 5 | 2% | - 0,5 |
| | >100 | 460 | 455 | 5 | 1% | - 1,6 |

TABLE 2

| | Puissance (Watts/m²) | Energie (Joules/cm²/Jour) | | a - b | $\frac{a-b}{b}$ (%) | Différence de Durée (Heures x 10) |
|---|---|---|---|---|---|---|
| | | Données 6 minutes ⓐ | Données horaires ⓑ | | | |
| JUILLET (11 jours) | >400 | 82 | 80 | 2 | 2% | + 3,3 |
| | >300 | 252 | 250 | 2 | 1% | + 1,4 |
| | >200 | 487 | 483 | 2 | 1% | - 1,1 |
| | >100 | 788 | 787 | 1 | 0,1% | - 5,1 |
| JANVIER (28 jours) | >700 | 131 | 120 | 11 | 9% | + 4,1 |
| | >500 | 419 | 401 | 18 | 4% | 2,2 |
| | >300 | 818 | 802 | 16 | 2% | - 2,9 |
| | >100 | 1309 | 1304 | 5 | 0,4% | - 4,0 |

TABLE 3

| | Puissance (Watts/m²) | Energie (Joules/cm²/Jour) | | a - b | $\frac{a-b}{b}$ (%) | Différence de Durée (Heures x 10) |
|---|---|---|---|---|---|---|
| | | Données 6 minutes ⓐ | Données horaires ⓑ | | | |
| JUILLET (11 jours) | >800 | 189 | 183 | 6 | 3% | + 4,6 |
| | >600 | 575 | 565 | 10 | 2% | + 2,0 |
| | >400 | 1108 | 1101 | 7 | 0,6% | + 6,7 |
| | >200 | 1780 | 1775 | 5 | 0,3% | + 5,1 |
| JANVIER (28 jours) | >800 | 60 | 53 | 7 | 13% . | + 0,2 |
| | >600 | 284 | 264 | 20 | 8% | - 1,6 |
| | >400 | 625 | 604 | 21 | 3% | - 0,9 |
| | >200 | 1064 | 1055 | 9 | 0,8% | - 0,5 |

TABLE 4

CARPENTRAS    MOIS : JANVIER

Nombre moyen de dixièmes d'heure par jour où la température de l'air est comprise entre les seuils $T_1$ et $T_2$ et le rayonnement solaire global entre les seuils $G_1$ et $G_2$

| $T_1/T_2$ \ $G_1/G_2$ | 0 / 40 | 40 / 80 | 80 / 120 | 120 / 160 | 160 / 200 | 200 / 240 | 240 / 280 | 280 / 320 | 320 / 360 | 360 / 400 |
|---|---|---|---|---|---|---|---|---|---|---|
| Inférieure à -4° | 0.4 | 0.1 | 0.2 | 0.1 | | | | | | |
| -4 / 0° | 3.0 | 1.2 | 1.0 | 0.5 | | | | | | |
| 0 / 4 | 4.4 | 3.2 | 1.5 | 1.7 | | | | | | |
| 4 / 8 | 10.1 | 6.0 | 5.8 | 4.5 | 0.3 | | | | | |
| 8 / 12 | 17.1 | 6.4 | 5.5 | 4.6 | 0.5 | | | | | |
| 12 / 16 | 4.7 | 2.9 | 2.3 | 1.6 | 0.5 | | | | | |
| 16 / 20 | | 0.1 | 0.1 | | | | | | | |
| 20 / 24 | | | | | | | | | | |
| 24 / 28 | | | | | | | | | | |
| 28 / 32 | | | | | | | | | | |
| 32 / 36 | | | | | | | | | | |

TABLE 5

CARPENTRAS    MOIS : JUILLET

Nombre moyen de dixièmes d'heure par jour où la température de l'air est comprise entre les seuils $T_1$ et $T_2$ et le rayonnement solaire global entre les seuils $G_1$ et $G_2$.

| $T_1 / T_2$ \ $G_1 / G_2$ | 0 / 40 | 40 / 80 | 80 / 120 | 120 / 160 | 160 / 200 | 200 / 240 | 240 / 280 | 280 / 320 | 320 / 360 | 360 / 400 |
|---|---|---|---|---|---|---|---|---|---|---|
| Inférieure à -4 | | | | | | | | | | |
| -4 / 0 | | | | | | | | | | |
| 0 / 4 | | | | | | | | | | |
| 4 / 8 | | | | | | | | | | |
| 8 / 12 | 0.1 | | | | | | | | | |
| 12 / 16 | 4.2 | 1.2 | 1.9 | 0.1 | | | | | | |
| 16 / 20 | 9.2 | 2.9 | 6.2 | 2.5 | 2.3 | 0.8 | 0.4 | 0.3 | 0.3 | |
| 20 / 24 | 5.0 | 2.3 | 3.2 | 3.1 | 3.8 | 6.6 | 3.0 | 2.7 | 2.4 | 0.1 |
| 24 / 28 | 5.8 | 2.0 | 3.6 | 2.0 | 2.5 | 5.1 | 6.0 | 7.7 | 5.6 | |
| 28 / 32 | 2.9 | 1.8 | 3.4 | 2.3 | 2.9 | 4.2 | 4.5 | 8.7 | 6.5 | |
| 32 / 36 | | 0.2 | 0.5 | 0.3 | 0.6 | 1.6 | 1.3 | 2.0 | 1.2 | |

($T_1$ et $T_2$ en °C; $G_1$ et $G_2$ en $J/cm^2/heure$

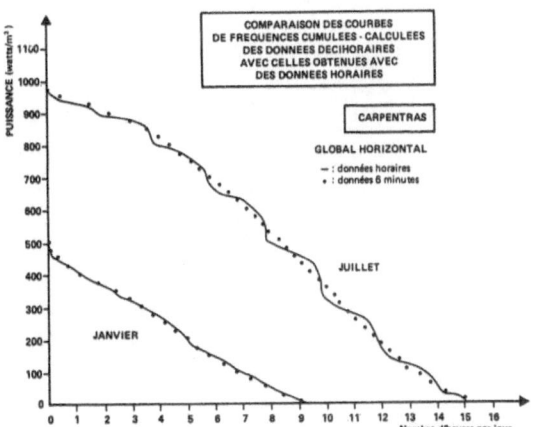

FIG. 1

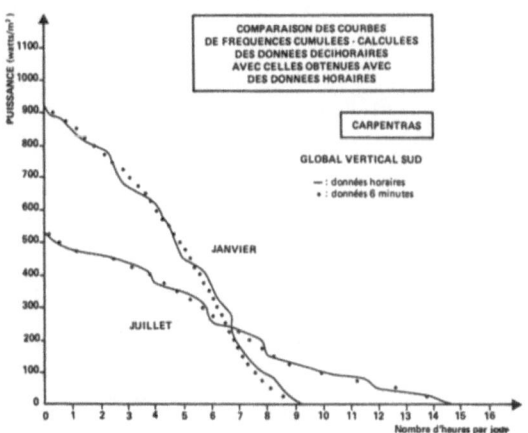

FIG. 2

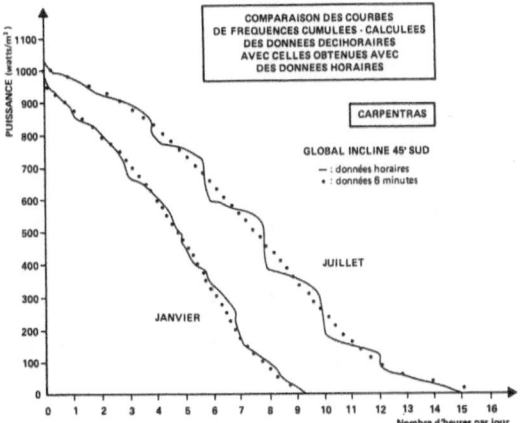

FIG. 3

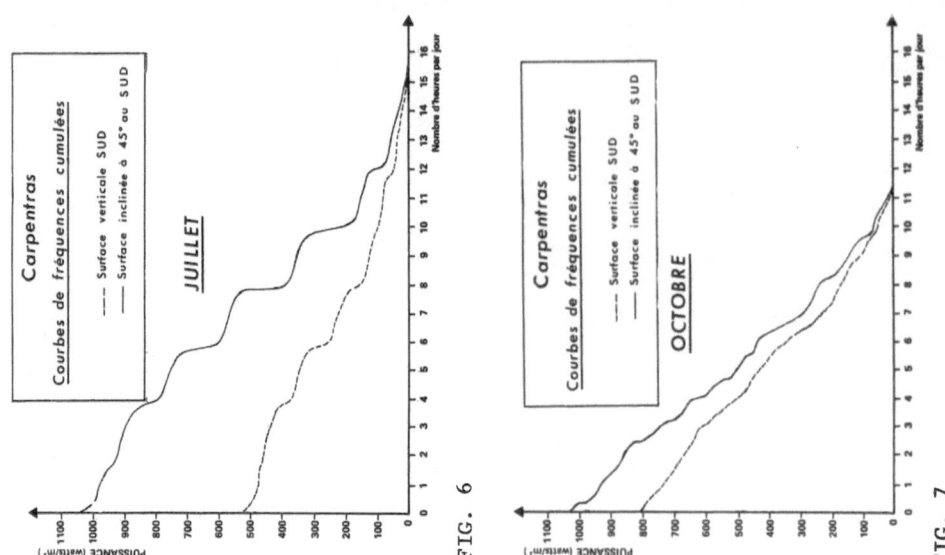

FIG. 6

FIG. 4

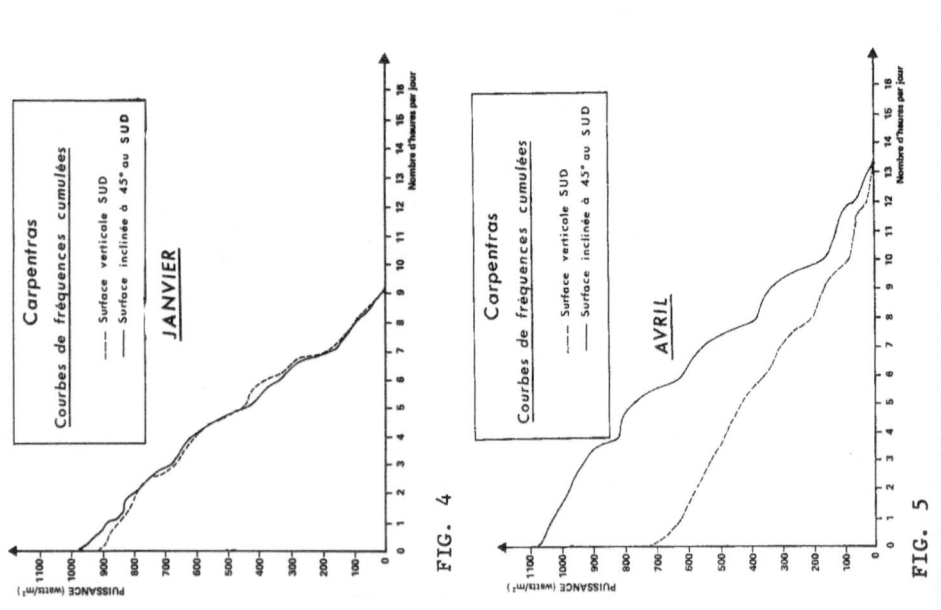

FIG. 7

FIG. 5

STATISTICAL DISTRIBUTION OF SOLAR RADIATION (HOURLY SUMS) : CUMULATIVE
FREQUENCY CURVES

Authors          : B. BOURGES, F. LASNIER
Contract Number  : ESF - 008 - F
Duration         : 39 months   1 April 80 - 30 june 83
Head of project  : B. BOURGES, Centre d'Energétique, Ecole des Mines de
                   Paris
Contractor       : ARMINES
Address          : Ecole des Mines de Paris - 60, bld St Michel
                   F-75272 - PARIS CEDEX 06

Summary
Cumulative Frequency Curves of solar irradiance are a suitable climatic
input for solar systems sizing simplified methods. From this study they
will be available for various european locations and types of surfaces
(orientation and tilting).
As far as horizontal plane is concerned, 30 locations-data have been analysed and practical formulas are given - Results are compared, from the
utilisability point of view, with US. methods.

For tilted planes, some available measured data are being processed. Various computation methods using measurements on horizontal planes as
input (Global and/or diffuse and/or direct) are being tested.

Agreement is satisfactory, as a first estimate, but more general and
accurate reconstitution algorithme should be found.

1 - AIMS AND METHODOLOGY
Frequency distribution of mean hourly global solar irradiance seems to be
a particularly well suited climatic input for solar system design methods[1],
even for some passive components [2].
The aim of this research is to study this distribution as Cumulative Frequency Curves (CFC) :
- to compute CFC's of global solar irradiance received by horizontal and
  tilted planes for various european locations (from measured data)
- from a statistical analysis of available data, to find general rules to
  get CFC's, in any location, even without measurements
- to prepare the results for publication, namely practical fitting formulas, preparation of atlas...

2 - CFC's OF SOLAR IRRADIANCE ON HORIZONTAL PLANE
Hourly data concerning 30 european meteorological stations (5 to 10 years
of data) have been processed on CFC's. A statistical analysis (Principal
Components Factor Analysis) of these curves has been performed. Conclusions
are the following : CFC's are characterized by 4 parameters :
- Maximal irradiance (solar noon, clear days) ($I_{max}$)
- Number of hours between sunrise and sunset ($d_j$)
- Mean daily sum of irradiance ($\bar{H}$)
- Secundary climatic and geographic factor ( $\varepsilon_2$)
Practical provisional formulas have been proposed [3], and are now being
improved. For instance, formulas are derived for France (Annex 1)

The utilisability function can be computed through integration of the CFC.
Two US Utilisability methods [4,5] have been compared with CFC-derived
values (figure 1) ; agreement is good for southern locations, but not
for others (particularly when high critical intensities)

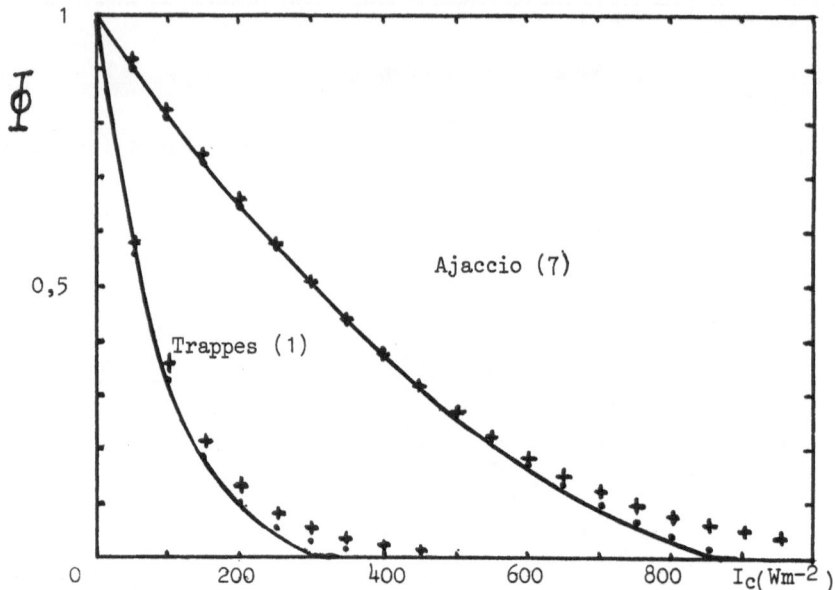

Fig. 1 : Utilisability $\Phi$ as function of critical intensity
+++ Klein (4)

... Collares-Pereira (5)

___ from CFC


## 3 - CFC's CURVES ON INCLINED SURFACES

For tilted surfaces not enough data are available for practical purposes ;
so it was necessary to establish an extrapolation from horizontal data to
reconstitute tilted surface frequency curves. First of all the presentation
of experimental frequency curves on tilted surfaces over several  years has
been performed. In second step several methods of correlation to separate
global horizontal irradiance into diffuse and direct componants have been
studied. Comparaison and validation with experimental data have been carried out for three vertical planes (East, South, West) exposed in U.K.
(Bracknell).

### 3.1 - Studied locations and weather data sets

| AREA | LOCATION | GVE | GVO | GVS | GV 60° |
|---|---|---|---|---|---|
| FRANCE | Odeillo | | | 72-74[+] | |
| IRELAND | Valentia | | | 76-77 | |
| SWITZERLAND | Locano | 58-71 | 58-71 | 58-71 | |
| ITALY | Ispra | | | | 79 |
| Danmark | Vaerlose-taes- | | | 59-73 | |
| | trop | | | | |
| U.K. | Bracknell | 67-76 | 67-76 | 67-76 | |

+ 1972-1974

(GVE - Global vertical East)

As an example the obtained experimental curves for Locarno to given in plot (2)

## 3.2 - Tilted planes computation algorithms

As we saw previously, measurements on inclined surfaces are not available on a wide-spread base.

So it is necessary, to get the CFC's, to compute them from more commonly measured parameters, namely global (G) and diffuse (D) radiation received by an horizontal plane and possibly direct normal radiation (I) or sunshine duration ($\sigma$).

For instantaneous (or mean hourly) values, global radiation $G_{\alpha,\beta}$ on a given plane (orientation $\alpha$; tilt angle :$\beta$ ) may be written

$$G_{\alpha,\beta} = I \cos \theta + D_{\alpha,\beta} + R_{\alpha,\beta}$$

where cos $\theta$ : sun-angle of incidence on the plane
$D_{\alpha,\beta}$ : diffuse radiation on the plane
$R_{\alpha,\beta}$ : ground reflected solar irradiance

The ground-reflected component is generally computed using an isotropic assumption and the albedo value $\rho$

$$R_{\alpha,\beta} = \rho \ \frac{(1 - \cos \beta)}{2} \ G$$

Methods may differ :
- by the way of computing the diffuse part $D_{\alpha,\beta}$ from the horizontal plane value D. Isotropic assumption is often made, but is discussed
- furthermore, when only G is known, by the way of making the separation between direct and diffuse components. Various correlations have been published

Tested methods are summarized in table (2)

| Method | Input data | Direct/Diffuse separation | $D_{\alpha,\beta}$ computation |
|---|---|---|---|
| 1 | G, D | | isotropic assumption |
| 2 | G, D | | § VI-3 |
| 3 | G, $\sigma$ | Franszen $\longrightarrow$ | § III.2.4 |
| | | | $\uparrow$ §III2.3 |
| 4 | G | Ruth and chant | |
| 5 | G | Bruno $\longrightarrow$ | §VI.3 et (ref6) |
| 6 | G | Orgill and Holland | §IV.2 |

Table 2

Comparison and validation with experimental data have been carried out for 3 vertical planes (East, West, South) exposed at BRACKNELL (U.K)over 9 years.

## 3.3 - Tested procedures

The analysis and comparison of the different procedures show that there are not big discrepancies in december between computed and experimental curves. In june the statistical relations (3 to 6 table 2) are very closed together and give in general case over estimated results regarding isotropic assumption- Plot (3) presents the comparison between isotropic, statistical and experimental curves for a south vertical plane in Bracknell The results are not affected for an east and west vertical plane exposed at Bracknell.

4 - CONCLUSION

From results obtained by processing Bracknell data on inclined surfaces, the following conclusions can be drawn :
- existing methods seem to be acceptable as a first estimate within a few percent (even the "isotropic" assumption), but show discrepancies in some cases
- it is, nevertheless, still premature to select one general method of reconstitution on tilted planes ; different locations, climates, and types of surfaces must be tested.

Then, a method will be choosen, (even it is provisional, until a more general and accurate is produced by Action 3.2).

This method will be applied to produce a representative sample of CFC's for surfaces facing due south with various tilt angle (30°, 60°, 90° and latitudes) for a set of european locations.
This CFC's can be analysed with a statistical methodology (simular to that used for horizontal plane CFC's) to get fitting formulas and, if possible, interpolation rules.
An atlas, with user's guide, will then be produced.

BIBLIOGRAPHIE
(1) J. Adnot, B. Bourges - "Simplified methods for the sizing of solar thermal plants" - E.E.C Contractors meeting (Project A), Meersburg1982
(2) W.A. Monsen, S.A. Klein and W.A. Beckman - "Prediction of direct gain solar heating system performance" - Solar Energy, vol. 27,1981
(3) B.Bourges - "Statistical distribution of solar radiation cumulative frequency curves" - E.E.C. Contractors meeting (project F), Brussels 1981
(4) S.A. Klein - "Calculation of Flat-Plate collector utilizability" Solar Energy, vol. 21, 1978
(5) M. Collares-Pereira and A. Rall - "Simple procedure for Predicting long term average performance of non-concentrating and of concentrating solar collectors" - Solar Energy, vol 23, 1979
(6) I.-G-J van den Brink-climatology of solar irradiance on inclined surfaces (february 82) n° 803 229 - IV.1 n° 803 229 - IV.2 (ESF-006-80 NLB)
(7) Liu, B.Y.H and R.C Jordan, The interrelationship and characteristic distribution of direct, diffuse and total solar radiation solar energy 4, 1 (1960)

ANNEX
CFC's of global solar irradiance on horizontal plane (formulas suitable for France)

Polynomial fitting of CFC
$$x = n_h/d_j = (1-y)(1+B_1y+B_2y^2+B_3y^3+B_4y^4)$$
where $y = I/I_{max}$

($n_h$ = number of hours during which solar irradiance has been greater than critical intensity I)

Reduced integral $\nu$
$$\nu = \overline{H}/(d_j \cdot I_{max})$$

INTERMEDIATE FACTORS

$$F_1 = -1.7707 + 5.6244 \, \nu - 1.5582 \, \nu^2$$

$$F_2 = \varepsilon_2 + 1.434 - 9.8645 \, \nu + 20.9105 \, \nu^2 - 13.1497 \, \nu^3$$

$$F_3 = 4.86858 + 3.13394 \, F_1 + 0.22875 \, F_2 - 13.91382 \, \nu$$


$B_i$ PARAMETERS


$$B_1 = -1.8003 + 1.73093 \, F_1 - 5.38431 \, F_2 - 10.78123 \, F_3$$

$$B_2 = 3.0744 + 0.15262 \, F_1 + 18.73507 \, F_2 + 71.17771 \, F_3$$

$$B_3 = -2.7150 - 2.65558 \, F_1 - 17.73181 \, F_2 - 130.92335 \, F_3$$

$$B_4 = 0.8853 + 1.70435 \, F_1 + 7.17479 \, F_2 + 70.19078 \, F_3$$

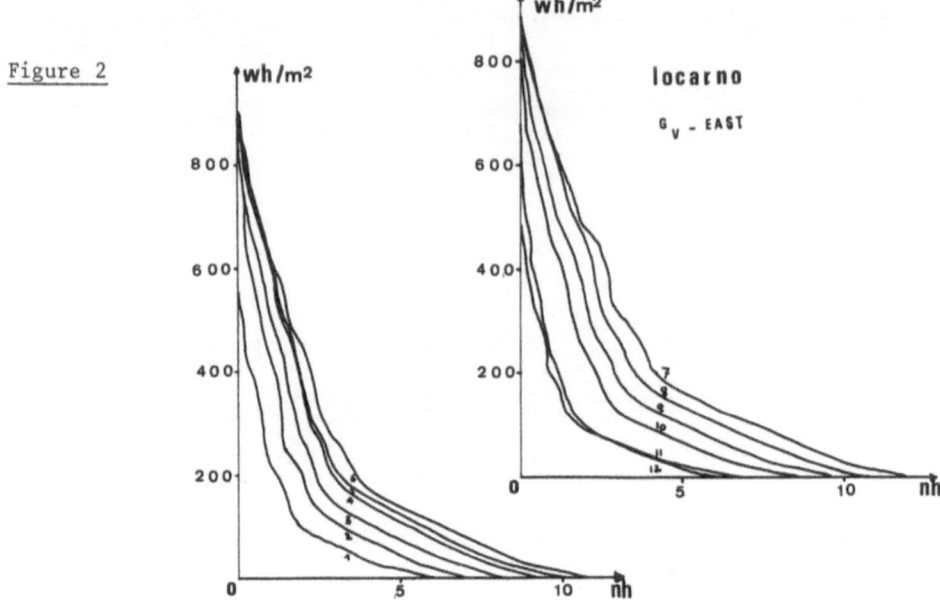

Figure 2

locarno

$G_V$ - EAST

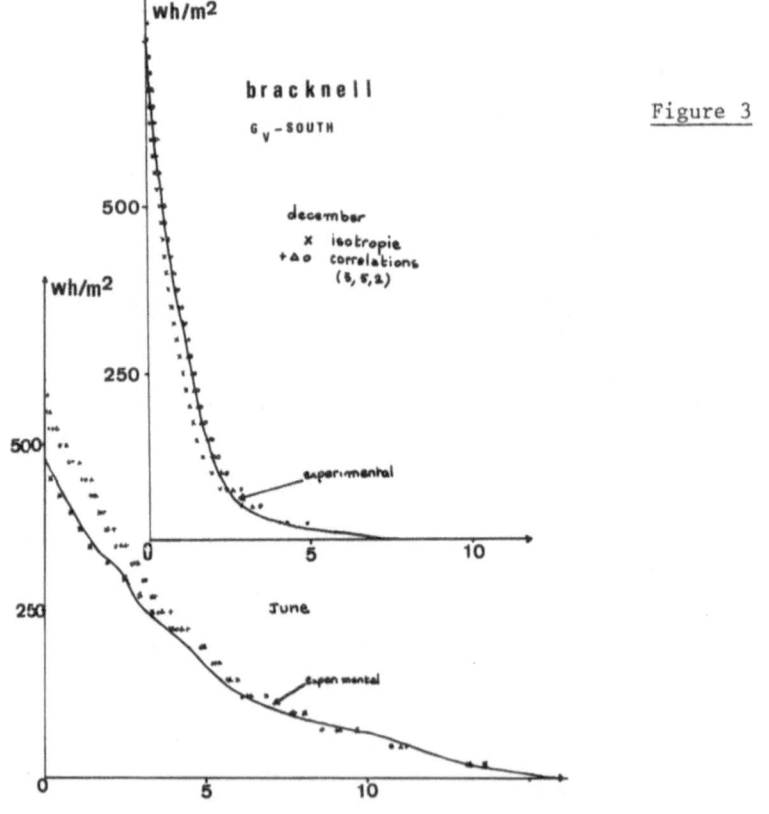

bracknell

$G_V$ - SOUTH

Figure 3

december

× isotropie

+∆o correlations

(s, s, z)

experimental

June

experimental

# STATISTICAL ANALYSIS OF SOLAR RADIATION DATA OF A
## HIGH ALTITUDE STATION

Contract number : ESF - 009 - F

Authors        : J.F. TRICAUD, Mrs LE PHAT VINH, Mrs PARRA

Duration       : 2 years        from July 1981 to June 1983

Total budget   : 71 kFF         CEC contribution : 35,5 kFF
                 (in national currency)

Head of project : A. VIALARON (ask J.F. TRICAUD)

Contracting organisation : CNRS

Address        : Laboratoire d'énergétique solaire

                 BP n° 5 Odeillo
                 66120 FONT ROMEU, France


## DESCRIPTION OF THE RESEARCH :

1) Statistical analysis of daily sums (production of cumulative frequency
   curves) for direct radiation and for global radiation on :
   - horizontal plane
   - vertical surfaces facing South, East and West
   - inclined South facing surface with slope equal to the latitude of
     Odeillo.

2) Analysis of time sequences of irradiance (tenth-hour, hourly and
   daily basis) for the same data.

## STATUS

Since October 1981 the data logging required for this work is realized
with a stored program computer. Data are registered on tapes and put on
a disk file in tenth-hour values (Solar time).

## FIRST RESULTS OBTAINED SO FAR

A first data processing has allowed to realize a second disk file in
hourly and daily values. Three examples of this disk file are given for
global radiation in April 1982 for vertical surfaces facing South, East
and West (see Tab. 1a, 1b and 1c respectively). Statistical analysis
require long period of data; in this way we have put on disk file data
for 1979, 1980 and 1981 (up to September). The control of those data is
running to day.

The cumulative frequency distribution of direct radiation in days per month
for daily suns in $kWh/m^2$ is given in Fig. 1 for January, February and
March 1982.

The number of consecutive days of direct radiation with daily sums above given thresholds is given in Fig. 2 for the first three monts of 1982.

Cumulative frequency distribution and number of given thresholds are given for the same months in Fig. 1-1, 1-2, 1-3 and Fig. 2-1, 2-2 and 2-3 respectively.

## FUTURE WORK UNTIL THE END OF THE CONTRACT

Statistical analysis of data for the period 1979 - 1982 :

- production of cumulation frequency curves for direct and global radiation;

- analysis of time sequences for tenth-hour, hourly and daily values for the same data.

ODEILLO　　RELEVE MENSUEL DE RAYONNEMENT GLOBAL VERTICAL EST

INSOLATION

!! PYRANOMETRE KIPP&ZONEN NO:5761 　!! HELIOGRAPHE: PYRHELIOMETRE 　!! ANNEE: 1982
!! ENREGISTREUR 　　　　　　　　　　　!! AVEC SEUIL: 110W/M2 　　　　　!!
!! INTEGRATEUR HP85 　　　　　　　　　!! UNITE EMPLOYEE: 1/10 HEURE 　!! MOIS: AVRIL
!! UNITE EMPLOYEE: J/CM2 RRN80

LATITUDE : 42 29' N
LONGITUDE: 02 07' E
ALTITUDE : 1580 M

INTERVALLES HORAIRES EN TEMPS SOLAIRE VRAI — INTERVALLES HORAIRES EN TEMPS SOLAIRE VRAI — "INSOLATION"

| J | -0-1 | -1-2 | -2-3 | -3-4 | -4-5 | -5-6 | -6-7 | -7-8 | -8-9 | -9-10 | -10-11 | -11-12 | MAT | R | -13-14 | -14-15 | -15-16 | -16-17 | -17-18 | -18-19 | -19-20 | -20-21 | -21-22 | -22-23 | -23-24 | SOIR | TOT | M | S | T |
|---|---|---|---|---|---|---|---|---|---|---|---|---|---|---|---|---|---|---|---|---|---|---|---|---|---|---|---|---|---|---|
| 91- | 0 | 0 | 0 | 0 | 0 | 0 | 1 | 11 | 9 | 50 | 95 | 122 | 299 | 0 | 45 | 34 | 30 | 49 | 32 | 14 | 1 | 0 | 0 | 0 | 0 | 206 | 505 | 0 | 1 | 1 |
| 92- | 0 | 0 | 0 | 0 | 0 | 3 | 45 | 294 | 322 | 293 | 219 | 122 | 1298 | 1 | 69 | 55 | 47 | 39 | 29 | 16 | 1 | 0 | 0 | 0 | 0 | 256 | 1554 | 52 | 60 | 112 |
| 93- | 0 | 0 | 0 | 0 | 23 | 158 | 254 | 285 | 218 | 166 | 107 | 211 | 1211 | 2 | 84 | 77 | 60 | 45 | 32 | 12 | 1 | 0 | 0 | 0 | 0 | 311 | 1522 | 61 | 38 | 99 |
| 94- | 0 | 0 | 0 | 3 | 17 | 88 | 236 | 172 | 111 | | | | 659 | 3 | 87 | 75 | 65 | 59 | 44 | 17 | 2 | 0 | 0 | 0 | 0 | 349 | 1008 | 32 | 44 | 76 |
| 95- | 0 | 0 | 0 | 2 | 29 | 42 | 99 | 108 | 134 | 99 | | | 513 | 4 | 62 | 56 | 49 | 42 | 26 | 15 | 2 | 0 | 0 | 0 | 0 | 252 | 765 | 26 | 45 | 71 |
| 96- | 0 | 0 | 0 | 6 | 113 | 214 | 180 | 177 | 158 | | | | 962 | 5 | 87 | 72 | 59 | 46 | 29 | 16 | 2 | 0 | 0 | 0 | 0 | 311 | 1273 | 52 | 38 | 90 |
| 97- | 0 | 0 | 0 | 10 | 128 | 279 | 260 | 178 | 163 | | | | 1108 | 6 | 54 | 51 | 28 | 31 | 27 | 15 | 2 | 0 | 0 | 0 | 0 | 208 | 1316 | 57 | 51 | 108 |
| 98- | 0 | 0 | 0 | 31 | 237 | 268 | 293 | 222 | 165 | | | | 1299 | 7 | 53 | 63 | 54 | 43 | 29 | 15 | 2 | 0 | 0 | 0 | 0 | 259 | 1558 | 62 | 35 | 97 |
| 99- | 0 | 0 | 0 | 40 | 218 | 290 | 283 | 249 | 171 | | | | 1338 | 8 | 58 | 52 | 50 | 42 | 28 | 14 | 2 | 0 | 0 | 0 | 0 | 246 | 1584 | 63 | 42 | 105 |
| 100- | 0 | 0 | 0 | 38 | 217 | 290 | 284 | 233 | 165 | | | | 1313 | 9/10 | 49 | 51 | 50 | 41 | 32 | 16 | 2 | 0 | 0 | 0 | 0 | 241 | 1554 | 64 | 57 | 121 |
| **TD** | 0 | | | | | 157 | 1173 | 2103 | 1974 | 1964 | 1608 | 1021 | 10000 | | 648 | 586 | 492 | 437 | 308 | 150 | 17 | 0 | | | | 2639 | 12639 | 469 | 412 | 881 |
| 101- | 0 | 0 | 0 | 0 | 0 | 37 | 192 | 260 | 213 | 154 | 96 | | 1222 | 11 | 71 | 56 | 43 | 34 | 31 | 15 | 2 | 0 | 0 | 0 | 0 | 252 | 1474 | 63 | 23 | 86 |
| 102- | 0 | 0 | 0 | 0 | 24 | 45 | 79 | 98 | 133 | 154 | 84 | | 617 | 12 | 77 | 63 | 30 | 20 | 21 | 8 | 0 | 0 | 0 | 0 | 0 | 219 | 836 | 23 | 15 | 38 |
| 103- | 0 | ? | ? | ? | 1 | 27 | 22 | 64 | 87 | 130 | 116 | | 447 | 13 | 84 | 62 | 58 | 51 | 35 | 17 | 3 | 0 | 0 | 0 | 0 | 310 | 867 | 12 | 33 | 45 |
| 104- | 0 | 0 | 0 | 0 | 56 | 258 | 283 | 181 | 69 | 143 | 102 | | 1092 | 14 | 78 | 52 | 63 | 50 | 31 | 14 | 3 | 0 | 0 | 0 | 0 | 291 | 1383 | 47 | 6 | 53 |
| 105- | 0 | 0 | 0 | 0 | 41 | 205 | 285 | 262 | 231 | 193 | 106 | | 1323 | 15 | 67 | 60 | 37 | 33 | 20 | 14 | 3 | 0 | 0 | 0 | 0 | 234 | 1557 | 59 | 8 | 67 |
| 106- | 0 | 0 | 0 | 0 | 2 | 8 | 13 | 35 | 49 | 86 | | | 228 | 16 | 112 | 58 | 48 | 25 | 21 | 4 | 0 | 0 | 0 | 0 | 0 | 294 | 522 | 0 | 0 | 0 |
| 107- | 0 | 0 | 0 | 0 | 25 | 23 | 39 | 222 | 249 | 177 | 102 | | 837 | 17 | 67 | 62 | 55 | 53 | 40 | 13 | 4 | 0 | 0 | 0 | 0 | 294 | 1131 | 37 | 36 | 73 |
| 108- | 0 | 0 | 0 | 0 | 56 | 219 | 282 | 274 | 214 | 96 | 66 | | 1207 | 18 | 66 | 52 | 31 | 14 | 7 | 3 | 2 | 0 | 0 | 0 | 0 | 175 | 1382 | 44 | 1 | 45 |
| 109- | 0 | 0 | 0 | 0 | 7 | 18 | 12 | 34 | 43 | 67 | 88 | | 269 | 19 | 86 | 81 | 42 | 0 | 0 | 0 | 0 | 0 | 0 | 0 | 0 | 209 | 478 | 6 | 14 | 20 |
| 110- | 0 | 0 | 0 | 0 | 0 | 43 | 159 | 255 | 237 | 169 | 121 | | 984 | 20 | 59 | 52 | 45 | 40 | 32 | 21 | 5 | 0 | 0 | 0 | 0 | 254 | 1238 | 44 | 61 | 105 |
| **TD** | 0 | | | | 249 | 1038 | 1685 | 1511 | 1332 | | | 967 | 8886 | | 767 | 598 | 452 | 320 | 243 | 126 | 26 | 0 | | | | 2532 | | 333 | 197 | 530 |
| 111- | 0 | 0 | 0 | 0 | 85 | 258 | 279 | 273 | 243 | 177 | 106 | | 1421 | 21 | 48 | 45 | 46 | 47 | 36 | 25 | 6 | 0 | 0 | 0 | 0 | 253 | 1674 | 65 | 63 | 128 |
| 112- | 0 | 0 | 0 | 0 | 72 | 257 | 264 | 256 | 229 | 185 | 93 | | 1356 | 22 | 49 | 46 | 43 | 41 | 33 | 22 | 6 | 0 | 0 | 0 | 0 | 240 | 1596 | 65 | 64 | 129 |
| 113- | 0 | 0 | 0 | 0 | 74 | 248 | 308 | 256 | 229 | 185 | 92 | | 1392 | 23 | 52 | 47 | 43 | 39 | 31 | 21 | 6 | 0 | 0 | 0 | 0 | 239 | 1631 | 65 | 62 | 127 |
| 114- | 0 | 0 | 0 | 0 | 60 | 210 | 284 | 281 | 239 | 181 | 106 | | 1361 | 24 | 68 | 66 | 66 | 18 | 36 | 21 | 7 | 0 | 0 | 0 | 0 | 249 | 1610 | 65 | 34 | 99 |
| 115- | 0 | 0 | 0 | 0 | 74 | 226 | 273 | 289 | 251 | 186 | 103 | | 1402 | 25 | 65 | 62 | 60 | 52 | 44 | 27 | 5 | 0 | 0 | 0 | 0 | 309 | 1711 | 65 | 54 | 119 |
| 116- | 0 | 0 | 0 | 0 | 53 | 201 | 230 | 276 | 213 | 175 | 112 | | 1260 | 26 | 75 | 65 | 66 | 64 | 37 | 22 | 6 | 0 | 0 | 0 | 0 | 335 | 1595 | 62 | 54 | 116 |
| 117- | 0 | 0 | 0 | 0 | 88 | 256 | 303 | 303 | 248 | 176 | 91 | | 1457 | 27 | 51 | 48 | 45 | 40 | 33 | 23 | 7 | 0 | 0 | 0 | 0 | 247 | 1704 | 66 | 64 | 130 |
| 118- | 0 | 0 | 0 | 0 | 100 | 269 | 315 | 299 | 234 | 163 | 91 | | 1472 | 28 | 60 | 47 | 38 | 32 | 23 | 20 | 8 | 0 | 0 | 0 | 0 | 250 | 1722 | 67 | 64 | 131 |
| 119- | 0 | 0 | 0 | 1 | 94 | 253 | 304 | 293 | 246 | 172 | 93 | | 1456 | 29 | 57 | 54 | 53 | 49 | 40 | 20 | 5 | 0 | 0 | 0 | 0 | 278 | 1734 | 66 | 46 | 112 |
| 120- | 0 | 0 | 0 | 0 | 2 | 5 | 13 | 16 | 35 | 79 | 96 | | 250 | 30 | 94 | 76 | 65 | 62 | 45 | 30 | 9 | 0 | 0 | 0 | 0 | 381 | 631 | 4 | 46 | 50 |
| **TD** | 2 | | | | 702 | 2187 | 2573 | 2534 | 2167 | 1679 | 983 | | 12827 | | 619 | 556 | 490 | 450 | 367 | 234 | 65 | 0 | | | | 2781 | 15608 | 590 | 549 | 1139 |
| **TM** | | 1108 | | | 4398 | 5991 | 6322 | 5642 | 4619 | 2971 | | | 24063 | | 2034 | 1740 | 1434 | 1207 | 918 | 510 | 108 | 0 | | | | 7952 | | 1394 | 1159 | 2553 |
| **MM** | 0 | 37 | | 147 | 200 | 211 | 188 | 154 | 99 | | | | 803 | MH | 68 | 58 | 48 | 40 | 31 | 17 | 4 | 0 | | | | 265 | | 46 | 39 | 85 |

Table 1a

- 185 -

ODEILLO

## RELEVE MENSUEL DE RAYONNEMENT GLOBAL VERTICAL SUD ET D'INSOLATION

| | | |
|---|---|---|
| PYRANOMETRE KIPPAZONEN  NO:2798 | HELIOGRAPHE: PYRHELIOMETRE — INSOLATION | ANNEE: 1982 |
| ENREGISTREUR | AVEC SEUIL: 110W/M2 | MOIS: AVRIL |
| INTEGRATEUR HP85 | UNITE EMPLOYEE: 1/10 HEURE | |
| UNITE EMPLOYEE: J/CM2 RRMBO | | |

LATITUDE : 42 29' N
LONGITUDE: 02 07' E
ALTITUDE : 1580 M

### INTERVALLES HORAIRES EN TEMPS SOLAIRE VRAI (matin)

Les colonnes -0-1 à -4-5 sont toutes à 0 pour tous les jours.

| JR | -5-6 | -6-7 | -7-8 | -8-9 | -9-10 | -10-11 | -11-12 | MAT |
|---|---|---|---|---|---|---|---|---|
| 91 | 0 | 8 | 9 | 6 | 41 | 85 | 102 | 251 |
| 92 | 1 | 15 | 98 | 167 | 225 | 263 | 274 | 1043 |
| 93 | 1 | 23 | 76 | 144 | 179 | 208 | 200 | 831 |
| 94 | 1 | 14 | 27 | 63 | 186 | 208 | 194 | 693 |
| 95 | 1 | 17 | 32 | 70 | 97 | 159 | 198 | 574 |
| 96 | 1 | 24 | 69 | 100 | 148 | 190 | 205 | 737 |
| 97 | 1 | 18 | 73 | 126 | 145 | 194 | 231 | 788 |
| 98 | 1 | 19 | 74 | 136 | 174 | 215 | 226 | 845 |
| 99 | 2 | 19 | 72 | 131 | 188 | 207 | 232 | 851 |
| 100 | 2 | 19 | 72 | 132 | 180 | 215 | 235 | 855 |
| TD | 11 | 176 | 602 | 1075 | 1563 | 1944 | 2097 | 7468 |
| 101 | 2 | 23 | 74 | 129 | 173 | 198 | 185 | 784 |
| 102 | 3 | 15 | 44 | 69 | 114 | 196 | 140 | 581 |
| 103 | 1 | 8 | 21 | 50 | 74 | 139 | 194 | 487 |
| 104 | 3 | 19 | 66 | 91 | 66 | 158 | 186 | 589 |
| 105 | 3 | 19 | 64 | 124 | 177 | 212 | 164 | 763 |
| 106 | 1 | 13 | 41 | 44 | 47 | 70 | 8 | 224 |
| 107 | 5 | 17 | 30 | 114 | 176 | 212 | 235 | 789 |
| 108 | 5 | 20 | 59 | 115 | 140 | 100 | 75 | 512 |
| 109 | 3 | 14 | 12 | 27 | 43 | 72 | 139 | 310 |
| 110 | 0 | 22 | 56 | 120 | 173 | 193 | 228 | 792 |
| TD | 24 | 165 | 439 | 880 | 1180 | 1527 | 1616 | 5831 |
| 111 | 4 | 19 | 55 | 116 | 166 | 201 | 219 | 780 |
| 112 | 4 | 19 | 54 | 116 | 164 | 197 | 213 | 767 |
| 113 | 5 | 22 | 57 | 116 | 163 | 195 | 214 | 772 |
| 114 | 5 | 23 | 57 | 113 | 160 | 201 | 215 | 774 |
| 115 | 5 | 22 | 52 | 113 | 168 | 202 | 215 | 777 |
| 116 | 5 | 26 | 59 | 112 | 141 | 192 | 216 | 752 |
| 117 | 5 | 22 | 52 | 111 | 160 | 193 | 210 | 753 |
| 118 | 5 | 22 | 52 | 111 | 150 | 183 | 205 | 729 |
| 119 | 6 | 24 | 52 | 108 | 155 | 187 | 204 | 736 |
| 120 | 3 | 11 | 15 | 20 | 39 | 88 | 114 | 290 |
| TD | 49 | 210 | 505 | 1036 | 1466 | 1839 | 2025 | 7130 |
| TM | 84 | 551 | 1546 | 2991 | 4209 | 5310 | 5738 | 20429 |
| MM | 3 | 18 | 52 | 100 | 140 | 177 | 191 | 681 |

### INTERVALLES HORAIRES EN TEMPS SOLAIRE VRAI (soir) — TOTAUX ET INSOLATION

| JOUR | -13-14 | -14-15 | -15-16 | -16-17 | -17-18 | -18-19 | -19-20 | SOIR | TOT | M | S | T |
|---|---|---|---|---|---|---|---|---|---|---|---|---|
| 1 | 37 | 28 | 26 | 52 | 36 | 16 | 1 | 195 | 446 | 0 | 1 | 1 |
| 2 | 263 | 238 | 198 | 132 | 83 | 22 | 1 | 937 | 1980 | 52 | 60 | 112 |
| 3 | 186 | 142 | 160 | 108 | 48 | 13 | 1 | 658 | 1489 | 61 | 38 | 99 |
| 4 | 147 | 180 | 166 | 128 | 76 | 19 | 2 | 718 | 1411 | 32 | 44 | 76 |
| 5 | 229 | 228 | 156 | 136 | 42 | 16 | 1 | 808 | 1382 | 26 | 45 | 71 |
| 6 | 204 | 187 | 143 | 98 | 38 | 20 | 1 | 692 | 1429 | 52 | 38 | 90 |
| 7 | 233 | 216 | 98 | 107 | 65 | 18 | 1 | 738 | 1526 | 57 | 51 | 108 |
| 8 | 226 | 185 | 120 | 67 | 44 | 19 | 1 | 662 | 1507 | 62 | 35 | 97 |
| 9 | 229 | 200 | 167 | 99 | 38 | 14 | 1 | 748 | 1599 | 63 | 42 | 105 |
| 10 | 216 | 215 | 159 | 114 | 69 | 19 | 1 | 793 | 1648 | 64 | 57 | 121 |
| TD | 1970 | 1819 | 1393 | 1041 | 539 | 176 | 11 | 6949 | 14417 | 469 | 412 | 881 |
| 11 | 200 | 155 | 90 | 51 | 45 | 16 | 1 | 558 | 1342 | 63 | 23 | 86 |
| 12 | 189 | 117 | 49 | 28 | 24 | 7 | 0 | 414 | 995 | 23 | 15 | 38 |
| 13 | 201 | 208 | 126 | 78 | 45 | 18 | 2 | 678 | 1165 | 12 | 33 | 45 |
| 14 | | | | | | | | 372 | 961 | 47 | 6 | 53 |
| 15 | | | | | | | | 352 | 1115 | 59 | 8 | 67 |
| 16 | | | | | | | | 307 | 531 | 0 | 0 | 0 |
| 17 | 231 | 186 | 120 | 108 | 59 | 18 | 5 | 727 | 1516 | 37 | 36 | 73 |
| 18 | 83 | 47 | 26 | 18 | 5 | 3 | 1 | 183 | 695 | 44 | 1 | 45 |
| 19 | 189 | 158 | 41 | 0 | 0 | 0 | 0 | 388 | 698 | 6 | 14 | 20 |
| 20 | 231 | 209 | 162 | 121 | 62 | 22 | 5 | 812 | 1604 | 44 | 61 | 105 |
| TD | 1698 | 1317 | 780 | 519 | 322 | 134 | 21 | 4791 | 10622 | 333 | 197 | 530 |
| 21 | 218 | 200 | 159 | 110 | 59 | 25 | 5 | 776 | 1556 | 65 | 63 | 128 |
| 22 | 216 | 197 | 163 | 117 | 60 | 24 | 6 | 783 | 1550 | 65 | 64 | 129 |
| 23 | 214 | 196 | 158 | 114 | 58 | 24 | 6 | 770 | 1542 | 65 | 62 | 127 |
| 24 | 215 | 159 | 77 | | | | | 498 | 1272 | 65 | 34 | 99 |
| 25 | 218 | 190 | 164 | 109 | 70 | 25 | 7 | 781 | 1558 | 65 | 54 | 119 |
| 26 | 173 | 165 | 128 | | | | | 696 | 1448 | 62 | 54 | 116 |
| 27 | 211 | 192 | 158 | 111 | 55 | 24 | 7 | 758 | 1511 | 66 | 64 | 130 |
| 28 | 209 | 148 | 155 | 107 | 51 | 23 | 7 | 700 | 1429 | 67 | 64 | 131 |
| 29 | 205 | 172 | 158 | 115 | 60 | 23 | 6 | 739 | 1475 | 66 | 46 | 112 |
| 30 | 161 | 101 | 93 | 54 | 32 | | | 586 | 876 | 4 | 46 | 50 |
| TD | 2040 | 1720 | 1413 | 1026 | 570 | 254 | 64 | 7087 | 14217 | 590 | 549 | 1139 |
| TM | 5708 | 4856 | 3586 | 2586 | 1431 | 564 | 96 | 18827 | 39256 | 1394 | 1159 | 2553 |
| MM | 190 | 162 | 120 | 86 | 48 | 19 | 3 | 628 | 1309 | 46 | 39 | 85 |

Table b

ODEILLO

RELEVE MENSUEL DE RAYONNEMENT GLOBAL VERTICAL OUEST

INSOLATION

LATITUDE : 42 29' N !! PYRANOMETRE KIPP&ZONEN NO:2050 !! HELIOGRAPHE: PYRHELIOMETRE !! ANNEE: 1982
LONGITUDE: 02 07' E !! ENREGISTREUR !! AVEC SEUIL: 110W/M2 !!
ALTITUDE : 1580 M !! INTEGRATEUR HP85 !! UNITE EMPLOYEE: 1/10 HEURE !! MOIS: AVRIL
!! UNITE EMPLOYEE: J/CM2 RRM80 !! !!

| JOUR | 0-1 | 1-2 | 2-3 | 3-4 | 4-5 | 5-6 | 6-7 | 7-8 | 8-9 | 9-10 | 10-11 | 11-12 | MAT | JOUR | 12-13 | 13-14 | 14-15 | 15-16 | 16-17 | 17-18 | 18-19 | 19-20 | 20-21 | 21-22 | 22-23 | 23-24 | SOIR | TOT | M | S | T |
|---|---|---|---|---|---|---|---|---|---|---|---|---|---|---|---|---|---|---|---|---|---|---|---|---|---|---|---|---|---|---|---|
| | INTERVALLES HORAIRES EN TEMPS SOLAIRE VRAI | | | | | | | | | | | | | | INTERVALLES HORAIRES EN TEMPS SOLAIRE VRAI | | | | | | | | | | | | INSOLATION | | | | |
| 91 | 0 | 0 | 0 | 0 | 0 | 1 | 9 | 10 | 7 | 45 | 75 | 82 | 229 | 1 | 31 | 24 | 23 | 56 | 43 | 25 | 1 | 0 | 0 | 0 | 0 | 0 | 203 | 432 | 0 | 1 | 1 |
| 92 | 0 | 0 | 0 | 0 | 0 | 1 | 15 | 48 | 61 | 69 | 75 | 67 | 336 | 2 | 152 | 197 | 239 | 247 | 255 | 177 | 7 | 0 | 0 | 0 | 0 | 0 | 1274 | 1610 | 52 | 60 | 112 |
| 93 | 0 | 0 | 0 | 0 | 0 | 2 | 18 | 31 | 45 | 64 | 76 | 93 | 329 | 3 | 125 | 137 | 196 | 172 | 89 | 17 | 2 | 0 | 0 | 0 | 0 | 0 | 738 | 1067 | 61 | 38 | 99 |
| 94 | 0 | 0 | 0 | 0 | 0 | 1 | 14 | 22 | 34 | 58 | 79 | 101 | 309 | 4 | 117 | 178 | 232 | 262 | 220 | 60 | 6 | 0 | 0 | 0 | 0 | 0 | 1075 | 1384 | 32 | 44 | 76 |
| 95 | 0 | 0 | 0 | 0 | 0 | 2 | 15 | 30 | 52 | 55 | 59 | 60 | 273 | 5 | 125 | 206 | 205 | 278 | 98 | 74 | 5 | 0 | 0 | 0 | 0 | 0 | 991 | 1264 | 26 | 45 | 71 |
| 96 | 0 | 0 | 0 | 0 | 0 | 2 | 20 | 35 | 53 | 63 | 70 | 85 | 328 | 6 | 126 | 165 | 187 | 168 | 53 | 80 | 10 | 0 | 0 | 0 | 0 | 0 | 789 | 1117 | 52 | 38 | 90 |
| 97 | 0 | 0 | 0 | 0 | 0 | 2 | 13 | 27 | 44 | 45 | 44 | 51 | 226 | 7 | 111 | 183 | 130 | 239 | 248 | 143 | 7 | 0 | 0 | 0 | 0 | 0 | 1061 | 1287 | 57 | 51 | 108 |
| 98 | 0 | 0 | 0 | 0 | 0 | 3 | 17 | 28 | 40 | 41 | 51 | 58 | 238 | 8 | 122 | 173 | 148 | 105 | 104 | 86 | 4 | 0 | 0 | 0 | 0 | 0 | 742 | 980 | 62 | 35 | 97 |
| 99 | 0 | 0 | 0 | 0 | 0 | 3 | 17 | 28 | 35 | 42 | 48 | 58 | 231 | 9 | 118 | 176 | 227 | 194 | 84 | 33 | 3 | 0 | 0 | 0 | 0 | 0 | 835 | 1066 | 63 | 42 | 105 |
| 100 | 0 | 0 | 0 | 0 | 0 | 3 | 17 | 28 | 35 | 40 | 45 | 53 | 221 | 10 | 110 | 191 | 219 | 231 | 254 | 139 | 5 | 0 | 0 | 0 | 0 | 0 | 1149 | 1370 | 64 | 57 | 121 |
| TD | | | | | 0 | 20 | 155 | 287 | 406 | 522 | 622 | 708 | 2720 | | 1137 | 1630 | 1806 | 1952 | 1448 | 834 | 50 | 0 | | | | | 8857 | 11577 | 469 | 412 | 881 |
| 101 | 0 | 0 | 0 | 0 | 0 | 4 | 19 | 31 | 41 | 51 | 61 | 79 | 286 | 11 | 123 | 140 | 109 | 68 | 92 | 30 | 3 | 0 | 0 | 0 | 0 | 0 | 565 | 851 | 63 | 23 | 86 |
| 102 | 0 | 0 | 0 | 0 | 0 | 4 | 15 | 32 | 47 | 57 | 66 | 76 | 297 | 12 | 131 | 109 | 52 | 18 | 27 | 6 | 0 | 0 | 0 | 0 | 0 | 0 | 343 | 640 | 23 | 15 | 38 |
| 103 | 0 | 0 | 0 | 0 | 0 | 1 | 8 | 21 | 44 | 52 | 79 | 90 | 295 | 13 | 130 | 193 | 164 | 121 | 90 | 43 | 6 | 0 | 0 | 0 | 0 | 0 | 747 | 1042 | 12 | 33 | 45 |
| 104 | 0 | 0 | 0 | 0 | 0 | 4 | 20 | 30 | 32 | 37 | 59 | 63 | 245 | 14 | 94 | 67 | 99 | 72 | 55 | 22 | 8 | 0 | 0 | 0 | 0 | 0 | 417 | 662 | 47 | 6 | 53 |
| 105 | 0 | 0 | 0 | 0 | 0 | 4 | 18 | 30 | 38 | 43 | 55 | 84 | 272 | 15 | 95 | 106 | 49 | 35 | 24 | 52 | 3 | 0 | 0 | 0 | 0 | 0 | 364 | 636 | 59 | 8 | 67 |
| 106 | 0 | 0 | 0 | 0 | 0 | 2 | 8 | 14 | 43 | 46 | 51 | 75 | 239 | 16 | 127 | 56 | 46 | 28 | 25 | 20 | 4 | 0 | 0 | 0 | 0 | 0 | 306 | 545 | 0 | 0 | 0 |
| 107 | 0 | 0 | 0 | 0 | 0 | 5 | 15 | 26 | 45 | 56 | 80 | 76 | 303 | 17 | 161 | 185 | 144 | 200 | 122 | 23 | 8 | 0 | 0 | 0 | 0 | 0 | 843 | 1146 | 37 | 36 | 73 |
| 108 | 0 | 0 | 0 | 0 | 0 | 5 | 19 | 28 | 36 | 49 | 61 | 59 | 257 | 18 | 74 | 47 | 25 | 13 | 7 | 4 | 2 | 0 | 0 | 0 | 0 | 0 | 172 | 429 | 44 | 1 | 45 |
| 109 | 0 | 0 | 0 | 0 | 0 | 4 | 14 | 11 | 26 | 40 | 62 | 82 | 239 | 19 | 126 | 155 | 41 | 0 | 0 | 0 | 0 | 0 | 0 | 0 | 0 | 0 | 322 | 561 | 6 | 14 | 20 |
| 110 | 0 | 0 | 0 | 0 | 0 | 0 | 22 | 36 | 45 | 50 | 49 | 57 | 259 | 20 | 118 | 193 | 240 | 294 | 303 | 236 | 33 | 0 | 0 | 0 | 0 | 0 | 1417 | 1676 | 44 | 61 | 105 |
| TD | | | | | 0 | 33 | 158 | 259 | 397 | 481 | 623 | 741 | 2692 | | 1179 | 1251 | 969 | 849 | 745 | 436 | 67 | 0 | | | | | 5496 | 8188 | 333 | 197 | 530 |
| 111 | 0 | 0 | 0 | 0 | 0 | 6 | 21 | 30 | 36 | 41 | 43 | 47 | 224 | 21 | 109 | 185 | 248 | 254 | 263 | 242 | 41 | 0 | 0 | 0 | 0 | 0 | 1342 | 1566 | 65 | 63 | 128 |
| 112 | 0 | 0 | 0 | 0 | 0 | 5 | 20 | 29 | 38 | 41 | 43 | 49 | 225 | 22 | 109 | 184 | 250 | 293 | 287 | 233 | 53 | 0 | 0 | 0 | 0 | 0 | 1409 | 1634 | 65 | 64 | 129 |
| 113 | 0 | 0 | 0 | 0 | 0 | 6 | 21 | 30 | 38 | 43 | 47 | 53 | 238 | 23 | 111 | 186 | 256 | 287 | 283 | 218 | 52 | 0 | 0 | 0 | 0 | 0 | 1393 | 1631 | 65 | 62 | 127 |
| 114 | 0 | 0 | 0 | 0 | 0 | 6 | 21 | 31 | 38 | 47 | 55 | 68 | 266 | 24 | 130 | 134 | 25 | 37 | 185 | 254 | 34 | 0 | 0 | 0 | 0 | 0 | 799 | 1065 | 65 | 34 | 99 |
| 115 | 0 | 0 | 0 | 0 | 0 | 7 | 22 | 31 | 39 | 46 | 53 | 63 | 261 | 25 | 121 | 183 | 244 | 233 | 253 | 118 | 16 | 0 | 0 | 0 | 0 | 0 | 1168 | 1429 | 65 | 54 | 119 |
| 116 | 0 | 0 | 0 | 0 | 0 | 7 | 23 | 34 | 42 | 51 | 55 | 76 | 288 | 26 | 120 | 138 | 248 | 282 | 247 | 183 | 35 | 0 | 0 | 0 | 0 | 0 | 1253 | 1541 | 62 | 54 | 116 |
| 117 | 0 | 0 | 0 | 0 | 0 | 7 | 22 | 31 | 38 | 42 | 46 | 53 | 239 | 27 | 114 | 190 | 253 | 294 | 296 | 235 | 60 | 0 | 0 | 0 | 0 | 0 | 1442 | 1681 | 66 | 64 | 130 |
| 118 | 0 | 0 | 0 | 0 | 0 | 8 | 22 | 32 | 39 | 52 | 57 | 52 | 262 | 28 | 116 | 144 | 254 | 299 | 313 | 267 | 75 | 0 | 0 | 0 | 0 | 0 | 1468 | 1730 | 67 | 64 | 131 |
| 119 | 0 | 0 | 0 | 0 | 0 | 8 | 23 | 32 | 38 | 42 | 46 | 53 | 242 | 29 | 116 | 165 | 250 | 294 | 203 | 33 | 9 | 0 | 0 | 0 | 0 | 0 | 1070 | 1312 | 66 | 46 | 112 |
| 120 | 0 | 0 | 0 | 0 | 0 | 2 | 8 | 11 | 15 | 30 | 68 | 89 | 223 | 30 | 131 | 145 | 151 | 185 | 187 | 151 | 13 | 0 | 0 | 0 | 0 | 0 | 963 | 1186 | 4 | 46 | 50 |
| TD | | | | | 0 | 62 | 203 | 291 | 361 | 435 | 513 | 603 | 2468 | | 1177 | 1654 | 2179 | 2458 | 2517 | 1934 | 388 | 0 | | | | | 12307 | 14775 | 590 | 549 | 1139 |
| TM | | | | | 0 | 115 | 516 | 837 | 1164 | 1438 | 1758 | 2052 | 7880. | | 3493 | 4535 | 4954 | 5259 | 4710 | 3204 | 505 | 0 | | | | | 26660. | 34540. | 1394 | 1159 | 2553 |
| MM | | | | | 0 | 4 | 17 | 28 | 39 | 48 | 59 | 68 | 263 | MM | 116 | 151 | 165 | 175 | 157 | 107 | 17 | 0 | | | | | 889 | 1151 | 46 | 39 | 85 |

Table c

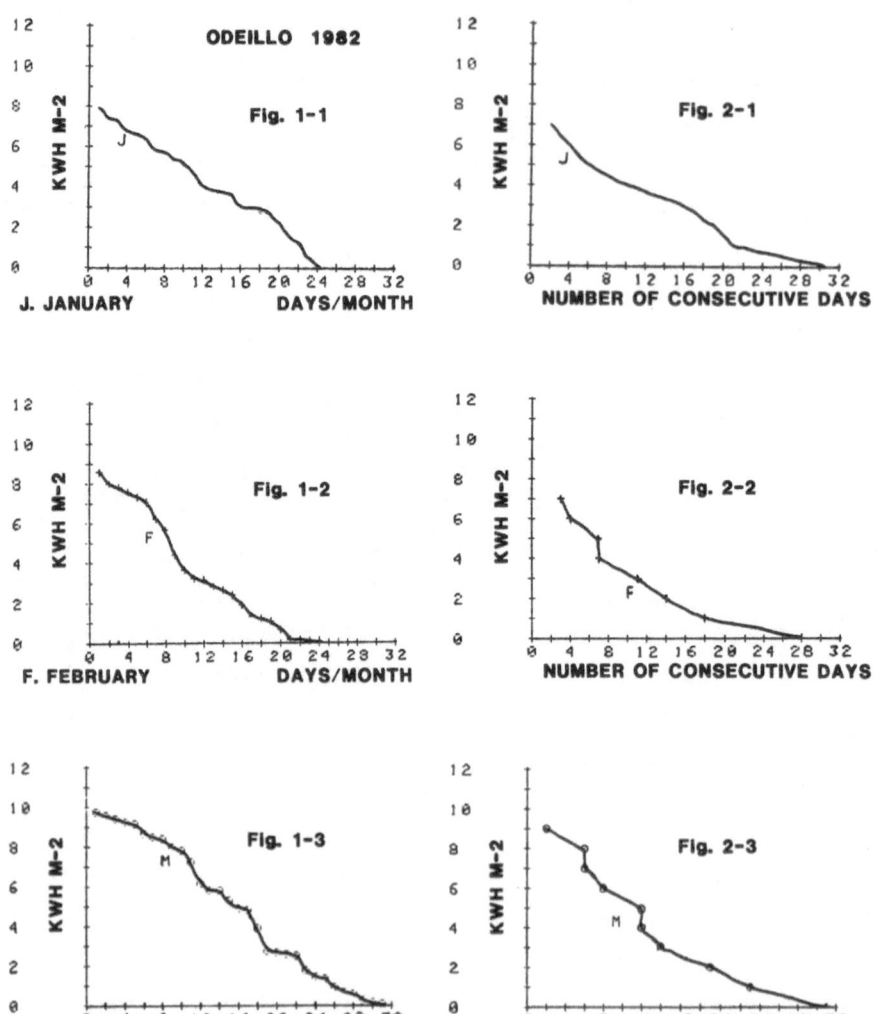

**Fig. 1-CUMULATIVE FREQUENCY CURVES OF DAILY SUMS OF DIRECT RADIATION**

**Fig. 2 -NUMBER OF CONSECUTIVE DAYS WITH DAILY SUMS OF DIRECT RADIATION ABOVE GIVEN THRESHOLDS**

ODEILLO 1982

Fig. 1-1

J. JANUARY

Fig. 2-1

NUMBER OF CONSECUTIVE DAYS

Fig. 1-2

F. FEBRUARY

Fig. 2-2

NUMBER OF CONSECUTIVE DAYS

Fig. 1-3

M. MARCH

Fig. 2-3

NUMBER OF CONSECUTIVE DAYS

# - ODEILLO - 1982

**KWH M-2**

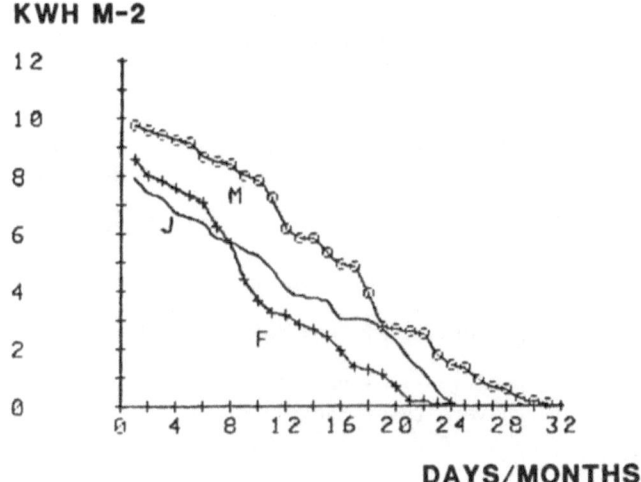

**DAYS/MONTHS**

Fig. 3 - Cumulative frequency curves of daily sums of direct radiation (1st of January to 31st of March)

**KWH M-2**

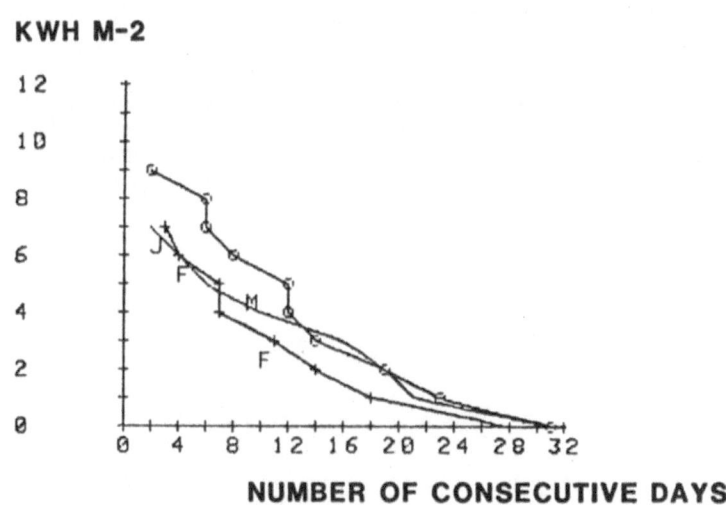

**NUMBER OF CONSECUTIVE DAYS**

Fig. 4 - Number of consecutive days with daily sums of direct radiation above given thresholds (1st of January to 31st of March)

# STUDIES OF RELATIONSHIPS BETWEEN INCLINED SURFACE IRRADIATION, TEMPERATURE AND WIND SPEED

Authors          : J.K. PAGE AND R.J. FLYNN

Contract number : ESF-033-81

Duration         : 18 months            Dates: 1st January, 1982 –
                                                30th June, 1983

Total budget     : £35,400              CEC Contribution: £16,700

Head of project : Professor J.K. Page, Department of Building Science

Contractor       : University of Sheffield, Sheffield, S10 2TN.

Address          : Department of Building Science,
                   University of Sheffield,
                   Sheffield.    S10 2TN.
                   UK.

Summary

    The key problem in this programme is the development of a satisfactory modelling methodology for estimating slope irradiation at the hourly level from a systematic set of observed hourly values of solar radiation on a horizontal surface.  A new mathematical model for estimating hourly values of slope irradiation has been completed and is now being used with the Bracknell observations of irradiation on vertical north, south, east and west surfaces to establish the validity or otherwise of the modelling process.  Special attention is being given to verifying the validity of the prediction of slope diffuse irradiation which forms a large component of the solar contribution in Northern Europe.

# 1.  Introduction

The key problem is the prediction of hourly values of the radiation on inclined planes from horizontal observed data.  Once accurate values can be predicted for inclined planes, it is simple to proceed to various forms of statistical analysis of the radiation climates of slopes, using the EEC solar radiation data tapes to generate the required data.  It was decided to use the Bracknell data as the basis of the studies of the accuracy of prediction of hourly values of solar radiation on inclined surfaces, because the available data set includes both hourly observations of global beam and diffuse radiation on a horizontal plane as well as radiation observations on four vertical surfaces facing north, south, east and west.  In addition the hourly sunshine record is also on the tape.

# 2.  Methodology

The hourly direct radiation can be estimated from such a record by subtraction of the hourly diffuse radiation from the hourly global radiation, provided accurate values of the shading ring correction are known.  Advantage is being taken of progress on the more accurate determination of shade ring corrections in the current Community programme to upgrade the existing corrections to provide a better estimate of the hourly direct beam irradiation.  It is simply a matter of trigonometry to convert the direct beam hourly irradiation to the corresponding slope value. Using the mid-hour solar altitude and the sunshine in that hour, one can estimate the clear sky direct irradiance normal to the beam by dividing the observed direct beam irradiance by the sunshine duration during that hour. One can thus attempt to derive an hourly value of the Linke Turbidity Factor.  The problem with this approach is that any errors in the sunshine record have a big influence on the derived turbidity factor.  This problem is especially important when the sunshine fraction is low.  In such circumstances it is likely to be necessary to use the monthly mean turbidity data rather than derived values of the turbidity otherwise serious problems of computational stability may arise.

The difficulties really lie in the prediction of slope diffuse irradiance.  It is important to achieve reasonable accuracy in this conversion as such a large proportion of the solar radiation in northern Europe is diffuse.  The project has built on the extensive work already carried out in connection with the prediction of irradiation on inclined planes for clear days, overcast days and monthly means.  The observed hourly diffuse radiation is split into two parts in proportion to the observed hourly sunshine, one part being attributed to the blue sky contribution and the other part to the overcast sky contribution.  The contributions of the two components on slopes are then estimated using the Moon and Spencer formula for the overcast sky component, and the new formula developed from the clear sky radiance model evolved for the production of inclined surface tables of radiation.  This radiance model is turbidity sensitive and for this reason it is necessary to estimate the Linke Turbidity Factor from the direct beam.

The hourly ground reflected component has normally to be estimated, but in the case of the Bracknell observations the vertical pyranometers were screened from the ground, so this complication was avoided, thus eliminating the uncertainties which would have arisen from lack of precise knowledge of ground albedo.

The hourly global radiation on each of the four vertical planes is thus estimated as the sum of the direct and diffuse components.  This

estimated value is then compared with the observed value and the modelling
reasons for the differences evaluated.

## 3. Progress to date

All the mathematical modelling necessary to carry out the procedures
outlined above has been completed, and the Bracknell EEC tape has been
reformatted to allow easier handling of the inter-relationships between the
7 hourly values involved. As the model is a complex one, it took a
considerable time to develop. It is now running very satisfactorily.

## 4. Clear day checks

Clear day checks on selected clear days have shown good agreement
between the theory and observation.

## 5. Current work on overcast day slope observations

Work is currently proceeding on the overcast day data. While the mean
slope predictions for overcast days are acceptable, currently the variance
of the errors is unacceptable and the hypothesis of the validity of the
Moon and Spencer formula for other than densely overcast conditions is
being re-examined. With thin cloud there is considerable directionality of
overcast radiation and a new model may be needed which produces results
which fall between the present clear sky and overcast sky predictions. The
work of Valko in Switzerland has shown by radiance observations that thin
overcast cloud produces exactly this situation. Once the overcast day
prediction problem is better understood, it will be possible to proceed to
the partially clouded problem on a better scientific basis.

## 6. Preparation of new Bracknell data tape

Work is to start shortly on punching the additional weather data for
Bracknell, so that cloud observations, temperature and wind data are on the
same tape as the radiation data. This will allow the project to proceed to
the final stage of examination of the statistical characteristics of slope
irradiation in relation to the statistical characteristics of the other
heat transfer variables.

## 7. Conclusion

Approaching the problem in this way will not only provide a proper
scientific basis for the study of the statistics of solar radiation on
inclined planes, but will also produce a verification or otherwise of the
techniques currently being used in the computer simulation of solar systems
to input solar radiation data falling on inclined surfaces using horizontal
observed radiation data as the source of hourly data generation.

ACTION 3.4

SENSITIVITY ANALYSIS OF THE USEFUL ENERGY OUTPUT

FROM SOLAR CONVERTERS AS A FUNCTION OF THE TYPE

AND TIME-STEP OF METEOROLOGICAL DATA

Report of action participant

# SENSITIVITY ANALYSIS OF THE USEFUL ENERGY OUTPUT FROM SOLAR CONVERTERS AS A FUNCTION OF THE TYPE AND TIME-STEP OF METEOROLOGICAL DATA

Authors : G. WATREMEZ - B. BOURGES

Contract number : ESF-023- f(G)

Duration : 19 months 1.06.80/31.12.81

Total budget : 250 000 FF    CEC Contribution : 50 %

Head of Project : B. BOURGES - Ecole des Mines de Paris

Contractor : Centre d'Energétique - ARMINES

Address : 60, bld St Michel - F 75272 - PARIS CEDEX 06

Summary :

This study is part of a European Concerted Action (Mr KOCH, KFA Jülich) and will help to answer two important questions : What si the accuracy of the useful energy output calculation according to the kind and time step of the meteorological data sets ? and subsequently:

How can these data be reduced to compute faster but accurately the useful energy output ?

For a set of meteorological data at short time steps as large as possible (Bourgoin Jallieu     , Paris = 44 days, Jülich = 270 days and other data from Trappes, Carpentras, Ukkel) a reference useful energy output is defined with MINERSOL for the DRU collector (used in the PTF) and an Evacuated Tubular Collector. It is compared to results derived with :
- either a simpler model (no inertia)
- or a smaller data basis (only averages over larger time steps or frequencies of values).

The comparison shows discrepancies up to 20 % due to the reduction in the model or in the data, largely climate-dependent : for the usual computation procedure and the usual hourly data basis both tendancies do compensate in Central Europe climates ; for other climates corrective terms are still to determine but the effects are under 10 % in most cases.

## 1 - Definitions

For the computation of the reference output, data obtained at time steps between 1 and 6 minutes are used (see § 2 here under). The program MINERSOL is then used. MINERSOL program is an accurate simulation model which predicts the thermal behaviour of solar converters. Its experimental validation has been performed in the course of other studies (1).

The reference output will be called in this paper E ref.

When we want to reduce the meteorological data basis, we keep only averages of the quantities over a period $\tau$ . With the use of MINERSOL, the computed energy output is $E_\tau$ ; for reasons to be found elsewhere (2) $E_\tau$ is always lower than $E_{ref}$.

The usual procedure is not to use MINERSOL or any other transient model : it is to neglect inertia and to write the useful power in steady-state (Hottel-Whillier-Bliss Equation) :

$$P^H = F_R \left[ (\tau\alpha)I - U_L(T_i - T_a) \right]^+$$

The sign + shows the influence of the control : the extracted power is considered nul if non-positive

The integration of $P^H$ over the initial data basis leads to an estimate of the output than we call $E^H_{réf}$ ; it is an over-estimator (1). When we move to averages over periods of a duration of $\tau$, $E^H_{ref}$ is reduced to $E^H_\tau$.

Two quantities are of particular interest : $E_{ref}$ and $E^H_{1h}$ ; the first one is the best estimator of the output (best model, best data basis), the second one is the result of the usual procedure (steady state model, hourly data).

## 2 - Available Meteorological Data and performed work

The quantities $E_{ref}$, $E_\tau$, $E_{ref}^H$ and $E_\tau^H$ have been computed for $\tau$ between 1 min and 60 min, for inlet temperatures of 30°C and 50°C in the case of the flat-plate DRU collector (used in the PTF), and for inlet temperatures of 100°C and 150°C in the case of an Evacuated Tubular Collector (close to the MAZDA collector). The results are available for part of our meteorological basis :

- Data Collected by ourselves in Sophia-Antipolis at 1 min-time-stepswere found uncorrect on the tilted planes during the treatment ; a new compaign is under way after a complete change in the apparatus (early in 1983) ;
- Some Data are available at 6 min. time-steps for Bourgoin-Jallieu (France, region of Grenoble) ; they have been provided by Prof KUHN (CNRS) after a first test over ten days ;
- 250 days of data were received from KFA Jülich at 10 min.time steps and not yet treated ;
- 44 days of data for the region of Paris were provided by Gaz de France and were used : they show up very clearly some phenomena because of their very small time step, both for temperature and radiation ;
- Data at 30 min.time steps for Belgium and at 6 min. time steps for Trappes and Carpentras were not yet treated.

Before reporting some results one should note the very large difference in the aspect of the radiation measurements between locations.

Here are represented first a "good" and a "bad" day in Paris (May) :

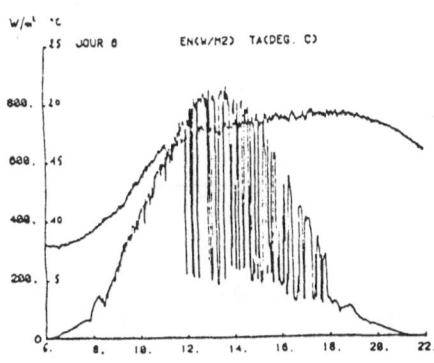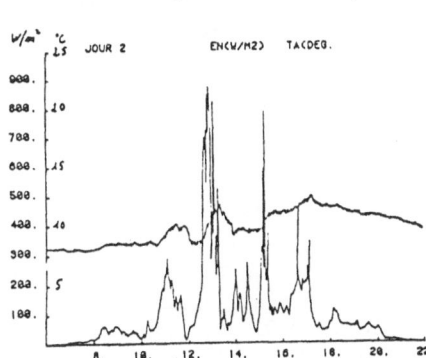

Then let us show the aspect of a good and a bad day in Bourgoin-Jallieu (september) :

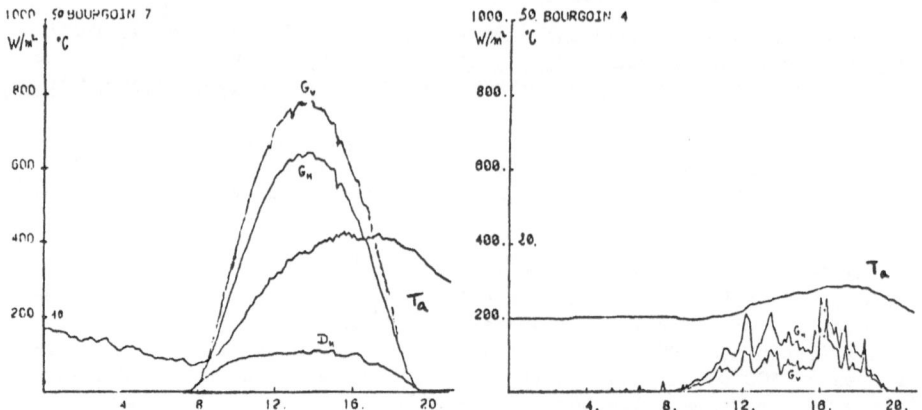

When we go South in France, the duration of transit of the clouds becomes larger and the difference between "good" and "bad" days stronger as may be seen in the existing statistical studies (2,3) and should be investigated more deeply.

## 3 - Influence of the Data Time-step with the reference model

The results are given here for two samples of data (Paris in May = P-M and Bourgoin in september = B-S) under the form of a normalised heat-balance (over the whole periods) : $E\tau/E_{ref}$. Dotted lines are for 30°C and full lines for 50°C, in the case of the flat-plate collector (left); on the right side the evacuated tubular collector is considered with 100°C as operating temperature (dotted line) or 150°C (full line) :

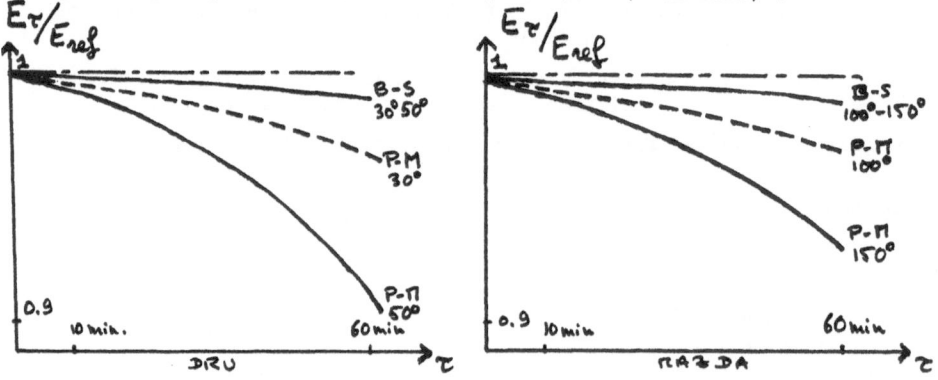

So the influence of the time step with a transient model is not negligible, at least in Paris (-10 % in the estimation of the output at 50°C). This is not yet an information for the usual model-builder who deals not with MINERSOL but with the Hottel-Whillier-Bliss equation and knows neither $E_{ref}$ nor $E_{1h}$ but only $E_{1H}^H$ (H.W.B Equation, Hourly data) For the meteorologist two informations are important :
- in some climates, meteorological data under 1 hour are not important for thermal conversion (photovoltaïcs and medium temperatures are still to be studied)
- in other climates they are meeded because the loss of information may reach 10 %, but 6/10 minutes integrals are enough.

These informations still lack generality : the treatment of the data basis is not terminated.

## 4 - Use of a simpler model (steady state)

As stated before the use of the steady state model leads to overestimations of the output of the collectors : $E^H_{ref}/E_{ref}$ exceeds largely unity (up to 1.3). But moving from small time steps to larger ones makes $E^H_\tau$ decrease and so get closer to $E_{ref}$ as may be seen from the following graphs (same conventions as previously) :

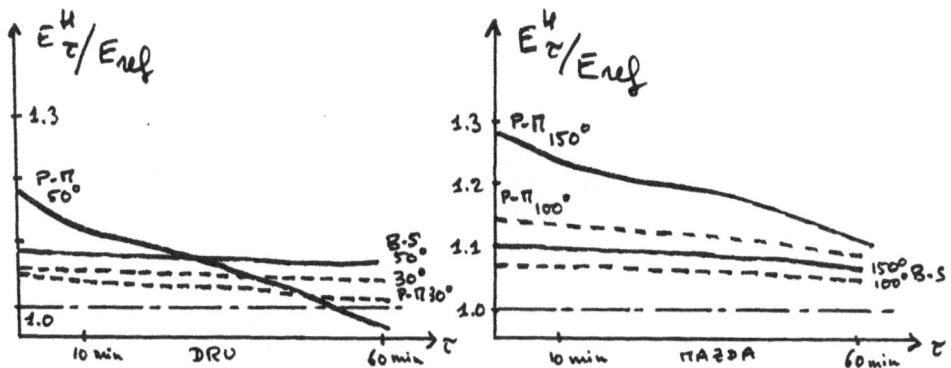

However almost in any case $E^H_{1h}/E_{ref}$ (the usual computation) remains higher than 1 : the difference is negligible in Paris and significant in Bourgoin-Jallieu (under 10 %).

This is an informatinn for the model-builder and not for the meteorologist: in Central Europe to forget completely about collectors'inertia and to use existing meteo-files may be correct, but could be misleading in Northern and Southern Climates. Generality of the result is not yet proved but on the other hand we want to stress the risk of using short time step data with the usual model (steady state) and to over-estimate largely the output of collectors...

## 5 - Use of Frequency Distributions

We know that system-designers need even more compressed data than hourly files (4,5). Part of our activity is the study of the reduction to averages or statistical distribution of the radiation values and/or outside temperatures.

The following table presents the deviations observed between the useful energy outputs computed by simple methods using the cumulative frequency curves. The considered collector is the DRU collector. The data set is the Paris sequence in May and reference is $E_{ref}$ determined previously

The table gives deviation in percent as compared to the reference value ; the five considered methods are :

Method 1 : simulation using Hottel Willier Bliss formula with one hour time meteorological data ($E^H_{1h}$)
Method 2 : The cumulative frequency curve for the period is employed and the mean air ambient temperature is considered for the whole period.
Method 3 : The cumulative frequency curve and a mean air temperature value calculated from 9h to 15h, are used
Method 4 : The calculation uses cumulative frequency curve and mean air temperature values calculated for each irradiation class.

Method 5 : CFC is used with two mean air temperature values :
  - daily mean value
  - nightly mean value

| INLET TEMPERATURE | $T_i = 30°C$ | $T_i = 50°C$ |
|---|---|---|
| - Method 1   H W B-1 hour | + 0,1 | - 3,8 |
| - Method 2   CFC, $\overline{T}_a$ | - 10 | -19,1 |
| - Method 3   CFC, $\overline{T}_a$ (9-15h) | - 4,2 | -13,7 |
| - Method 4   CFC, $\overline{T}_a$ for each class of irradiation | - 0,4 | - 4,5 |
| - Method 5   CFC, $T_a$ for the day, $\overline{T}_a$ for the Night | - 6,1 | -15,5 |

(deviations in %)

By using a mean of the air temperature calculated from 9h to 15h (daily mean) or mean values of Tam calculated for each class of radiation, the results are more accurate than those obtained by using a mean value of Tam for all the period.

6 - Discussion

After the completion of the data treatment (dec 1982) the following months and locations will be covered :
- Sophia Antipolis : uncorrect data
- Bourgoin Jallieu : september only for the present stage
- Jülich : january, april, july
- Paris GdF : march, april, may
- Trappes : january, july
- Carpentras : 12 may-17 june, january, april.

An extension to mediterranean climates (south France, Greece, Italy) and Northern European Climates (U.K. Ireland, Denmark, ...) is still necessary. The results for regions with latitudes between 45° and 50° may be however considered as ready for publication at the end of the present work.

REFERENCES

(1) "Experimental validation of a transient program simulating the dynamic behaviour of Solar Energy Collection Loops", G. Watremez, et alii, Bryghton 1981

(2) "Contribution à l'étude théorique et expérimentale des boucles de captage thermique de l'Energie Solaire", J. Adnot, Thesis, 1981

(3) "Etude des liaisons entre l'irradiation journalière et les variables météorologiques courantes", M. J. Mejon, thesis, 1979

(4) "Statistical Distribution of Solar Radiation (Hourly Sums) : Cumulative Frequency Curves", F. Lasnier and B. Bourges, same meeting

(5) "Simplified Methods for the sizing of Solar Thermal Plants", J. Adnot and B. Bourges, CEC contractors meeting prog.A, Meersburg, 1982.

ACTION 4.1

IMPROVEMENT OF MEASUREMENTS IN NATIONAL NETWORKS OF GLOBAL,
DIRECT, DIFFUSE IRRADIANCE AND IRRADIANCE ON   INCLINED
SURFACES


Action leaders's progress report
J.W. GRUETER, Kernforschungsanlage Jülich GmbH, Federal
Republic of Germany


Reports of action participants :

- Variability in space of solar radiation

- Improvement of measurement of diffuse solar radiation

- Progress toward establishing a comprehensive solar
  radiation data and calibration center in Greece

- Solar radiation in Italy

- Improvement of radiation network

- Investigation of the accuracy of shading ring corrections

- The extension and improvement of the UK radiation network

## ACTION 4.1:
## IMPROVEMENT OF MEASUREMENTS IN NATIONAL NETWORKS
## OF GLOBAL, DIRECT, DIFFUSE IRRADIANCE
## AND IRRADIANCE ON INCLINED SURFACES


Action leader:        J.W. GRÜTER
                      Kernforschungsanlage Jülich GmbH
                      STE/Postfach 1913
                      D-5170 JÜLICH

Participants:         Met DK      Meteorologisk Institut Copenhagen
                                  (Frydendahl)

                      DWD         Meteorologisches Observatorium Hamburg
                                  (Dehne)

                      Met GR      Meteorological Institute,
                                  National Observatory, Athens
                                  (Lalas)

                      IFA         Istituto de fisica dell'atmosfera,
                                  Rome
                                  (Guerrini)

                      Met IR      Valentia Observatory, Cahirceveen
                                  (Murphy, McWilliams)

                      Met NL      Koninklijk Nederlands Meteorologisch
                                  Instituut, De Bilt
                                  (Slob)

                      Met UK      Meteorological Office, Bracknell
                                  (Durbin)

Tasks:                - Improvement of instrumentation
                        (in particular for diffuse radiation)

                      - Densification of networks for global, direct,
                        and diffuse irradiance

                      - Measurements on inclined surfaces

## 1. Abstract

Action 4.1 of Project F deals with the radiation networks of the member states of the European Community. New equipment and measuring techniques for networks are assessed or tested and eventually installed by the national services. Action 4.1 cooperates with action 1 in assessing calibration procedures for pyranometers.

Since the last Project F survey meeting 1981, the following steps have been reached:

- The densification of networks as foreseen by the programmes is terminated. There are still administrative problems in fulfilling contractors work in Italy. Greece has started in installing a primary station in Athens. A Greek data bank for solar energy purposes is founded.
- A new approach on assessing the spatial distribution of global radiation with data from the Danish network gave good results. The method could be applied for the assessment of the spatial density of radiation stations in Europe, before going ahead with a further densification of the radiation networks.
  The method of calibrating network pyranometers by substandards was found to be adequate with an accuracy of $\leq 5$ % (combined action 1, 4.1).
- The correction formulas for diffuse radiation measurements by shadow band techniques has been improved leading to an overall accuracy of $\pm 10$ W/m$^2$. They may be used for shadow-bands with b/r    0.3. This result is based on three independent analyses in Hamburg, De Bilt, and Valentia. A unification into a "universal correction formula for Europe" seems to be possible.
- The final report of the KNMI is available on the climatology of solar irradiance on inclined surfaces part IV measurements.
- "Solar weeks", a data collection campaign for meteorological data of the first fortnight in November 1981, have been organized. The data collection showed substantial differences in the data dessimination methods of the national services.
- Data of solar irradiation, hourly and daily values, measured at the radiation stations of the networks, have been collected for the month April 1982, in order to supply action 4.3 with ground reference data.

## 2. General Outline of Action 4.1

Radiation mapping, like the European Solar Radiation Atlas, is based on measurements obtained from radiation networks. These networks are maintained worldwide by meteorological services on a national basis. By the two Solar Energy Programmes of the Commission of the European Communities the services of the Member States received substantial support in upgrading their networks. Because of the different status of these networks and the differing interests in the Member States, the services accepted contracts according to their individual needs. Belgium and France did not participate in action 4.1, Belgium because of its sufficient network, and France since the national service had already its own improvement programme.

The objectives of action 4.1, commenced already in 1978, in the present Solar Energy Programme are

- Improvement of instrumentation,
- Densification of networks for global, direct, and diffuse irradiance,
- Measurements on inclined surfaces.

These objectives should be reached in two different ways:

A. The contracts covered mainly the installation of new or improved equipment for individual services, or the contracts covered the installation and maintenance of additional equipment like pyrano-meters on inclined surfaces or pyrheliometers.

B. The contracts contained the assessment and improvement of certain measuring techniques, generally used in the European Community or studies of common interest for the participants or the goals of the Solar Energy Programme.

It is obvious that the work according to mode A cannot result into extensive scientific reports, in spite of its great importance for the scientific infrastructure in the European Communities. Most of the work in action 4.1 was done by individual contractors, but also common actions were performed. The title of the contracts are listed in table 1.

Table 1:
Title of contracts in action 4.1

| Country | Contract | Title |
|---------|----------|-------|
| DK | ESF-025-DK | Improvement of the Danish radiation network |
| D | ESF-004-80-D | Improvement of measurement of diffuse solar radiation |
| GR | ESF-027-GR | Installation of a radiation station in Athens Initiation of a data bank for Greece |
| I | ESF-017-I | Development of a network for measurement of direct solar radiation and data elaboration |
| EIR | ESF-019-EIR | Investigation of the accuracy of shading ring corrections applied in the measurements of diffuse solar radiation |
|  | ESF-018-EIR | Improvement of radiation network |
| NL | ESF-006-80-NL | Climatology of solar irradiance on inclined surfaces - IV; for action 4.1 measurements |
| UK | ESF-024-80-UK | Extension and improvement of the Meteorological Office network of radiation stations in the United Kingdom |

The achievements of the individual contracts during the period 10/81 - 10/82 are briefly described in section 3.

Besides the work performed through individual contracts, the discussion and cooperation between the contractors were very importants. The calibration procedures, recommended now by action 1, e.g. are based on contributions, given by the participants of action 4.1. Therefore action 1 and 4.1 worked in close cooperation. All meetings were held together.

Three concerted actions were taken outside contractors' duties, based on the cooperative effort of the participants:

- Data source catalogue for solar researchers.
  A comprehensive description of the radiation networks in the European Communities, not terminated yet due to necessary discussions prior to further work, to be put into this topic.
- "Solar weeks".
  A collection of data throughout the Community relevant for solar systems, measured during the period Nov. 1-14, 1981.
- Ground reference data for action 4.3.
  A collection of radiation data, measured by the national services during the month April 1982.

## 3. Executive Summary of period Nov 1981 - Oct 1982

### 3.1 Individual contractors' work

#### 3.1.1 Radiometric networks

Contract ESF-025-DK   1.7.79 - 30.6.83
The Danish Meteorological Institute
responsible: K. Frydendahl

The goals of the work in Denmark under this second contract are to run the stations established during the first contract and to build a programme system to handle this bulk of data and present them in a form suitable for different purposes. For the moment this programme system is able to produce data, but still needs to be improved which will be done within this year.

Contract ESF-018-EIR   1.1.80 - 31.12.82
The Irish Meteorological Service
responsible: E.J. Murphy

The Irish Meteorological Service network of solar radiation recording stations has been extended by the addition of two new stations - Clones and Malin Head.

Global and diffuse radiation data are available for both stations from 1st June 1981.

Contract ESF-027-GR
National Observatory, Meteorological Institute of the
National Observatory of Athens
responsible: D.P. Lalas

Several instruments have been ordered for installation in the vicinity of Athens. A data bank for existing data has been established. Analytic work for calculation of solar energy potentials in Greece have been performed and will be published soon.

Contract ESF-017-I   1.1.80 - 30.6.83
Istituto di Fisica dell'Atmosfera
responsible: A. Guerrini

Due to continuing difficulties the progress in improving the existing instrumentation is slow. The present status is as follows.

9 stations are by now working, namely: Pisa, Grosseto, Genova, Capo Mele, Bologna, Cosenza, Castrovillarisibari, Vigna di Valle, Napoli.

By the end of 1982, it is planned to have a total of 25 stations. Equipment for 28 stations exists at present. No progress was made in the installation of equipment for direct radiation. The responsibility of following-up the contractors' work is taken now directly by the Commission.

Contract 1: ESF-024-80-UK    1.1.80 - 30.6.83
Contract 2: ESF-038-81-UK    1.7.81 - 30.6.83
Meteorological Office
responsible: W.G. Durbin until 15.9.82/G. Jenkins
        Contract 1: Global (G) and diffuse (D) radiation.
        Measurements of G and D commenced at Hemsby in April 1981 and at Camborne in September 1981.
        Works Services on renovating the building chosen for radiation equipment at Stornoway are almost complete and it is expected that the station will commence making measurements during the autumn of 1982.
        Shanwell is in operation since Dec 1981.
        Measurements of direct radiation (I) are now made at Eskaldemuir using Eppley pyrheliometers.
        Contract 2: Measurements of G and D commenced in Aughton in Nov 1981.
        Works Services have commenced at Finningley and should be completed by the autumn of 1982.
        Agreement has been reached to establish a meteorological office at Aviemore. The logal formalities have been dealt with and it is expected that Works Services will commence during this summer of 1982.
(i)     An Eko Sunphotometer was received and proved satisfactory in acceptance tests (albeit after a little trouble initially). It has been transferred to Eskdalemuir where routine turbidity measurements have commenced.
(ii)    A second Eko Sunphotometer has been ordered for use at Bracknell. Delivery is expected late September 1982. Shading ring corrections are under consideration.

3.1.2 Spatial correlation of irradiation

Contract ESF-025-DK    1.7.79 - 30.6.83
The Danish Meteorological Institute
responsible: Knud Frydendahl
        Investigations on spatial distributions have been performed using the dense Danish network. A new approach on the representation of the established correlations has been made.
        The statistical structures in data have been investigated at the aim of revealing properties which can be exploited especially for the later study of spatial correlations. Both hourly and daily values of global radiation has been considered such that the various months have been analysed separately. It has become clear from these analyses that tradional calculations of measures of correlation do not to a satisfactory degree compensate for certain variance structures in data when spatial correlation is of concern. In fact, increasing variance with increasing level of global radiation can be shown to be analysed properly by means of a log-transformation of data. Hereby correlations between measurements only depending on interdistances can be obtained and this transformation at the same time facilitates the formalization of a statistical model. The report outlines the structure of the model which will be used for further analyses. It is an important characteristic of the model that it specifies the (logarithmic) mean values as composed by a site-specific

parameter and a time-specific parameter, hence allowing the variation in global radiation over time to be systematic rather than stochastic. All correlations are defined in the light of this model.

### 3.1.3 Diffuse sky radiation

Contract ESF-004-80 D    1.1.80 - 31.12.82
Deutscher Wetterdienst
responsible: F. Kasten
    To improve the accuracy of diffuse solar radiation data gained by ring shaded pyranometers, a new correction formular for losses of the Hamburg ring has been derived. This formula has been confirmed by data sets of different periods and of different pyranometer types, whereby the constants were lightly modified. The correlation coefficient is always near to .80. The intercorrelations between the parameters has been determined. Details of the new approach have been published earlier in the last survey meeting.
    For further improvements of the correction formula the influence of the distribution of sky radiance must be taken into account. The so-called sky scanner was therefore equipped for broad band measurements, and a proper data acquisition system has been established using a micro-processor to sample and store the mean values in a register with 1440 places according to a fine subdivision of the sky hemisphere.
    The main reference data of diffuse solar radiation delivers a sun-following shade disk device with a shaded angle of $9.6^{\circ}$. To control the transfer of the improved correction formula to different kinds of shade ring devices in national networks, a compact reference shade disk instrument has been built.
Figure 2 shows the influence of the improved formula.
Contract ES-F-019-EIR    1.1.81 - 31.12.82
responsible: S. McWilliams
    The diffuse sky radiation as measured with a solarimeter and shading ring must be corrected to sompensate for the sky radiation cut-off by the ring. This correction is generally computed on the basis of the geometric area of the sky cut-off and assuming isotropic sky. This assumption results by applying a geometrical correction factor $k_{geo}$, e.g. after research paper 895 of the British air ministry, in values of $D_R$ = $k_{geo} \cdot D_m$, where $D_m$ is the measured value of the shaded pyranometer. $D_R$ differs from the true value $D_c$ by a factor "k" = $D_c/D_R$. Hourly values of the factor k as found at Valentia are analysed.

### Results
o    The overall mean values were found to be k = 1.053.
o    There appears to be no significant dependence of k on Solar Elevation.
o    Variations in k with solar declination are limited to about ± 1.6 %.
o    The most significant variations of k are associated with $D_R/G$.
The following equation was found to fit the data very closely:

$$k = 1.1431 - 0.1451 \ (D_R/G)^3 - 0.000682 \cdot \delta$$

81.7 % of the data are brought within ± 5 % of the true value and only 4.5 % have a residual error in excess of ± 10 %. The results will be checked by extending the analysed data to include the additional year 1980. Fig. 3 shows the result of the improved formula.

Contract ESF-006-80-NL    1.7.79 - 31.12.81
responsible: W.H. Slob
    The diffuse solar irradiance can be measured with a shading ring

pyranometer while the diffuse irradiance can also be calculated from the global horizontal irradiance G and the direct irradiance I measured with a pyrheliometer. The pyrheliometer measurement was used to derive a correction factor for the shading ring pyranometer (Eppley SBS). It appears that such a correction factor is dependent on the hourly clearness index defined as;

$$K_T = G/G_o$$

in which G is the mean hourly global irradiance and $G_o$ is the mean hourly extra terrestrial irradiance, both on a horizontal plane. The relation can be brought into formulae which results in the following equation.

$$f = 0.981 + 0.389 \ K_T + 0.049 \ K_T^2; \text{ ring dimensions } b/r = 0.24 \quad (A)$$

The above derived equation is tested for another period together with a constant correction factor and a relation for the correction factor as it is derived by contractor ESF-004-80-D. This equation is

$$f = A + B \ (D_m/G)^3 + C. \ \delta + D/\tau \qquad (B)$$

with

$$\tau = \ln. \ I_o \sin h/(G-D_m)$$
$\delta$ = declination of the sun, north positive degrees
A = 1.2486    C = -.00066
B = -.1927    D = -.06007

Parameters A...D fitted by ESF-004-80-D for ring dimensions b/r = 0.169. The result of this comparison is given in table 2.

Table 2:
Comparison of different shading ring correction methods
based on hourly values
$D_m$ = measured uncorrected shade ring value
$D_c$ = calculated from G and I

| CORRECTION EQUATION | $\dfrac{D_m \text{ corrected}}{D_c}$ | STANDARD DEVIATION |
|---|---|---|
| f = 1.118 | 1.000 | 0.102 |
| f = equation (A) | 1.005 | 0.073 |
| f = equation (B) | 0.991 | 0.075 |

It appears that the constant correction factor method is improved by applying a correction factor function as can be seen in the reduction of the standard deviation. Both correction factor functions give the same precision.

Conclusions for the shade ring corrections

    As conclusions from the three independent studies one can derive
o  there is a strong relation between the clearness index or the ratio $D_m/G$ and the correction factor,
o  there is a weak explicit relation between the declination and the correction factor.

The contractors should treat their data base in direction of a unification of their results.

## 1. Proposal
An absolute dependence on the ring dimensions should be put into the formula along the lines of the Irish participant.

## 2. Proposal
The residual variations should be correlated
A. to the clearness index $G/G_o$
B. to the ratio $D_m/G$
resulting into a correction formula

$$D = D_m \cdot k_{geo} \cdot f \qquad\qquad k_{geo} \text{ evaluated for isotropic sky conditions}$$

$$f = \sum a_i (G/G_o)^i$$

or
$$f = a_o + a_1 (D_m/G)^3 + (\alpha_2 \cdot \delta) \text{ to be proved if last term is necessary.}$$

where the fit parameters $a_i$ should be universal in Europe for the same kind of instrument set.

If with this relatively simple corrections the accuracy can be pushed below $\pm 10$ W/m$^2$, no further improvement is necessary.

The formula should be tested then on a European data basis for all kinds of cloudiness and turbidity conditions before recommendation.

### 3.1.4 Radiation on Inclined Surfaces

Contract ESF-006-80-NL    1.1.80 - 1.7.82
Koninklijk Nederlands Meteorologisch Instituut
responsible: W.H. Slob

Abstract of the final report, available from the KNMI, named "Climatology of Solar Irradiance on Inclined Surfaces - IV, Part I Measurements".

This is one of the final reports of the research work carried out by the Royal Dutch Meteorological Institute (KNMI) under contract no. ESF-006-80 NL(B) for the commission of the European Communities (DG XII) and for the Dutch Solar Energy Programme under project no. 4.341, job no. 3.1.4 and 3.1.5.

The work has been done in close cooperation with the Institute of Applied Physics TNO-TH (TDP).

The complete final reporting consists of three separate reports. These are:

Part I:     Measurements
Part II:    Validation of calculation models
Part III:   Climatology of solar irradiance on the horizontal and inclined surfaces in the Bilt.

This abstract describes the results of Part I; Measurements.

In Cabauw (51.96 N, 4.93 E) in the Netherlands, the following input variables were measured from March 1979 till October 1981:

- The global solar irradiance, with Eppley pyranometers, on the following orientations:  . 90$^o$ east, south, west, north
  . 67.5$^o$ south
  .. 45$^o$ east, south east, south, south west, west
  . 0  horizontal
- The direct solar irradiance on a surface perpendicular to the sunbeam

with an Eppley pyrheliometer.
- The ground reflected solar irradiance with an Eppley pyranometer in the inverted position.

The 6 minute averages of all these data have been recorded and presented in this report in the form of hourly, daily and monthly totals as well as in frequency distributions.

All hourly and 6 minute data are also available on magnetic tape. In order to present a reliable data set there has been carried out a so-called "cleaning" procedure for the 6 minute averaged values. The hourly irradiation data are based on these "cleaned" 6 minute values. During the whole measuring period, a reference pyranometer has been circulating along all orientations. In this way we have achieved a reliable control on the measured irradiance data. It appeared that all pyranometers gave results within 2 % of the reference pyranometer measurements.

There has also been carried out a statistical analysis of the global irradiance over an hourly period. It appears that the relative standard deviation almost linearly decreases with increasing global irradiance. Also the relation between the relative sunshine duration and the hourly clearness index is shown.

During the measuring period a number of special measurements are carried out. The influence of a 0.53 µm filter is shown for the horizontal and the south-45 orientation as well as the direct irradiance. It appears that this influence is rather constant. The comparison of three Eppley PSP with three Kipp CM 10 pyranometers is shown for different orientations. It appears that the Eppley PSP gives higher irradiance values for irradiances below 200 - 300 $W/m^2$. There has been derived a new equation for the correction factor of a shading ring pyranometer which gives reliable results. At last there is described the operation and results of the all-sky camera system.

It should be mentioned that a special magnetic tape will be available around June 1982 which contains not only the hourly radiation data but also the most relevant meteorological parameters.

The former contractor could not attend the action meeting because the contract is terminated.

## 3.2    Concerted actions

### 3.2.1 Data source catalogue

On the meeting in Athens, November 1981, the action leader collected the available information on the national radiation networks. Looking closer into this information, the action leader came to the conclusion that the conversion of the available material needs more than a month to a suitable and  informative publication.

#### Status of the information
There exist well written reports on the French, Danish, Greek, Irish, and UK networks. The reports are listed in table 3.

The list of radiation stations in the European Community contains 167 stations. The distribution over the landscape of the European Community is shown in figure 1.

The collection procedures are differing from country to country, one collecting data minute by minute, others measuring averages of 1/2 or 1 h with scanning periods of <20 sec. In some countries climatological parameters are measured separately on different recording devices, in some countries automatic stations for all parameters are installed.

Available Descriptions of National Radiation Networks

| Country | Published by | Title (year of publication) |
|---------|-------------|----------------------------|
| Denmark | Meteorologish Institut, Copenhagen | The Danish Radiation Network (1981) |
| France | Direction de la Météorologie, Paris | 1. Energie solaire: Catalogue des Données Disponibles (1981)<br>2. Le gisement solaire en France (1980) |
| Greece | Meteorological Institute University of Athens, Athens | Meteorological data for applications of solar and wind energy (1979) (in Greek) |
| Ireland | Meteorological Service, Dublin | Solar Radiation Observations 1976 (1977) |
| United Kingdom | Meteorological Office, Bracknell | Solar Radiation Data Available from the Meteorological Office (1981) |

In most countries radiation is recorded on the time base of local apparent time (LAT), in spite of the fact that the climatological data are measured or observed at GMT. Solar researchers need for their purposes a data set of radiation and climatological data, measured at the same time or time interval. This requires interpolation routines, which have to be developed.

At some stations only radiation values are recorded and not climatological data. In Germany radiation and climatological data are stored at different branches of the weather service. The format of the data storage differs from country to country.

In conclusion: the access to the data sources for solar researchers needs further improvement, before a comprehensive European publication has great value to the solar community. Furthermore, the existing data source catalogue, prepared by the IEA, is available which with some updates would give good information on the radiation networks. Due to these considerations the action leader postponed the data source catalogue. He proposes that the Commission has a closer look to this topic also in view of future activities of the improvement programme within Project F.

### 3.2.2 Solar weeks November 1-14, 1981

The design or simulation of solar systems needs as primary input data meteorological information. In the past years many groups have identified the most important weather elements which influence the performance of solar systems. Meteorological measurements are performed in many institutions; but the governmental meteorological services only are responsible for reliable measurements on a national basis.

Since 1977 supports the European Commission by action 4.1 the meteorological services of the member states in upgrading their radiation networks. This is done in view of the fact that
- radiation is the driving meteorological parameter in the performance
  of any solar system,
- the radiation networks in many member states were in an inadequate
  status for solar energy application studies,
- the networks of climatological parameters were due to statements of the
  services in a good shape.

Now by the end of the second 4 years Solar Energy Programme it is
of some interest to assess, if the necessary data are easily to be obtained and/or if further investments into meteorological instrumentation
throughout the Community is useful. The action leader started therefore
a data collection campaign, by announcing the period November 1-14, 1981,
as "Solar weeks" on the action meeting Nov. 10-12, 1981, held in Athens,
Greece.

Organisation
- 1st Request on December 15, 1981
  The institutions, listed in table 3, column 2, were addressed with a
  request to transmit radiation and climatological data to the action
  leader. The parameters are selected data sets with high probability
  that they can be obtained as complete as possible. Therefore long wave
  radiation and humidity were not requested. Direct normal incidence
  radiation and diffuse radiation were regarded as complementary. The
  list of stations were taken from Dogniaux's survey made for the WMO.
- Repeated Request in April 82 to those institutions who did not return
  the answer form.
- Campaign closed on July 31, 1982.

Results
        The results of the campaign are shown in table 4. One finds the
address of the organisations to whom the requests are forwarded, their
responses and the structure, as the data were obtained. In the 9 member
countries out of the 187 stations reporting global radiation, 73 stations
report global radiation and at least the climatological data, some 10
in addition direct radiation, ca. 50 stations global diffuse and climatological data. The reporting format is differing from country to country.
Mostly the climatological data are reported at GMT which radiation data
are reported at LAT.
        Only four participants have sent or were willing to send the data
on one magnetic tape. The tape format density and work differs strongly
from 800 to 16000 BPI, 7 and 9 track tape, ASCII code and EBCDIC.
        Therefore it is very time-consuming to prepare an integrated data
set, regardless from which country the data were. It is not due to the
action leader to assess the information, one can derive from table 4.
One statement may be permitted: Some cooperative effort should be made
to facilitate the access for European solar researchers to the necessary
meteorological data base.

# Table 4: Information tableau of the solar weeks

| | Address | dates | | | stations and available information | | | | | | | | | |
|---|---|---|---|---|---|---|---|---|---|---|---|---|---|---|
| | of 1. request/origin of data | a. b. | c. d. | e. | data available | instruments as reported units time base of measurements data storage | | | | | | | | |
| | | | | | G D SS CC T A V[f) I | G | D | I | SS | CC | T[g) | A[h) | V | |

| | | | | | | | | | | | | | | |
|---|---|---|---|---|---|---|---|---|---|---|---|---|---|---|
| B | R. Dogniaux Institut Royal Météorologique de Belgique 3 Av Circulaire Bruxelles | 21.12. 81 19.1. 82 | 3 7 | 1 2 | x x x  x  x x x x  x  x x x | Pyranom. J/cm² | NIP | C.St. min | WMO[i) Oct. | 0.1 | 0.1 | kn | | |
| | | | | | LAT 1/2 sums | | | | GMT every half h | | | | | |
| | | | | | | | | | inst. value (j) | | | | | |
| | | | | all data on one tape | | | | | | | | | | |
| DK | K. Frydendahl Meteorologisk Institut Lyngbyvej 100 2100 København | 23.12. 81 11.7. 82 | 16 3 3 19 1 | 9 3 3 1 | x              x x    x       x x    x | Pyr. J/cm² | | C.St. .1h | | 0.1 | | | | |
| | | | | | all data at civil time hourly values global rad. and temp. on tape 1 SS on papertape synoptic measurements (different stat.) on tape 2 | | | | | | | | | |
| D | F. Kasten (only radiation data) Meteorologisches Observatorium Deutscher Wetterdienst Frahmredder 95 2000 Hamburg 65 | 16.12. 81 15.3. 82 | 26 12 26 | 14 12 | x x x | Pyranom. J/cm² | NIP | --------------------------- | | | | | | |
| | | | | | LAT hourly sums | | | | | | | | | |
| | | | | radiation data on 1 tape from MOH | | | | | | | | | | |
| | For the same stations the climatological data were available only from the Zentralamt des Deutschen Wetterdienstes, division K5, PB 185, D-6050 Offenbach. No negotiations due to costs. | | | | | synoptic measurements at GMT | | | | | | | | |
| F | P. Fournier Ministère des Transports Direction de la Météorologie 2 Av Rapp 75340   Paris CEDEX 07 | 4.1. 82 10.2. | 31 20 | 11 1 19 | x x x  x  x x x x x x x      x  x x x | Pyranom. J/cm² | NIP | C.St. 0.1h | WMO Oct | 0.1 | 10 | m/s | | |
| | | | | | LAT hourly sums | | | | G M T synoptic measurements | | | | | |
| | | | | all data on 1 tape | | | | | | | | | | |
| G | Lalas[1) Department of Meteorology University of Athens Ippokratous 33 Athens 144 | 9.1. 82 | | | (1 x x x  x  x x x) | no information but at least one station in Athens | | | | | | | | |
| | | no further contact until 31st July | | | | | | | | | | | | |
| | The national meteorological service, Glyfada, Post Office TT5, Athens, was not contacted, because Prof. Lalas is contractor of action 4.1. | | | | | | | | | | | | | |
| I | A. Guerrini[1) Istituto di fisica atmosfera P. le L. Sturzo 31 00144 Roma | | | no answer/at least 9 stations with CM5 pyranometers for global and diffuse radiation running | | | | | | | | | | |
| | S. Palmieri Servizio Meteorologico Aeronautica Piazzale Archivi 00144 Roma | 2.3. 82 | | at least 31 stations with bimetallic pyranometers running; no further contact until 31st July | | | | | | | | | | |
| IRL | E.J. Murphy Valentia Observatory Cahirciveen/Co Kerry | 21.12. 81 8.2. 82 | 6 7 | 4 2 | x x x  x  x x x x      x  x x x | Pyranom. J/cm² | NIP | C.St. 0.1h | total Oct. | 0.1 | 0.1 | kn | set of synoptic data delivered | |
| | | | | | LAT hourly sums | | | | G M T hourly inst. values | | | | | |
| | | | | | on punched cd and on magn. tape | | | | on magnetic tape | | | | | |
| NL | C.W. v. Scherpenzeel Koninklijk Nederlands Meteorologisch Instituut P.O. Box 201 3230 AE de Bilt | 8.6. 81 | 5 10 | 5 | x    x  x  x x x | information to be requested offer of data/not ordered due to costs all data on 1 m. tape/4 weeks after ord. | | | | | | | | |
| UK | J.B. Elms/MetOlc(2) room t.12 Meteorological Office Headquarters annex Bracknell/Bershire RG12 2UR | 4.12. 81 28.4. 82 | 27 5 42 | 8 5 12 2 | x x x  x  x x x x x (daily totals) x x x | Pyranom. Wh/m² | NIP | C.St. 0.1h | Oct. | 0.1 | 1 | kn | | |
| | | | | | LAT hourly sums | | | | G M T synoptic measurements | | | | | |
| | | | | all data on 1 magnetic tape | | | | | | | | | | |

a: date of 1st answer; b: date of delivery of data; c: number of stations reported; d: number of stations after Dogniaux's list; e: number of stations with instrumentation as indicated in next column: G = global radiation, D = diffuse radiation; I = direct radiation; SS = sunshine duration; CC = cloud cover; T = dry air temperature; A = wind direction; V = wind speed; f: instantaneous values in 10 m height, sometimes stations for wind and radiation have a distance of up to 10 km; g: degree Centigrade; h: degree; i: WMO code; j: 10 min average before reading time.

### 3.2.3 Ground reference data for action 4.3

The action leader was asked by the Commission to organize an intercomparison between radiation data produced by satellite image processing
and ground reference data measured in the radiation networks. For this
purpose the month April 82 was selected. The national services were
warned on the Athens meeting November 82, that a request on radiation
data only will be sent to them after April. The action leader gave a
second announcement to those who participated in the solar week. A formal
request was circulated on July 1. The status on Sept. 10, 82, was as
follows: Daily sums from all countries but Italy have been received or
are announced. Hourly sums have been received from the United Kingdom
and Germany and are announced by Belgium, Denmark, Greece, Ireland, and
France. The data reformation and transmission to the participants of
action 4.3 will continue during September. This exercise was very successful except that only 95 out of 187 stations can be used for ground truth
intercomparisons.

### 4. Conclusions and suggestions for future activities

In spite of certain difficulties the action "Improvement of measurement in national networks of global direct, diffuse irradiance and irradiance on inclined surfaces" was successful. The instrumentation could be
upgraded substantially; measurement techniques are better understood;
calibration procedures have been unified to obtain comparable results
throughout the European Community.
Some effort is still needed to unify the results in shade ring measurements of diffuse radiation.
The Community certainly should take action to improve the access
to meteorological data as they are needed for simulation and design purposes in solar energy applications.
The action leader stimulated on the action meeting in Copenhagen
a round table discussion on possible future activities in the framework
of action 4.1 and the next solar energy programme of the Commission.
Besides technical details, which one may include into future research
areas, the participants gave priority to the following items in the order
listed below:
o A working group on network instrumentation including all meteorological
  parameters should continue the objectives of action 4.1 and 1.
o Before going ahead with additional equipment assessment studies should
  be performed on
  - the improvement achieved so far,
  - the possible complement of sattelite image processing and ground mea-
    surements,
  - the density of the networks.
o Correlation of network data to small time scale and short distance
  data (micro-to-macroscale correlations).
o The forecast of radiation regimes for solar applications in a small
  time-scale (hours or days).
o Cloud cover measurements in networks.
The participants gave a high rank to the international cooperation
in the EC, coordination with other international bodies like IEA, WMO
should be considered.

References:
1. IAEA Solar Radiation Data Source Catalogue; edited by: W. Josefsson, M.L. Westerberg, SMHI, Norrköping, Sweden, Oct. 1980
2. Inventory of Radiation Measurements, Region VI (Europe), edited by R. Dogniaux, World Meteorological Organization, May 1978.
3. Solar Radiation Data, Project F Survey, in: "Solar Energy R&D in the European Community", Series F, Volume 1, ed. by W. Palz.

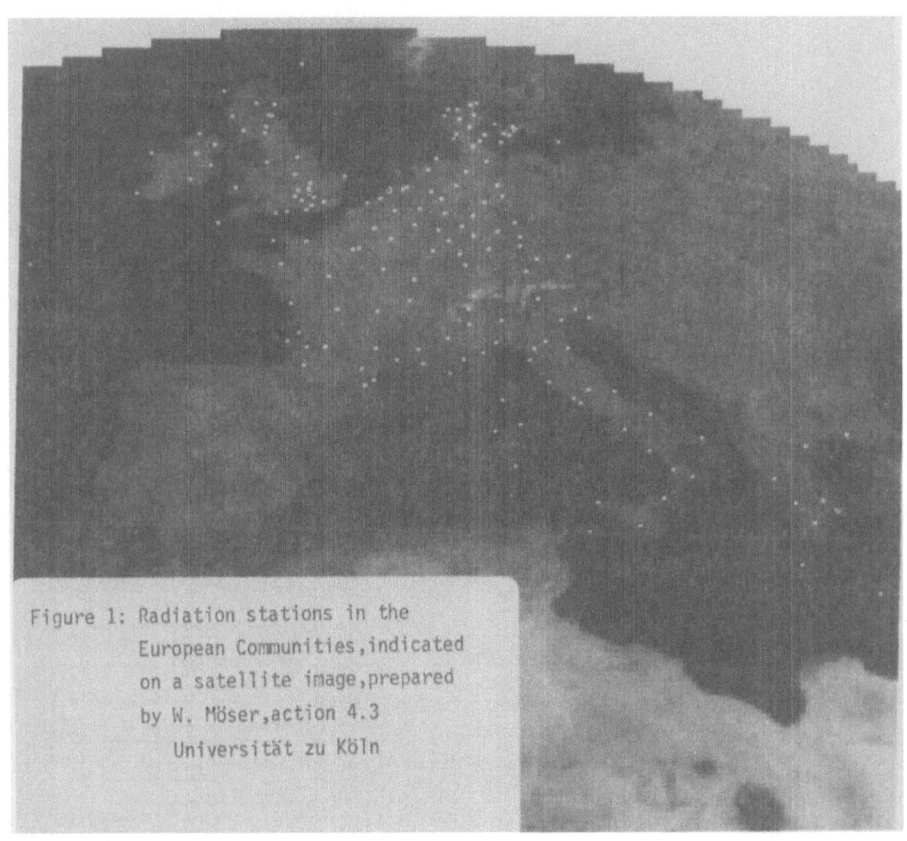

Figure 1: Radiation stations in the European Communities, indicated on a satellite image, prepared by W. Möser, action 4.3 Universität zu Köln

Frequency distribution of deviations between true and corrected measured values of diffuse radiation based on hourly periods of measurement

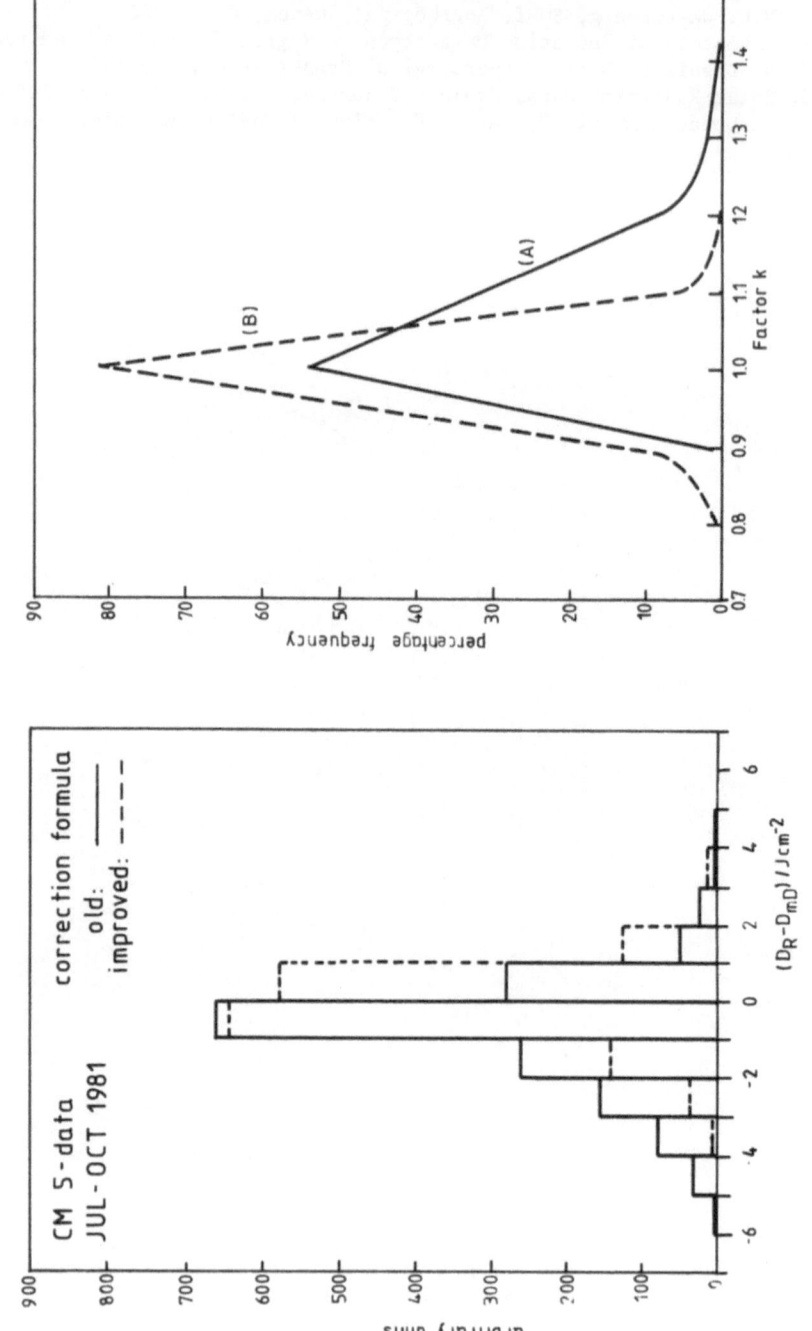

Fig. 2: Results of Hamburg

Fig. 3: Results of Valentia

# VARIABILITY IN SPACE OF SOLAR RADIATION

Authors           : P. Allerup, K. Frydendahl

Contract number : ESF-25-DK

Duration          : 48 months          1 July 1979 - 30 June 1983

Total budget     : 300 500 D Kr       CEC contribution : 147 000 D Kr

Head of project : Knud Frydendahl

Contractor       : Danish Meteorological Institute

Address           : Lyngbyvej 100
                    DK 2100 København Ø
                    Denmark

## Summary

Data from six out of 19 existing stations measuring global radiation have
been prepared for a statistical analysis. Series of simultaneous measurements from the period January 1980 to August 1981 have been apt to an automatic error-finding procedure and have been  corrected due to systematic
fluctuations over time. The statistical structures in data have been investigated at the aim of revealing properties which can be exploited especially for the study of spatial correlations. It has become clear that tra-
ditional calculations of measures of correlation do not to a satisfactory
degree compensate for certain variance structures in data. In fact, a log-
transformation of data is shown to be necessary. The report outlines the
structure of a statistical model which will be used for further analyses.
It is an important characteristic of the model that it specifies the (logaritmic) mean values as composed by a site-specific parameter and a time-
specific parameter, hence allowing the variation in global radiation over
time to be systematic rather than stochastic. All correlations are defined
in the light of this model.

## 1.1  Introduction.

The goals of the work in Denmark under this second contract are to run the stations established during the first contract and to build a programme system to handle this bulk of data and present them in a form suitable for different purposes. For the moment this programme system is able to produce data, but still needs to be improved which will be done within this year. Further we want to utilize the dense danish network to analyse spatial correlation structure of global radiation.

## 1.2  Statistical analyses - data description.

19 stations are measuring global radiation in Denmark. Data from six stations have been analysed in this report. The last 13 stations will be studied later. Raw data from the period January 1980 until August 1981 from each of the six stations were at first subject to an error finding procedure implemented on a computer. The six series of observations were pooled monthwise such that only simultaneous measurements of ½ hour values of global radiation on all six stations are included in the resulting data base. Much effort has been put into the work of removing beforehand known systematic errors because a later statistical modelling of data otherwise may over-estimate the stochastic errors. The six series of simultaneous measurements (½ hour values) have been accompagnied by two series of simultaneous 10 minute measurements taken at the same station i.e. with interdistance zero.

## 1.2.1  Outline  of the statistical analyses.

It seems that not much work has been done in the field of statistical modelling of solar data. Hay and Suckling (1976) and (1979) and Valko (1980) all present analyses carried out by means of statistical techniques rather than developing a formal mathematical statistical description of the data. In fact, application of the technical tools assumes the existence of a statistical model. However, formulation and test of fit of such models seldom occurs and results obtained by means of the technical approach may therefore be non-valid, since no justification for the use of the technique has been stated.
        For the study of spatial correlations the presence of a statistical model becomes crucial since calculations of ordinary correlation coefficients have no meaning outside the normal distribution and even within the range of this distribution do have interpretations depending on the specification of the mean value structure. How this specification will be done depends  among other things on structures found in data and is therefore an empirical problem to be dealt with, when formulation of a statistical model takes place.
        Another wellknown tool when analysing solar data is the so-called coefficient of variation (Hay 1979, Latimer and Won 1980). Being the ratio of standard deviation on differences over aritmetic mean (see below) for two stations, it is at first  sight an easy interpretable measure of extrapolation error when the spatial representativity of a station is under consideration.
        However, the use of this tool implicitly assumes certain structures to be found in data - or assumes the existence of a specific statistical model fitting the data. This will be discussed in the sequel.
        The objective point for the danish research work is detailed analyses of the spatial correlation structure of global radiation taking into account the dense network. Great emphasis will therefore be put on empirical data analysis in order to create the basis for a statistical model which in turn provides the necessary technical tools.

## 1.3 Empirical findings.

Relations between the measurements from the six stations can be analysed graphically by pairwise comparions. This is done in fig. 1 - fig. 2 where simultaneous measurements from a fixed month are plotted for different periods: ½ hour and one hour. The shown examples represent naturally but a fraction of all possible combinations and are selected to demonstrate relations with short and (relatively) long interdistance.

It is obvious from the figures that ½-hour values are hardly correlated at all. This might be expected. It is, however a bit surprising that low degree of correlation can be found by the daily average relations too. A common property of all figures is that much of the correlation - this being low or high - is owing to the two clusters representing "total cloudiness" and "clear sky". Analysis of these (and the many others) graphs, furthermore, reveals an allready wellknown structure. In fact, the dispersion of points increases with increasing level of measurement. In other words the standard deviation on the difference between measurements increases with increasing level of measurement. This functional relationship need not be simple as the example giving in fig. 3 shows. The graph illustrates for all combinations of stations the standard deviation on differences (ordinate) and mean value for sum of measurements (abscissa). The points are referring to hourly values in a fixed month. It seems clear that the functional relationship is not in any way linear and hence the ratio - i.e. the variation coefficient - of standard deviation over mean value does not compensate for dependency on the average level of measurement. Therefore this quantity is less able to be used satisfactory when studying spatial relations since the average level influences relationship between coefficients of variability and distance. Furthermore, the distribution of differences (for given level of average measurement) seems not to follow the normal distribution, which adds to the problem of interpretating the standard deviation - and in turn the coefficient of variability, see fig. 4 for an example. A less problematic way of dealing with the basic structure in the figures fig. 1 - 2 is to seek for a simple transformation of the observations such that the increasing variance with increasing measurement level disappears. Or more correctly remains constant, independent of the measurement level.

In the figures fig. 5 - fig. 6 the effect of a log - transformation of the observations is demonstrated. The presented graphs are referring to the same situations given in fig. 1 - fig. 2. Uncorrelated structures of course remain uncorrelated after the transformation, but it is clear, however, that patterns of increasing dispersion now show up to be rather constant. A more convenient way of presenting the effect of log-transformation is given in fig. 7, where the logaritmic difference (ordinate) is plotted against the sum of logaritmic measurements (abscissa). The graph at the same time provides the distribution of logaritmic difference shown as frequency distribution along the ordinate axis.

## 1.4 Formal description of empirical structures.

We will summarize shortly the most important characteristics found in the empirical relation between logaritmic observations. It may be stressed that these summaries are based on a bulk of empirical relations of which only a fraction has been shown in this report. All structures are found within fixed months to avoid the systematic influence, at this stage of investigations, done by the change of potential radiation over the year.

(1) The increasing pattern of dispersion found in the direct measurements can be described satisfactory by a logaritmic transformation of data.

(2) The logaritmic difference between measurements has distribution
whose variance (standard deviation) is independent of the level
of measurements. Fixed month, fixed time period (½ hour, hour or
day) and fixed distance between the stations.

(3) The variance on the logaritmic difference increases with increasing distance  etween the stations and decreases with increasing
sample period (½ hour, hour and day).

(4) The distribution of logaritmic differences can be approximately
described by the normal distribution - fixed month, fixed time-
period and fixed distance between the stations.

### 1.4.1  Stating a statistical model - future work.

Let $X_{it}$ be the logaritmic radiation at station No. i (i=1,...,6) and at
time period No. t ( the t'$^{th}$ ½ hour, hour or day). The empirical structures
can now be represented in the following statistical formalization, which
will be the point of attack for the further work.

(5)  $E ( X_{it} ) = \alpha_i + \beta_t$

(6)  $V ( X_{it} ) = \sigma_i^2$

(7)  $Cov ( X_{it}, X_{jt} ) = \sigma_{ij}$

(8)  $X_{it} \sim$ normal distribution

In fact, (5) states the mean value of radiation to be composed, additive
by two characteristic parameters, $\alpha_i$ characterizing the specific station
and $\beta_t$ characterizing the systematic change of solar radiation with time t.
Immediately we get from (5) and (6).

(9)  $E ( X_{it} - X_{jt} ) = \alpha_i - \alpha_j$

(10) $V ( X_{it} - X_{jt} ) = \sigma_i^2 + \sigma_j^2 - 2\sigma_{ij}$

which corresponds to the shown structure in fig. 7 stating the independence
of logaritmic difference $X_{it} - X_{jt}$ by measurement level.
    It is important to stress that the model specification (5) - (8) is
restricted to the seperate months and the notation for this effect is from
practical reasons omitted.
    The covariance structure(7) allows for a description by interdistance
Dist (i,j) between the stations. This will of course be of central interest
for the further work.
    The test of fit of the model specification will continue and the two
long series of measurements with interdistance zero will play an important
role in this respect. These data will at the same time be useful when estimating the lowest attainable level of inaccuracy for extrapolations, since
no spatial effects are present here. A single example from the analysis of
these data is given in fig. 8, where the logaritmic difference (ordinate)
is plotted against the level of ( not logaritmic transformed) measurement
(abscissa). The points indicate - as expected by model - a constant standard deviation of approximately 0,04 corresponding to deviations in the
original scale of approximately 6%. This special case with interdistance
zero is in agreement with Hay and Suckling (1977).

References:

Latimer, J.R. and Won, T.K. (1980): Recommendations concerning meteorological networks for solar energy applications. (IEA Programme) Swedish meteorological and hydrological institute.

Suckling, P.W. and Hay, J.E (1976): The spatial variability of daily values of solar radiation for British Columbia and Alberta, Canada. Climatological bulletin No. 20.

Hay, J.E. and Suckling, P.W. (1979): An asessment of the network for measuring and modelling solar radiation in British Columbia and adjacent areas of Western Canada. Canadian Geographer XXIII.3.

Hay, J.E. and Suckling, P.W. (1977): Problems associated with the determination of solar radiation fluxes in Western Canada. Presented at the anual conference of the Association of American Geographers, Salt Lake City.

Valko, P. (1980): Some empirical properties of solar radiation and related parameters. Swiss Meteorological Institute, Zurich.

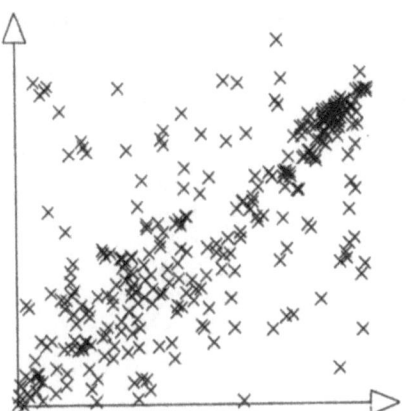

Fig.1. ½ hour measurements from 30225 and 30383 (interdistance 11km). July 1980.

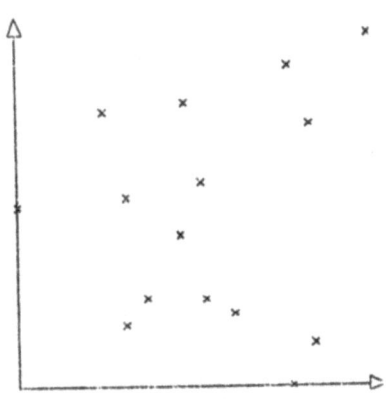

Fig.2. 1 hour measurements from 29011 and 29255 (interdistance 93km). October 1980.

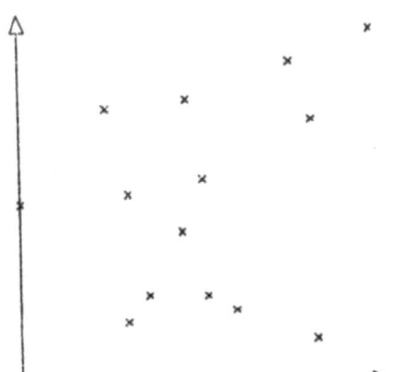

Fig.3. Standard deviation (ordinate) on differences versus average mean value (abscissa). 1 hour measurements. 15 pairwise combinations of six stations. June 1981.

Fig.4. Distribution of differences between 29011 and 29255 1 hour measurements. November 1980.

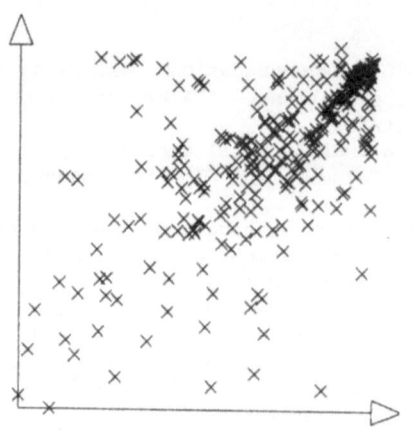

Fig.5. ½ hour logaritmic measurements from 30225 and 30383 (interdistance 11 km). June 1980.

Fig.6. 1 hour logaritmic measurements from 29011 and 29255 (interdistance 93 km). October 1980.

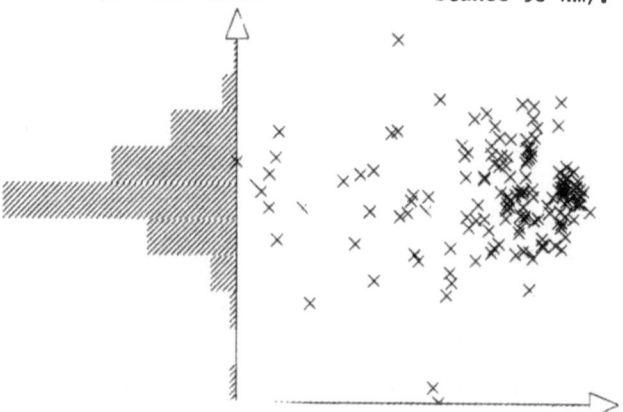

Fig.7. Logaritmic differences (ordinate) versus logaritmic sum (abscissa) of 1 hour measurements from 30188 and 30383. July 1981.

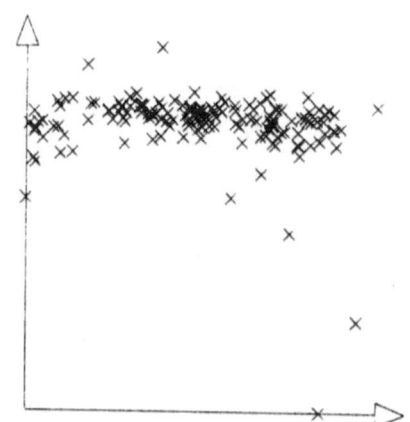

Fig.8. Logaritmic differences (ordinate) of 1 hour measurements taken at interdistance zero, versus sum of measurements (abscissa). July 1981.

# IMPROVEMENT OF MEASUREMENT OF DIFFUSE SOLAR RADIATION

Authors            : F. Kasten, K. Dehne, W. Brettschneider

Contract number : ESF-004-80 D (B)

Duration           : 3 Years          January 1980 - December 1982

Head of project : F. Kasten, Deutscher Wetterdienst
                   Meteorologisches Observatorium Hamburg

Contractor        : Deutscher Wetterdienst, Zentralamt

Address            : Frankfurter Strasse 135
                     6050 Offenbach am Main

## Summary

To improve the accuracy of diffuse solar radiation data
gained by ring shaded pyranometers, a new formula correcting
for losses caused by the Hamburg ring has been derived. The
formula has been validated by data sets of different time
periods and from different pyranometer types, the latter with
slightly modified constants. The correlation coefficients are
always close to 0.80. For further improving the correction
formula, the influence of the sky radiance distribution is to
be taken into account. The so-called sky scanner was equipped
for broad band measurements, and an appropriate data acquisition system was designed which uses a microprocessor to sample
and store up to 1440 mean radiance values of the sky hemisphere. The reference data of diffuse solar radiation are delivered by a sun-following shade disk device with a shaded
angle of 9.6°. A compact reference shade disk instrument has
been built to be used for comparison with different kinds of
shade ring devices used in the national networks in order to
validate the transfered correction formula.

## 1. Introduction

The use of a shade ring device is a convenient method to measure diffuse solar radiation. Up to now, the inherent problem of correcting the undesired ring losses of diffuse solar radiation is insufficiently solved by simple correction formulae. The aim of the present study is a) to derive an improved correction formula depending on parameters which are easily available for a selected shade ring device, b) to transfer this correction formula to other types of shade ring devices used in EC countries.

## 2. Methods
### 2.1 Derivation of the correction formula

As standard device, the "Hamburg ring" was selected whose width to radius ratio $b/r = 0.169$ represents a medium value within the group of ring devices. Its shaded view angle of $9.6°$ corresponds to the view angle of the Linke-Feussner-Actinometer.

The hourly sums $D_{mR}$ obtained with this device were compared to reference data gained with a shade disk device $D_D$ of $9.6°$ shade angle or calculated from measured data of direct solar radiation I by $D_C = G - I \cdot \sin\gamma$ (G = global radiation; $\gamma$ = solar elevation). The correction formula was to be represented by a function of the type $f = D_D/D_{mR} = f(D_{mR}, G, \delta)$.

For further improving the formula, the spatial distribution of sky radiances is to be evaluated.

### 2.2 Transfer of the correction formula

The correction formula of the standard device can be transfered to shade ring devices with different ratios $b/r$ by a theoretical transfer formula.

To validate the transfered correction formula, simultaneous measurements of diffuse solar radiation with the reference device are planned preferably at the places of the operational routine in order to take account of the particular cloudiness and dust conditions.

## 3. Results
### 3.1 Listing of previous results obtained in the project

a) Statistical correlation applied to the data sets gained from both a CM 10 pyranometer with shading device and a Normal Incidence Pyrheliometer (NIP), delivered the following correction formula

$$f = A + B \cdot (D_{mR}/G)^3 + C \cdot \delta + D/\tau'$$ (1)

where $\tau = \ln [I_0 \cdot \sin\gamma/(G - D_{mR})]$,

$\delta$ = declination of sun,

$I_0$ = extraterrestrial I.

b) Using the data sets of the period January 1981 - May 1981, the f-values derived from both types of reference data differed by about 1-2 %. The correlation coefficients amount to 0.78 and 0.80, respectively. The magnitude of f is dominated by the terms $A + B \cdot (D_{mR}/G)^3$.

c) The formula for transfering the correction factor from the standard device to shading ring devices with other (b/r)-ratios is given by

$$f = 0.169 \cdot f_{St}/[0.169 \cdot f_{St} - (f_{St} - 1) \cdot b/r]. \qquad (2)$$

d) Theoretical correction formulae derived for clear sky and overcast sky by assuming ideal spatial distributions of sky radiances have been compared with the statistically derived formula.

## 3.2 New results
### 3.2.1 Improved correction formula

The evaluation of data gained by a CM 10 pyranometer in the period January 1981 - October 1981 yielded the same correlation coefficient as before; the constants of equation (1) showed but a small modification, see following table.

|  | period | A | B | C | D | Corr. coeff. |
|---|---|---|---|---|---|---|
| $f_{III}$ | Jan-Oct 1981 | 1.161 | -.112 | 0.0009 | -0.0246 | .78 |
| $f_{II}$ | Jan-May 1981 | 1.162 | -.108 | 0.0017 | -0.0311 | .78 |

If equation (1) is used for correcting the data gained by a pyranometer of type CM 5, the constant A has to be changed to 1.135.

The effect of the improved correction formula becomes evident in Fig.1 which shows uncorrected and corrected values on three days with different sky conditions.

For comparing the improved correction formula to the old correction formula of MetObs Hamburg, the frequency distributions of the deviations of the corrected data $D_R$ from the reference data $D_D$ are shown in Fig. 2.     Using the improved formula, the classes of deviations less than $\pm$ 1 J cm$^{-2}$ and $\pm$ 2 J cm$^{-2}$ have relative class frequencies of 0.78 and 0.95, respectively. In case of the old formula, the relative frequencies of the same classes are 0.60 and 0.80. The variance of the deviations was found to decrease from 1.99 (J cm$^{-2}$)$^2$ for the old formula to 0.98 (J cm$^{-2}$)$^2$ for the improved formula.

### 3.2.2 Special investigations on further improvements

To study the structure of the improved formula, the correlation coefficients of the correlations between $f_{III}$ and the single parameters, and of the intercorrelations between the parameters were calculated:

| $f_{III} \wedge (D/G)^3$ | $f_{III} \wedge \delta$ | $f_{III} \wedge \tau'$ | $(D/G)^3 \wedge \delta$ | $(D/G)^3 \wedge \tau'$ | $\tau' \wedge \delta$ |
|---|---|---|---|---|---|
| -.717 | .474 | .598 | -.281 | -.908 | .201 |

The figures demonstrate the high degree of correlation between $f_{III}$ and $(D/G)^3$ and between $(D/G)^3$ and $\tau'$.

To measure the real distribution of sky radiances, a sky scanner was equipped with a filtered broad band silicon sensor and broad band flexible fiber optics. The data acquisition system was modified by including a microprocessor to sample and store the means of the measured radiances into a register of 72 x 20 places corresponding to a hemispherical resolution of $5^\circ$ azimuth and of $4.5^\circ$ elevation angle. The stored values are read out on a cassette tape from which they are transfered by data line to a computer centre for further evaluation.

### 3.2.3 Transfer of the correction formula

An improved shade disk device has been developed which is to be used as reference instrument for validating the transfer of the correction formula. The device, see Fig. 3, is of simple and compact design. A strong rod which points to the pole bears the mount for the pyranometer and the axis of the motor-driven perch with the sun-following disk.

### 4. Conclusions

a) The frequency of deviations larger than $\pm$ 1 $J\cdot cm^{-2}$ of the corrected values from the reference values is reduced from 40 % to 20 % if the improved correction formula is used.

b) Further improvements require special case studies of the distribution of sky radiances which will lead to a more complex structure of the correction formula.

c) Comparisons of shading ring devices used in national networks with the reference shade disk device are to be performed in order to validate the improved correction method.

Fig 1 – Comparison of uncorrected $(D_{mR})$ and corrected $(D_R)$ hourly sums of diffuse solar radiation with the corresponding reference values $D_D$ on three days with different sky conditions. Explanation of the symbols :

| Sky | $D_{mR}$ | $D_R$ |
|---|---|---|
| overcast | □ | ■ |
| cloudy | ▽ | ▼ |
| clear | o | ● |

Fig. 2 – Frequency distributions of the deviations of corrected hourly sums $D_R$ of diffuse solar radiation from the reference values $D_D$. CM5-pyranometer with Hamburg ring, July – October 1981.

Dashed line : corrected with the old formula
Solid line : corrected with the improved formula.

Fig. 3 – Shade disk device, new compact design to be used as reference instrument at different sites.

PROGRESS TOWARD ESTABLISHING A COMPREHENSIVE
SOLAR RADIATION DATA AND CALIBRATION CENTER IN GREECE


Author            :  D.P. LALAS

Contract number :  ESF.027.G(B)

Duration          :  24 months          Start:  No signed contract yet

Total budget    :  Drs. 6,735,028    EEC Contribution:  Drs. 3,367,514

Head of project :  Prof. D.P. Lalas, Meteorological Institute

Contractor       :  Meteorological Institute
                      National Observatory of Athens
Address          :  National Observatory of Athens
                      Thesion, Athens
                      Greece

Summary

Progress in establishing a complete station and calibration center for all
parameters of solar insolation in Greece is reported.  This includes the
installation of new instruments, the calibration of substandards and the
improvement of data logging.  Efforts to collect and archive the existing
solar radiation data for Greece are discussed along with the subsequent
development of algorithms for the calculation of available solar energy.

During the first six months of this project, even though no contract was signed mainly because of internal administrative difficulties at the University of Athens, the following were accomplished:

## 1. MEASUREMENTS

1.1. <u>Intercalibration</u>. A previously unused precision pyranometer (Model PSP-Eppley Labs Serial No. 14899F3 with Schott WF 245 clear glass cover) was sent to Carpentras, France twice (on 21 May 1981 and 17 December 1981) to be calibrated as the substandard for the Greek network.

1.2. <u>Diffuse radiation measurements</u>. A shadow band (Model SBS, Eppley Labs) has been installed to continuously measure diffuse radiation on a horizontal plane. Average hourly values have been recorded since 1 September 1981 and have been utilized in the theoretical studies described in paragraph 3 below.

1.3. <u>Total solar radiation at 1000m altitude</u>. A total solar radiation sensor (Model Star, Eppley Labs) has been in place since 1 September 1981 on top of Mt. Hymettus at an altitude of 1020m above MSL and less than 4km from the main National Observatory of Athens (NOA), which is at an altitude of 107m above MSL in downtown Athens. Some problems have developed because of adverse environmental conditions at the top of the mountain, and the sensor reliability was questionable. It was replaced by a Lamdba Instruments Corp. (USA) pyranometer, Model LI-200S on 13 July 1982. In addition, smoke particles and total suspended particulates are being monitored at NOA to provide a data base for the estimation of the radiation absorption.

1.4. <u>Direct solar radiation measurements</u>. A solar tracker (Eppley Labs) was installed at NOA on 1 January 1982. Direct solar radiation and direct solar radiation at various frequency ranges measured by cut-off filters OG1 (525-2800nm), RG2 (630-2800nm) and RG8 (710-2700nm) have been recorded continuously since. In addition, independent measurements of the same quantities are taken four times a day during the regular meteorological observation times (0800, 1100, 1400, and 1700 LST) by trained observers (a Kipp & Zonen pyrheliometer is used).

1.5. <u>Ordering instruments</u>. To cover the additional measurements that have been proposed, two more pyranometers have been purchased and will be installed shortly. In addition, a data logger capable of recording all the solar radiation inputs has been obtained and is now being tested.

## 2. DATA BANK

A collection of all available data for Greece that refers to solar energy has been initiated. The results to date include the following:

2.1. Direct and total insolation instantaneous values, as well as values of direct radiation with OG1, RG2, RG8 and quartz cut-off at 0800, 1100, 1400 and 1700 LST along with pressure at NOA for the period 1966 to 1981, have been collected and archived.

2.2. Monthly sunshine duration values for the period 1900-1981 at NOA have been obtained and recorded.

2.3. Monthly sunshine duration values for the last 30 years at 28 stations in Greece have been obtained and digitized.

2.4. Hourly values of total solar radiation at NOA from 1977 to 1981 have been recorded on tape. It is hoped that this time series will be expanded to include the years 1956 to 1977 within the next six months.

2.5. The hourly values of total solar radiation at Patras for the years 1975 to 1981 have been obtained.

2.6.  Hourly values of total solar radiation for 1 year at various
locations (Rodos, Crete, Zakynthos, etc.) measured by the Greek Public
Power Corporation (PPC) have been obtained.

The data collection and archiving has progressed to the point that by
the end of the year it will be 90% complete.  Auxiliary data, such as
cloudiness, will be collected then to make the archive complete.

## 3.  ANALYTICAL WORK

Since only a limited number of measurement stations are economically
feasible, there will always be locations where solar energy potential will
have to be calculated.  Because solar energy, unlike wind energy, is
continuously distributed various algorithms based on available
meteorological data can be utilized.  An effort to generate such algorithms
for Greece has been initiated. The first results will appear in two papers:

3.1.  "Estimation of global, direct and diffuse solar radiation in
Athens under clear sky conditions."  This paper has been accepted for
publication in Archiv für Meteorologie.  It compares three well-known
models with data for Athens and comes to the conclusion that for cloudless
days the Bird model produces the best estimates of solar irradiance in
Athens.

3.2.  "Methods for the calculation of solar radiation intensity on an
inclined surface in Greece."  This report (in Greek) utilizes the Bird
model discussed above and a new partition method for cloudiness to
calculate direct total and diffuse radiation at an inclined surface.
Tables for a surface tilted at 30°, 45° and 60° are given to assist solar
system designers and will appear in the J. Greek Engineering Society.  In
addition, an article currently in preparation based on this report will be
submitted to Solar Energy.

In both publications, the assistance of EEC under this contract is
acknowledged.

## SOLAR RADIATION IN ITALY

Author                      : A. LAVAGNINI

Contract number             : ESF/071/I - (S)

Duration                    : 42 months from 1/180 to 30/6/83

Total budget                : 112 000 000 LIT   CEC contribution : 56 000 000 LIT
                              (in national currency)

Head of project             : Dr. Angelo GUERRINI

Contracting organisation:   Istituto di Fisica dell'Atmosfera
                            Consiglio Nazionale delle Ricerche

Address                     : P.le L. Sturzo, 31 - 00144 ROME (Italy)

## 1. INTRODUCTION

In Italy the global radiation is systematically detected by the Air Force Meteorological Service in 30 localities starting from 1958 as daily
totals by means of Fluess-Robitzsch pyranometers During this period the
instruments were not periodically calibrated. Obviously through the years
a loss of sensitivity was realized in most of them. Since the data detected
in such a way were considered unreliable, a "correction" was applied to
them and this will be shown afterwards. The modified data, relative to 10
stations were sent to the EC in order to compile the "European Atlas of
Solar radiation".

For all the 30 stations relative to a period of twenty years 1958-1977
were then obtained:

a) the cumulative frequency distributions of the sunshine hours and of
   global radiation;
b) the sunshine hours - global radiation correlation;
c) the analysis of the sequences relative to the days with the lowest values
   of global radiation.

Moreover an accurate study was carried out on the cloud cover and on the
solar radiation at Adrano (Sicily) in consideration of the presence of the
EC solar plant (1,2).

## 2. DATA MODIFICATION OF THE DAILY GLOBAL RADIATION

To the lack of periodical calibration surveys of the pyranometers, an
attempt was made to obviate through the purchasing, the intercalibration
and the installation (made in 1972) of 30 new instruments of the same Ful-
ness-Mobitzsh type. For each station during the first operational year of
the new equipment, were selected those days whose diagrams of the daily
global radiation have presented a trend such as to consider them as clear
sky days and without any particular atmospheric turbidity. Such values were
reported on the graph, and an example of which is shown in fig. 1 and the
interpolating curve was drawn, by band. Similar curves were produced for
each year from 1958 to 1971. Since it was considered that such a curve

should be characteristic of the measurement locality, if no remarkable modifications occur in the surrounding territory, the difference between the relative ordinate for 1972 and that relative to each year, was attributed to the decay of the instrument. Therefore all the curves relative to the years from 1958 to 1971 were "normalized" to that of 1972.

This method gives rise to many perplexities. For three stations (Genoa, Vigna di Valle and Adrano), having at disposal other measurements carried out by means of electrical pyranometers, a comparison (3) was effected with the data of the Robitzsh pyranometers modified as previously described. It resulted that the values modified are statistically higher of a quantity varying from 3 to 15% according to the station examined and to the season of the year.

A more accurate verification may be done when the 30 new stations which are being installed near those already operating, will have carried out measurements for a sufficient period of time.

## 3. PROCESSING CARRIED OUT

### 3.1. Frequency distributions

Frequency distributions of global radiation and sunshine hours were produced for each month of the year. The relative tables may be sent upon request.

These results were utilized to produce maps showing, in each Italian area, with which frequency given thresholds of daily global radiation may be exceeded (3).

In figs 2a and 2b are shown two examples relative to winter and autumn seasons.

### 3.2. Global radiation-sunshine hours correlation

It was effected applying the Angstrom's formula

$$\frac{G}{G_o} = A \frac{N}{N_o} + B$$

where
- G and N are the daily totals of global radiation and sunshine hours respectively;
- $G_o$ and $N_o$ are the extratmospheric radiation and the day duration.

In table 1 are reported the A and B coefficients of the straight line, the r correlation coefficient and the standard deviations.

### 3.3. Persistence analysis

It has been moreover carried out an analysis showing the day sequences with global radiation lower than a prefixed threshold. In figure 3 it is shown, as an example, the graph relative to Vigna di Valle for the autumn season. In abscissae are reported the consecutive days, in ordinate the per cent of the days and on the top of each broken line, the threshold considered (in MJ/m$^2$·day).

The longest sequences of days with lower levels of total energy occur, as expected, in December in the Po valley and in the North Adriatic sea with a few exception (in Trieste and Udine in January) (3).

To give an idea, the probability to have 4 consecutive days, in each of which the daily total energy incident on the horizontal plane is lower than 2 MJ/m$^2$ is: in Milan 10%, Bologna 5.7% and Ancona 6.2%.

BIBLIOGRAPHY

1) A. LAVAGNINI, A.M. SEMPREVIVA.
   Total cloudiness and high clouds frequency on eastern Sicily.
   IFA 82/08, July 1982.

2) F. BARBALISCIA, A. LAVAGNINI
   Analysis of solar radiation caracteristics of the solar station of
   Adrano. IFA 82/19, September 1982.

3) A. LAVAGNINI
   Solar radiation in Italy
   IFA 82/22

4) C. STANGHELLINI, A. LAVAGNINI
   Prediction of daily sunshine hours from cloud cover: role of climate
   and latitude.
   Submitted to "Solar Energy".

TABLE 1

| STATIONS | A | B | r | $\sigma^2_N$ | $\sigma^2_G$ |
|----------|------|------|------|--------|--------|
| ALGHERO  | 0.57 | 0.24 | 0.88 | 0.0671 | 0.0283 |
| ANCONA   | 0.49 | 0.30 | 0.84 | 0.0776 | 0.0264 |
| BOLOGNA  | 0.47 | 0.31 | 0.87 | 0.0649 | 0.0195 |
| BOLZANO  | 0.50 | 0.29 | 0.79 | 0.0491 | 0.0198 |
| BRINDISI | 0.51 | 0.30 | 0.85 | 0.0671 | 0.0242 |
| CAGLIARI | 0.48 | 0.29 | 0.83 | 0.0659 | 0.0222 |
| C.MELE   | 0.50 | 0.29 | 0.87 | 0.0704 | 0.0228 |
| C.PALIN  | 0.38 | 0.35 | 0.78 | 0.1007 | 0.0232 |
| CROTONE  | 0.51 | 0.30 | 0.82 | 0.0574 | 0.0223 |
| FOGGIA   | 0.51 | 0.28 | 0.89 | 0.0718 | 0.0234 |
| GENOVA   | 0.48 | 0.27 | 0.87 | 0.0778 | 0.0235 |
| M.CIMONE | 0.65 | 0.20 | 0.85 | 0.0848 | 0.0489 |
| MESSINA  | 0.56 | 0.26 | 0.90 | 0.0656 | 0.0258 |
| MILANO   | 0.45 | 0.31 | 0.85 | 0.0691 | 0.0195 |
| M.TERMIN | 0.57 | 0.22 | 0.88 | 0.0827 | 0.0346 |
| NAPOLI   | 0.50 | 0.29 | 0.85 | 0.0667 | 0.0236 |
| OLBIA    | 0.50 | 0.26 | 0.86 | 0.0708 | 0.0243 |
| PANTEL   | 0.54 | 0.27 | 0.86 | 0.0620 | 0.0243 |
| PESCARA  | 0.46 | 0.32 | 0.84 | 0.0749 | 0.0224 |
| PIANOSA  | 0.45 | 0.33 | 0.81 | 0.0757 | 0.0234 |
| PISA     | 0.52 | 0.25 | 0.88 | 0.0745 | 0.0258 |
| PROSA    | 0.41 | 0.48 | 0.74 | 0.0763 | 0.0230 |
| R.CIAMP  | 0.50 | 0.28 | 0.87 | 0.0714 | 0.0233 |
| TORINO   | 0.47 | 0.31 | 0.84 | 0.0661 | 0.0206 |
| TRIESTE  | 0.46 | 0.28 | 0.85 | 0.0838 | 0.0245 |
| UDINE    | 0.41 | 0.31 | 0.83 | 0.0841 | 0.0202 |
| USTICA   | 0.47 | 0.33 | 0.83 | 0.0674 | 0.0216 |
| VENEZIA  | 0.46 | 0.31 | 0.83 | 0.0667 | 0.0205 |
| VIGNA    | 0.53 | 0.25 | 0.90 | 0.0727 | 0.0246 |

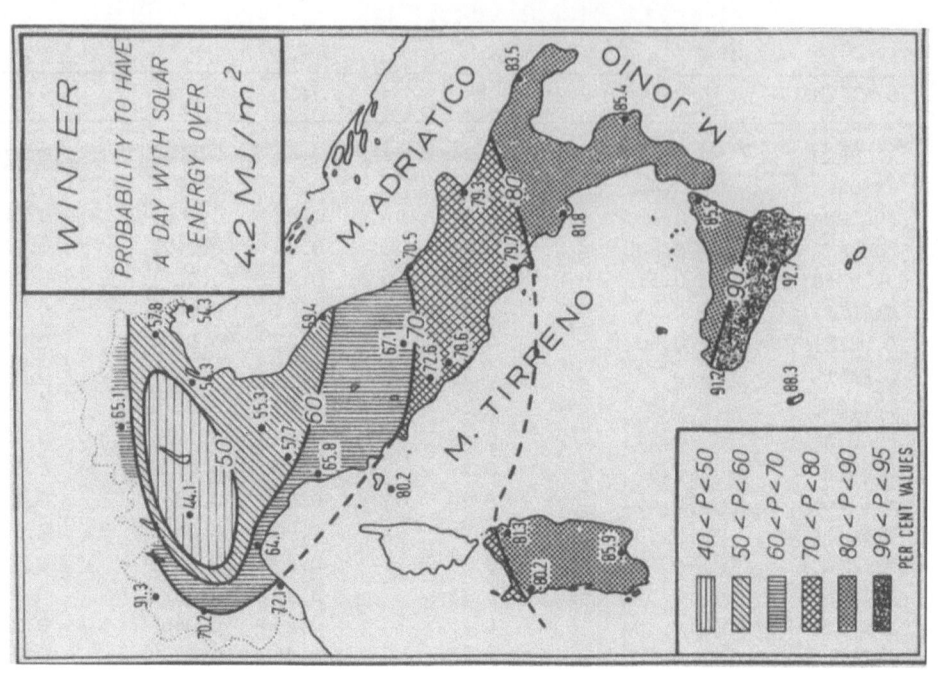

WINTER

PROBABILITY TO HAVE
A DAY WITH SOLAR
ENERGY OVER

4.2 MJ/m²

M. ADRIATICO

M. JONIO

M. TIRRENO

40 < P < 50
50 < P < 60
60 < P < 70
70 < P < 80
80 < P < 90
90 < P < 95

PER CENT VALUES

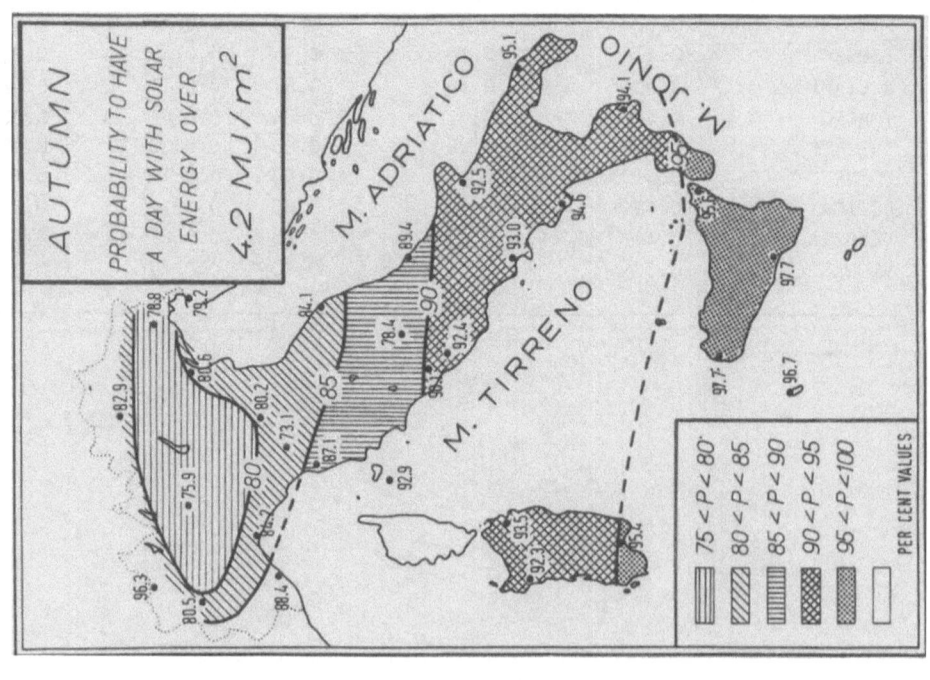

AUTUMN

PROBABILITY TO HAVE
A DAY WITH SOLAR
ENERGY OVER

4.2 MJ/m²

M. ADRIATICO

M. JONIO

M. TIRRENO

75 < P < 80
80 < P < 85
85 < P < 90
90 < P < 95
95 < P < 100

PER CENT VALUES

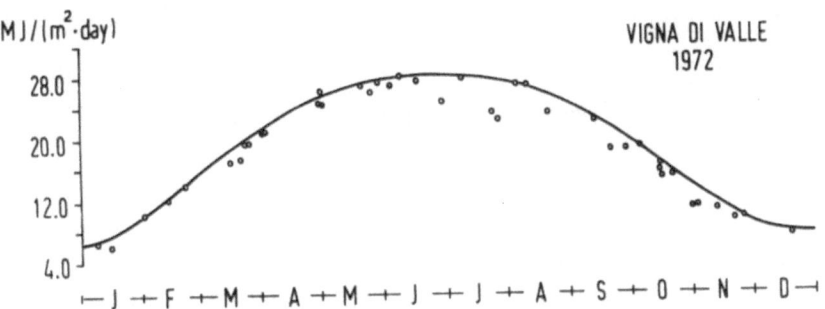

Daily global solar radiation of the Vigna di Valle station
in the year 1972, measured in clear sky and limpid air

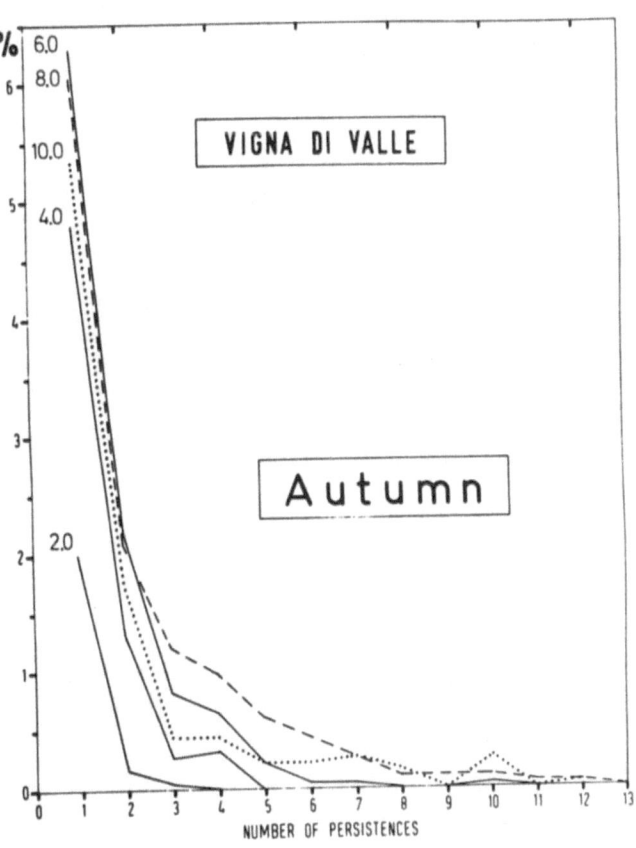

# IMPROVEMENT OF RADIATION NETWORK

Author                : E.J. Murphy

Contract number       : ESF-018-EIR(G)

Duration              : 36 months     1 January 1980 - 31 December 1982

Total budget          : IR£12,994     CEC contribution : IR£6,497

Head of project       : E.J. Murphy

Contractor            : Irish Meteorological Service

Address               : Glasnevin Hill
                        Dublin 9
                        IRELAND

## Summary

The Irish Meteorological Service network of solar radiation recording stations has been extended by the addition of two new stations - Clones and Malin Head.  Global and diffuse radiation data are available for both stations from 1st June 1981.

## 1.1 Introduction

The Irish Meteorological Service network of solar radiation stations consisted of 4 stations measuring global and diffuse radiation and one station which measured global radiation only.

Under Action 4.1 in the present contract period it was planned to extend this network by the addition of two new stations - Clones and Malin Head (see Figure 1). Both stations are synoptic weather stations with long term records of hourly observations including duration of bright sunshine.

The new equipment was delivered to Valentia Observatory in mid-March 1981, tested and calibrated during April 1981 and installed at the new stations in May 1981.

## 1.2 Measuring equipment

The equipment for each station consisted of 2 Kipp and Zonen CM6 pyranometers (global and diffuse), 2 Kipp and Zonen BD8 flat-bed recorders, 1 Eppley electronic integrator with Digitec printer (global) and 1 shading-ring.

## 1.3 Results obtained

Global and diffuse data are available for both stations from 1st June 1981. A small number of hours data for Malin Head are not available due to instrument malfunction. Interruptions to the power supply at Clones which, unlike Malin Head, does not have a standby electrical generator account for the loss of some 20 hours global and diffuse data. Instrument malfunctions at Clones was responsible for the loss of an approximate total of 4 weeks diffuse data.

## 1.4 Future work

Future work consists of continuous data recording at both stations, supervision and maintenance of equipment, examination of data and preparation of same for publication.

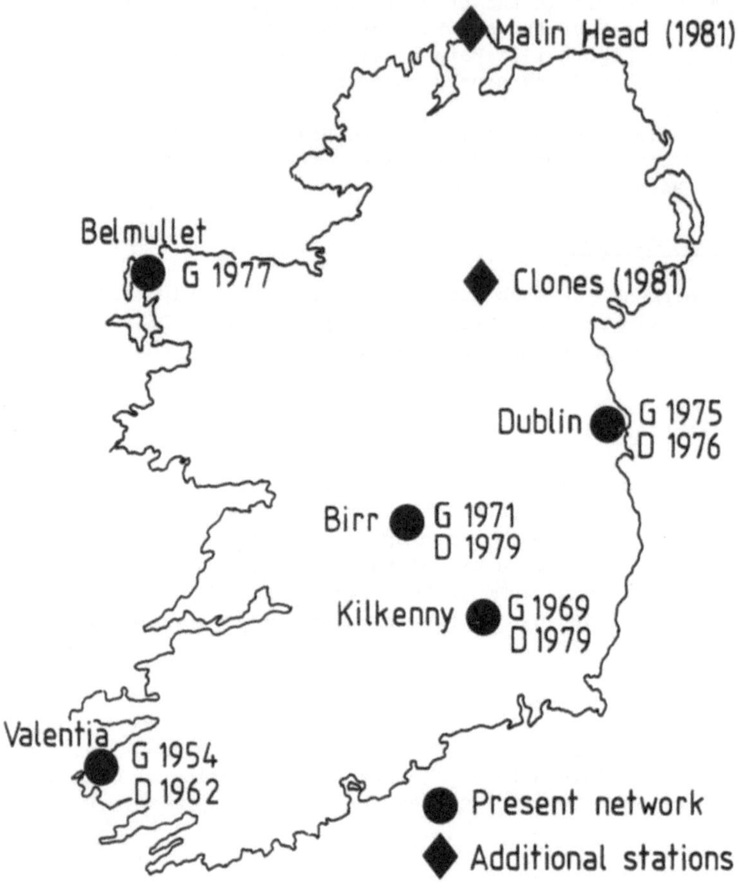

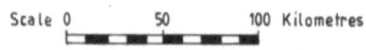

Malin Head (1981)

Belmullet
● G 1977

◆ Clones (1981)

Dublin ● G 1975
D 1976

Birr ● G 1971
D 1979

Kilkenny ● G 1969
D 1979

Valentia ● G 1954
D 1962

● Present network
◆ Additional stations

IRISH METEOROLOGICAL SERVICE
SOLAR RADIATION STATION NETWORK

Scale 0      50      100 Kilometres

Fig. 1

# INVESTIGATION OF THE ACCURACY OF SHADING RING CORRECTIONS APPLIED IN THE MEASUREMENT OF DIFFUSE SOLAR RADIATION

Author : S. McWilliams

Contract number : ES-F-019-EIR(N)

Duration : 24 months   1st January 1981 - 31st December 1982

Total budget : IR£11,450   CEC contribution   IR£5,725

Head of project : S. McWilliams

Contractor : S. McWilliams

Address : c/o Valentia Observatory
Cahirciveen, Co. Kerry,
Ireland.

## Summary

The Diffuse Sky Radiation as measured with a pyranometer and shading ring must be corrected to compensate for sky radiation cut off by the ring. This correction is generally computed on the basis of the geometric area of the sky cut off and assuming isotropic sky. This assumption results in values of $D_R$ which differ from the true value $D$ by a factor "K" = $D_c/D_r$.
The frequency distribution of k for the year 1979 was:

| k = | 0.9 | 1.0 | 1.1 | 1.2 | 1.3 | 1.4 | 1.5 |
|---|---|---|---|---|---|---|---|
| frequency % = | 2.4 | 54.7 | 32.0 | 8.7 | 1.7 | 0.4 | 0.1 |

There was no significant dependence of k on Solar Elevation and only a very slight dependence on Solar Declination ($\delta$).
The most significant variations of k are associated with $D_R/G$ and the following equation was found to fit the data closely:

$$k = 1.1431 - 0.1451(D_R/G)^3 - 0.0006826$$

When the observed data are adjusted by applying the factor k as derived from this formula 81.7% of the data are brought within $\pm$ 5% of the true value and only 4.5% have a residual error in excess of $\pm$ 10%.

# 1. Project description

## 1.1 Introduction

The measurement of Diffuse Sky Radiation is most frequently done by means of a pyranometer plus shading ring. The measured values must be corrected to compensate for the sky radiation cut off by the ring. This correction is generally computed on the basis of the geometric area of the sky cut off and assuming isotropic sky.

There is considerable doubt as to the accuracy of the shading ring correction based on isotropic sky.

The purpose of this project is to assess the error involved in this method of applying shading ring corrections and analyse it with respect to Solar Declination and Elevation, Duration of bright Sunshine, Cloud Amount and (D/G).

## 1.2 Procedure

At Valentia Observatory a pyrheliometer on solar tracker has been measuring the Direct Sun at normal incidence (I) since November 1978. The Global Solar Radiation on a horizontal surface (G) has also been measured.

The Diffuse Sky Radiation on the horizontal surface ($D_c$) will be computed from the relation

$$D_c = G - I SinE \dots\dots\dots\dots\dots\dots\dots\dots\dots\dots (1)$$

where E is the solar elevation.

The Diffuse radiation is also measured with shading ring. These measurements will be corrected for the geometric area of the sky cut off by the shading ring assuming isotropic conditions giving $D_R$.

$$\text{Thus } D_c = G - I SinE = kD_R \dots\dots\dots\dots\dots\dots (2)$$

where k is the factor required to reduce the data obtained by means of the solarimeter and shading ring to the true value.

The factor k will be computed for hourly values over a 2-year period and it will be analysed in relation to Solar Declination (D), Elevation (E) and $D_R/G$

# 2. Results obtained

The value of "k" has now been computed for each hour of the year 1979. Those hours when the solar elevation was below $10°$ were omitted from the analysis.

## 2.1 Frequency distribution

The overall mean value was found to be k = 1.053. The frequency distribution of k values for the full year was as follows:

| k = 0.9 | 1.0 | 1.1 | 1.2 | 1.3 | 1.4 | 1.5 |
|---------|-----|-----|-----|-----|-----|-----|
| Frequency % = 2.4 | 54.7 | 32.0 | 8.7 | 1.7 | 0.4 | 0.1 |

This means that on overall average the Diffuse radiation data as generally derived from the shading ring plus geometric correction based on isotropic sky are underestimated by 5.3%. About 42.9% of

the hourly values are underestimated by 10% or more and 10.9% of the
data are undervalued by 20% or more.

## 2.2 Variation with Solar Elevation (E)

The data were assembled according to Solar Elevation (E) for
ranges of $5^o$ and the mean value of k found for each range as
follows:

| $E^o$ | 10-14.9 | 15-19.9 | 20-24.9 | 25-29.9 | 30-34.9 |
|---|---|---|---|---|---|
| k | 1.053 | 1.043 | 1.048 | 1.050 | 1.057 |
| $E^o$ | 35-39.9 | 40-44.9 | 45-49.9 | 50-54.9 | 55-59.9 |
| k | 1.051 | 1.054 | 1.064 | 1.057 | 1.060 |

The results seem to indicate no significant dependence of k on
Solar Elevation.

## 2.3 Variation of k with Solar Declination (D)

The data were next assembled for $5^o$ ranges of Solar declination
centred at the values indicated in the table and mean values
obtained as follows:

| Declination$^o$ | $21.7^o$ | $17.5^o$ | $12.5^o$ | $7.5^o$ | $2.5^o$ |
|---|---|---|---|---|---|
| k | 1.053 | 1.041 | 1.065 | 1.070 | 1.062 |
| Declination$^o$ | $-2.5^o$ | $-7.5^o$ | $-12.5^o$ | $-17.5^o$ | $-21.7^o$ |
| k | 1.049 | 1.052 | 1.060 | 1.050 | 1.037 |

Since the overall mean value of k is 1.053 it would appear that the
contribution of Solar Declination to the variation is limited to
about $\pm$ 1.6%.

## 2.4 Variation of k with $D_R/G$

The values of $D_R/G$ were computed for each hour and the
corresponding values of k assembled in band widths of 0.1 for $D_R/G$
except that due to the smaller number of occasions with $D_R/G$ less
than 0.3 these values were grouped together in the 0.3 band width.
The results were as follows:

| D/G | 0.3 | 0.4 | 0.5 | 0.6 | 0.7 | 0.8 | 0.9 | 1.0(or greater) |
|---|---|---|---|---|---|---|---|---|
| k | 1.131 | 1.135 | 1.127 | 1.101 | 1.091 | 1.073 | 1.045 | 0.993 |

These values would indicate that the value of $D_R/G$ was the most
important factor in the variation of "k" values and shows a definite
trend of increasing k with decreasing values of $D_R/G$ while the
Diffuse radiation data as at present derived would be accurate when
the skies are overcast or completely clouded.
   In view of the importance of the value of $D_R/G$ in the
determination of the k values an effort was made to fit an equation
to represent the variation of k with $D_R/G$. The Declination (D) was
also included as a variable even though, as stated above, it's
contribution was small.
   The following equation was found to fit the data very closely:

$$k = 1.1431 - 0.1451(D_R/G)^3 - 0.000682\delta\ldots\ldots\ldots\ldots\ldots(1)$$

The fit of the equation to the actual observed values may be seen from the following table for the annual means (ignoring the Declination contribution which is negligible)

| $D_R/G$ | 0.3 | 0.4 | 0.5 | 0.6 | 0.7 | 0.8 | 0.9 | 1.0 |
|---|---|---|---|---|---|---|---|---|
| k from observed values | 1.131 | 1.135 | 1.127 | 1.101 | 1.091 | 1.073 | 1.045 | 0.993 |
| k computed from formula | 1.139 | 1.134 | 1.125 | 1.112 | 1.093 | 1.069 | 1.037 | 0.998 |

## 3. Conclusion

When the observed data are adjusted by applying the factor k as derived from the above equation to obtain $k_1$ (adjusted k value) the frequency distribution becomes:

| $k_1$ = | 0.8 | 0.9 | 1.0 | 1.1 | 1.2 | 1.3 |
|---|---|---|---|---|---|---|
| frequency % = | 0.4 | 11.2 | 81.7 | 6.0 | 0.5 | 0.2 |

The frequency distribution of k (for observed data) and $k_1$ (for adjusted data)are shown on Fig. 1.

It can be seen that by adjusting the data by means of equation (1) the accuracy is improved to the extent that 81.7% of the data are within +5% of the true value and only 4.5% have a residual error in excess of $\pm$ 10%.

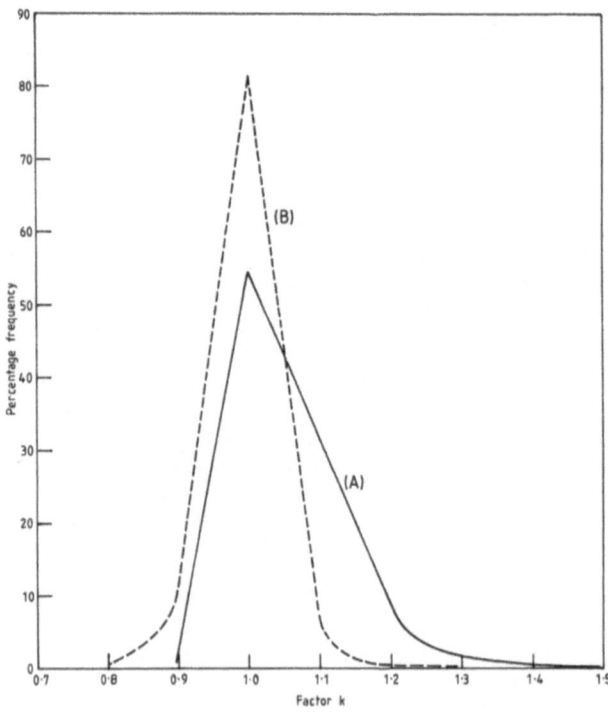

Fig. 1 - Frequency distribution of k for (A) original data and (B) data after adjustment by means of equation (1).

THE EXTENSION AND IMPROVEMENT OF THE U.K. RADIATION NETWORK

Author                  :   J. H. SEYMOUR

Contract Number   :   ESF—024—80—UK(H)     Contract 1
                          ESF—038—81—UK(H)         "      2

Duration                :   42 months 1/1/1980—30/6/1983   —1
                          24 months 1/7/1981—30/6/1983   —2

Total Budget         :   107 KECU   CEC Contribution:   47 KECU —1
                            58 KECU      "            "      :   29 KECU —2

Head of Project    :   (from September 1982) Dr G. J. Jenkins,
                          Meteorological Office, Bracknell.

Contractor            :   U.K. Secretary of State for Defence, represented by the
                          Secretary of the Meteorological Office, Mr F. R. Howell.

Address                 :   Secretary Meteorological Office
                          Meteorological Office
                          London Road
                          Bracknell
                          Berkshire   RG12 2SZ

Summary

    With CEC support the Meteorological Office radiation network has
been extended to improve the spatial distribution of stations within
the United Kingdom.  Additional observations of turbidity using
sunphotometers and of direct radiation have been included to increase
the range of measurements made.  An investigation into the shade ring
corrections required in the case of a non—isotropic distribution of
diffuse radiation has been held up by hardware problems but will be
commencing shortly.

# 1. Background

At the start of the four year period the Meteorological Office maintained eight stations at which routine measurements of solar radiation were made. Since then two have been lost — Kew Observatory and Cardington — but with support from the CEC the network has been extended to eleven stations, with three more due to open by the end of the contract period. All stations measure (or will measure) global (G) and diffuse (D) radiation on a horizontal surface, while additional measurements are made at certain stations (Table I, Figure I).

# 2. Proposals

2.1 It was proposed under Contract 1
(i)  to extend measurements of G and D to the four stations Stornoway, Shanwell, Hemsby and Camborne.
(ii)  to commence measurements of the direct component (I) at Eskdalemuir and to continue such measurements at Easthampstead but with an improved pyrheliometer.
(iii) to investigate possible improvements to the corrections made to D measurements to account for over-shading by the ring used.

2.2 It was proposed under Contract 2
(i)  to extend measurements of G and D to the three stations Aughton, Aviemore and Finningley.
(ii)  to commence turbidity measurements at Easthampstead and Eskdalemuir.

# 3. Extensions to the Network

3.1  The locations of the stations currently operating (at September 1982) and of stations which will be operational by the end of the contract period are shown in Table I and Figure I.

## TABLE I

### The U.K. Meteorological Office Radiation Network (9/82)

| Station | Lat. | Long. | G | D | I | V | L | B | T |
|---|---|---|---|---|---|---|---|---|---|
| Aberporth | 52.1 | 4.6 W | O | O | | | | | |
| Aldergrove | 54.7 | 6.2 W | O | O | | | | O | |
| Aughton | 53.5 | 2.9 W | O | O | | | | | |
| Aviemore | 57.2 | 3.8 W | P2 | P2 | | | | | |
| Camborne | 50.2 | 5.3 W | O | O | | | | | |
| Crawley | 51.1 | 0.2 W | O | O | | | | | |
| Easthampstead | 51.4 | 0.8 W | O | O | O | O | O | | P2 |
| Eskdalemuir | 55.3 | 3.2 W | O | O | O | | | O | O |
| Finningley | 53.5 | 1.0 W | P2 | P2 | | | | | |
| Hemsby | 52.9 | 1.7 E | O | O | | | | | |
| Lerwick | 60.1 | 1.2 W | O | O | O | O | O | O | |
| London W.C. | 51.5 | 0.1 W | O | O | | | | | |
| Shanwell | 56.4 | 2.9 W | O | O | | | | | |
| Stornoway | 58.2 | 6.3 W | P1 | P1 | | | | | |

Key O = operational  P1,2  proposed under contract 1, 2 but not yet operational.

G,D = Global, Diffuse radiation on a horizontal surface
  I = direct radiation   V = radiation falling on vertical surfaces facing
                               north, south, east and west.
  L = radiation falling on a south facing surface inclined at the latitude
      angle   B = net radiation   T = turbidity.

3.2   Of the planned contract 1 stations Stornoway and Camborne will
complete the chain of stations (Lerwick-Stornoway-Malin Head (Irish
proposal) - Valentia - Camborne - Rennes - La Rochelle) defining the
northwestern boundary of the CEC region and supplying hourly G and D
data.  Hemsby and Shanwell together with the contract 2 additions will
improve the spatial distribution of stations within the United
Kingdom.
Measurements of I at Eskdalemuir help to fill the latitude gap between
Lerwick and Easthampstead.

## 4.   Instrumentation

Kipp CM5 pyranometers are used for all G and D measurements, Eppley
pyrheliometers on sun-following mounts for I.  The D pyranometer is shaded
by a ring made to a standard Meteorological Office pattern (radius:
width = 0.2) and geometric corrections are applied to the measured data on
the assumption of the isotropic distribution of diffuse radiation.  Data
are recorded on magnetic tape at intervals of one minute and these data, as
well as the derived hourly mean irradiations are archived at Bracknell.
Instruments are calibrated at Easthampstead and calibrations are traceable
to the World Radiometric Reference (WRR) through the TMI cavity radiometer
of the National Radiation Centre.
Sunphotometers made by Eko Instruments (model MS120) are being used for
turbidity measurements under Contract 2.

## 5.   Progress Made

### 5.1   Contract 1 - Network Extension
(i)      Measurements of G and D commenced at Hemsby in April 1981,
at Camborne in September 1981 and at Shanwell in December 1981.
(ii)     Work on renovating the building to house the equipment
at Stornoway is complete and it is planned that the station will
commence measurements in October 1982.
(iii)    Measurements of I are now being made as routine at
Eskdalemuir and Easthampstead using Eppley NIPs.
### 5.2   Contract 2
(i)      Measurements of G and D commenced at Aughton in November
1981.
(ii)     Preparatory work has been completed at Finningley and it
is planned to install equipment there by the end of 1982.
(iii)    A synoptic observing station is being established at
Aviemore.  It is hoped that the installation of radiation equipment will occur during the Spring of 1983.
(iv)     After an initial electronics problem was overcome
turbidity measurements with the sunphotometer started at
Eskdalemuir in December 1981.  Since then atmospheric conditions
have resulted in a disappointingly low number of observations.
A second instrument is due for delivery shortly to be used at
Easthampstead.

## 6. Improvements to Shade Ring Corrections (Contract 1)

### 6.1 Current Practice
At present the "Kew" pattern shading ring is used at all Meteorological Office stations for D measurements which are then corrected for overshading by invoking the normal assumption of an isotropic distribution for diffuse radiation. Several workers, eg (1), have demonstrated the need for further corrections in the non-isotropic case which is more normal in the U.K. Accurate values of D are required when the direct radiation I is to be deduced from G and D as, for example, when estimating radiation on inclined surfaces.

### 6.2 Proposal
It was planned to develop an occulting disc mechanism and to make continuous measurements of diffuse radiation $D_d$ which could then be related to those made with the normal shading ring $D_r$.

### 6.3 Progress
A device was made but found to be unsatisfactory in performance. Plans for modification were produced but the cost of carrying them out or of rebuilding the equipment was such that in the present financial climate it was decided not to proceed with either. Instead earlier measurements made by Painter (1), using equipment normally used for the calibration of standard pyranometers, will be repeated but this time at two locations – Easthampstead and Eskdalemuir – and with improved data recording facilities. Additional disc mechanisms have now been built but problems were encountered in locating suitable drive motors. These are due for delivery shortly and measurements will commence in Autumn 1982. The Easthampstead results will be examined for consistency with the earlier data and then will be compared with the results from Eskdalemuir to see if there is a significant latitude dependence associated with the corrections required.

References

(1) H. E. Painter, 1981 The Shade Ring Correction for Diffuse Irradiance Measurements. Solar Energy, Vol 26 pp 361-363, 1981.

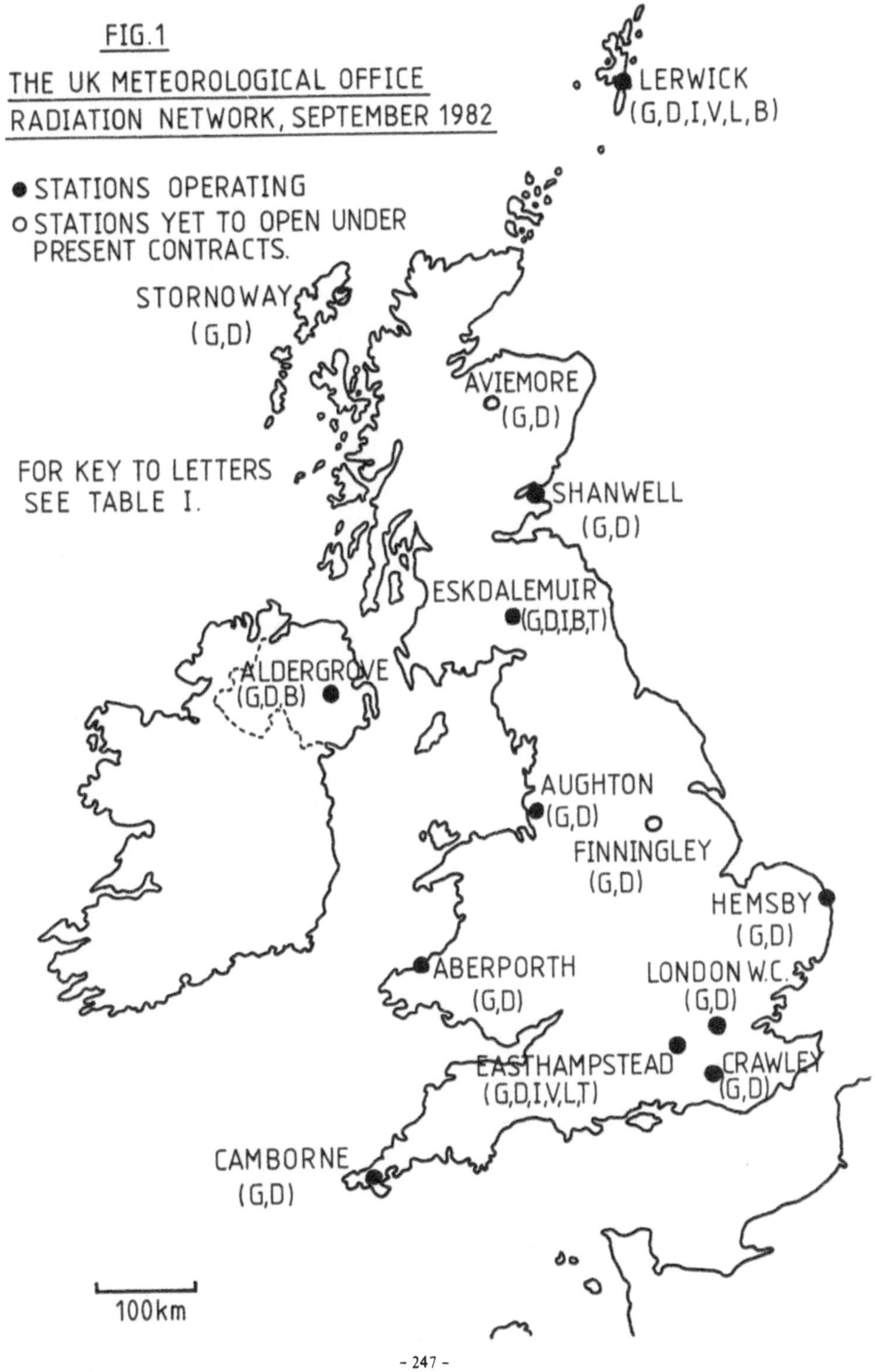

FIG.1

THE UK METEOROLOGICAL OFFICE
RADIATION NETWORK, SEPTEMBER 1982

● STATIONS OPERATING
○ STATIONS YET TO OPEN UNDER
  PRESENT CONTRACTS.

LERWICK
(G,D,I,V,L,B)

STORNOWAY
(G,D)

AVIEMORE
(G,D)

FOR KEY TO LETTERS
SEE TABLE I.

SHANWELL
(G,D)

ESKDALEMUIR
(G,D,I,B,T)

ALDERGROVE
(G,D,B)

AUGHTON
(G,D)

FINNINGLEY
(G,D)

HEMSBY
(G,D)

ABERPORTH
(G,D)

LONDON W.C.
(G,D)

EASTHAMPSTEAD
(G,D,I,V,L,T)

CRAWLEY
(G,D)

CAMBORNE
(G,D)

100km

ACTION 4.2

SPECIAL MEASUREMENTS

Action leaders's progress report
    J.L. PLAZY, Agence française pour la maîtrise de
    l'énergie, Sophia Antipolis, Valbonne, France

Reports of action participants :

  - Study of I.R. radiation from the sky

  - The flux density of radiation energy originating from the
    circumsolar sky measured at groundbased stations

  - Special measurements of radiation in high altitude station

  - Installation of equipment to measure the separate com-
    ponents of global solar radiation on a vertical surface

## ACTION 4.2 - SPECIAL MEASUREMENTS

Action budget (CEC contribution)  :   138,2 kUC

Action Leader  :   J.-L. PLAZY
                   Agence Française pour la Maîtrise
                   de l'Energie
                   SOPHIA-ANTIPOLIS
                   F-06565 VALBONNE


Participants  :

- Direction de la Météorologie - PARIS
  Contract Nr  ESF-005-F
  J.A. BEDEL

- Institut für Meteorologie
  Johannes Gutenberg Universität - MAINZ
  Contract Nr ESF-015-D
  R. EIDEN

- Laboratoire d'Energétique Solaire - CNRS - ODEILLO
  Contract Nr ESF-009-F
  J.-F. TRICAUD

- Institut Royal Météorologique de Belgique - UCCLE
  Contract Nr ESF-003-B
  R. DOGNIAUX

- Irish Meteorological Service - DUBLIN
  Contract Nr   ESF-018 E
  E.J. MURPHY


Task :

Development of instrumentation for specific needs in
solar energy.

Atmospheric turbidity, solar aureole, spectral and long
wave radiation, distribution of the luminance.

SUMMARY

The action 4.2 is related to special measurements for the
solar energy application. It covers as well specific instrumentation
development like a solar aureole radiance measuring device, a solar
spectrum distribution selective pyranometer, a reflected radiation
pyranometer, than statistic and climatologic studies on the
atmospheric turbidity and infrared long wave radiation.

In the state of development of the action 4.2 some results
can be considered as reached, mainly :

. one year of complete spectral measurements of global solar radiation
  was performed at Uccle for height wavelenghts ;

. complementation of the radiation network special measurements in
  Valentia and Odeillo for reflected radiation ;

. more than 200 instantaneous measurements of the circumsolar sky
  irradiance in three locations : Mainz (FRG), Deuselbach (FRG)
  and Adrano (Sicily).

. a statistical formula to calculate the long wave atmospheric
  radiation (the equivalent sky temperature) wathever the
  meteorological conditions on a daily and hourly basis.

The work to be done before the end of the programme will be to
appreciate the winning and losses of circumsolar sky radiance
according to the turbidity variation and the cloudy conditions
to check the efficiency of the focusing devices, and to make a
climatological study of the atmospheric turbidity using the
available data of Trappes and Carpentras where measurements are
regularly performed.

1 - GENERAL TASK OF THE ACTION 4.2

The solar energy applications need not only solar radiation
data relating to the available energy but other informations
about the quality of this energy and about various other parameters
which interface with the solar devices functioning. This is the
reason why the C.E.C solar R & D programme includes an action
for developing new specific instruments and methods in view
to obtain and analyse the required complementary data.

Among the various possible subjects of interest the
1979-1982 programme approachs the following aspects :

. Circumsolar sky radiation measurements. This point is treated
  by the Johannes Gutenberg University of Mainz, to obtain a
  better knowledge of the radiant energy flux provided by the
  circumsolar sky and to allow better calculation of the efficiency
  of focusing devices such as the European solar electrical plant
  of Adrano located in Sicily ;

. Spectral single wavelenghts measurements of the global solar
radiation reaching the ground. The Institut Royal Météorologique
de Belgique owns an equipment which works at eight wavelenghts
allowing to reconstitute the spectral distribution of the
global radiation ;

. Special measurements of reflected radiation. The Irish
Meteorological Service and the Centre National de la Recherche
Scientifique planed to install pyranometer to measure respectively
in Valentia and Odeillo the radiation reflected by the ground
in view to allow the improvement of calculation methods giving
the incoming radiation on tilted plane.

. Statistical analysis of long wave radiation and atmospheric
turbidity collected data in Trappes and Carpentras. The Direction
de la Météorologie have to propose calculation formula to deduce
hourly and daily sums of the infra-red atmospheric radiation
and to analyse the temporal atmospheric turbidity factor in
these two locations.

As indicated in the strategy paper of the R & D programme reports
should be provided by the contractors on infra-red radiation,
turbidity, spectral and circumsolar radiation. A comprehensive
report over the whole research made in this action will be
prepared by the action leader at the end of the period.

## 2 - STATE OF THE WORK

### 2.1 - Spectral distribution of the global solar radiation

The first step of the I.R.M.B. was to adapt proper
methods of calibration to spectral measurements. Two methods
were selected :

. the first method was based on the use of an integrating
sphere in which the spectral flux of the penetrating
radiation was measured with a reference cell and compared
with the signal delivered by the spectropyranometers. This
method needs very stable irradiance conditions.

. the second method is based on the comparison between the
spectropyranometers and a monochromotor.

The spectropyranometers operate at eight wavelenghts included
in the range 300 to 1000 nm. These wavelenghts are :

315 - 400 - 446 - 545 - 646 - 730 - 816 and 914 nm.

A complete year of measurements (1980) at UCCLE was practiced.
The data are now published by the IRMB. Misc. Série B Nr 52and
53 under the title "Distribution Spectrale du Rayonnement
Solaire à UCCLE". Exemples of such data are given in volume 1
Serie F. Solar Radiation Data published by D. REIDEL
PUBLISHING COMPANY for the Commission of the European
Communities.

## 2.2 - Circumsolar sky radiation

In order to determine the order of magnitude of the circumsolar irradiance according to the absorption and scattering
process caused by the atmospheric molecules, aerosols and
cloud particles the Johannes Gutenberg Universität of Mainz
has developped a special photometer which allows the sky
circumsolar irradiance measurement from 1 to 10 degrees
distance from the center of the sun.

A silicon photovoltaic cell (UV 444 B, EG and G) is used
as detector, it is associated with a flat filter which give
it a quite (± 4 %) flat sensitivity response according to
the wavelenght within the range 460 to 975 nm. The photometer
is calibrated in absolute radiometry.

In parallel with the circumsolar irradiance measurements the
Linke Turbidity factor was measured with an other photometer
which uses too a photovoltaic sensor. The turbidity factor
is calculated over the range 400-1030 nm.

About 200 runs  of measurements have been practiced in three
locations :

MAINZ (FRG)        polluted atmosphere

DEUSELBACH (FRG)   clean continental atmosphere

ADRANO (Sicily)    clean maritime atmosphere

As far as the statistical analysis allows it on a such sample
some concurrents can be proposed :

On clear sky conditions the highest values of circumsolar
irradiance are found in the polluted area and the lowest in
the maritime one.

In general the circumsolar irradiance does not exeed 10 % of
the direct solar radiation in a disc of 10° solar distance
around the sun exepted in highly polluted atmosphere.

For the same Linke turbidity factor the circumsolar irradiance
is higher in cloudy conditions than in cloudless one as
shown in the following exemple :

Linke turbidity factor   :        4,65

optical air mass         :        1,7

Direct normal irradiance :   410   W  $m^{-2}$

Circumsolar irradiance

    cloudy case        :     19   W  $m^{-2}$
    cloudness case     :     1,5 W  $m^{-2}$

Aperture angle           :        10°

The table I obtained from various measurements gives an idea
of the circumsolar irradiance level according to the turbidity
factor value and the cloud presence or missing. It appears
then for the same turbidity factor the circumsolar irradiance
is always higher when clouds are present. The ratio between
the circumsolar irradiance at an angular distance of 5° and
of 10° is lower on cloudy conditions than in a cloudless
atmosphere.

TABLE I

| Turbidity factor | Optimal mass | Direct solar irradiance $W. m^{-2}$ | Weather conditions C = CLOUDY NC = CLOUDYLESS | Circumsolar irradiance $W. m^{-2}$ $\varphi = 5°$ | $\varphi = 10°$ |
|---|---|---|---|---|---|
| 6.2 - 6.3 | 1.65 | 330 | C | 85 | 100 |
|  |  |  | NC | 8 | 23 |
| 5.1 - 5.3 | 1.8 | 355 | C | 28 | 42 |
|  |  |  | NC | 8 | 21 |
| 4.6 - 4.7 | 1.7 | 410 | C | 20 | 35 |
|  |  |  | NC | 6 | 17 |
| 3.1 - 3.2 | 1.6 | 533 | C | 12 | 22 |
|  |  |  | NC | 3.5 | 11 |
| 2.2 - 2.5 | 1.4 | 615 | C | 5.8 | 13 |
|  |  |  | NC | 3.2 | 8.5 |

However the obtained data do not allow to establish a
relationship between the circumsolar irradiance and the
angular distance for various values of Turbidity factor. This
shows the necessity of more investigation in such measurements
results.

2.3 - Reflected radiation measurements

The Centre National de la Recherche Scientifique (CNRS) has
installed in Odeillo equipments for measuring the reflected
radiation on a vertical plane and a pyrheliometer to measure
both the direct radiation and the time duration during the
direct radiation is above the level of 110, 200, 500, 600
and 800 W. $m^{-2}$. At the present time this equipment is operating
and the data will be available on hourly and daily basis
according to the format used by the Direction de la Météorologie.

The Irish Meteorological Service has purchased and installed in Valentia Observatory equipments to measure the solar radiation on a vertical south-facing surface. The data collection on hourly basis begun service May 1981.

Table II gives an exemple of such measurements and a comparison between calculated and measured data.

Under overcast sky conditions the measured and calculated values are closed ones to the others, during sunny periods an asymetry can appear probably due to local factors. This statement shows the necessity to include in the calculation models entrance data taking in account the local conditions and not only the albedo values.

|  | 5 to 6 | 6 to 7 | 7 to 8 | 8 to 9 | 9 to 10 | 10 to 11 | 11 to 12 | 12 to 13 | 13 to 14 | 14 to 15 | 15 to 16 | 16 to 17 | 17 to 18 | 18 to 19 |
|---|---|---|---|---|---|---|---|---|---|---|---|---|---|---|
| **Ist JUNE 1981 – OVERCAST DAY** | | | | | | | | | | | | | | |
| Measured values | 3 | 7 | 9 | 4 | 27 | 23 | 20 | 11 | 5 | 4 | 4 | 2 | 4 | |
| Theoretical values | 3 | 6 | 10 | 6 | 18 | 21 | 20 | 10 | 6 | 4 | 4 | 3 | 3 | |
| **21st JUNE 1981 SUNNY DAY** | | | | | | | | | | | | | | |
| Measured values | 9 | 16 | 30 | 40 | 43 | 41 | 35 | 28 | 21 | 13 | 8 | 5 | 14 | 10 |
| Theorical values | 7 | 13 | 20 | 23 | 27 | 30 | 31 | 31 | 31 | 27 | 25 | 20 | 17 | 9 |
| **18th DECEMBER 1981 – SUNNY DAY** | | | | | | | | | | | | | | |
| Measured values | | | | 1 | 16 | 22 | 16 | 10 | 3 | 1 | 1 | | | |
| Theorical values | | | | 1 | 3 | 8 | 10 | 10 | 9 | 4 | 1 | | | |

### TABLE II

Exemples of measured and calculated values of ground relected solar irradiation on a vertical surface in Valentia Observatory.

### 2.4 – Atmospheric longwave radiation

This parameter is measured continuously in France in two locations : Trappes since 1979 and Carpentras since 1976. The research programme managed by the Direction de la Météorologie was conducted towards threes axes :

### 2.4.1 – a statistical study

For Carpentras the climatological parameters obtained are :

hourly mean value       :       $0,325$  kW.h. $m^{-2}$
mean sky temperature :       $+ 2°C$
hourly standard deviation : $0,043$ kW.h.$m^{-2}$
sky temperature deviation : $\pm$ $9°$
maximum  hourly value :  $0,453$ kW.h.$m^{-2}$ (+ 25°8 C)
minimum hourly value :   $0,203$ kW.h.$m^{-2}$ (– 28°6 C)

The study of the atmospheric radiation variations during the year conduces to separate the year in two periods a stable flat radiation period from November to April during which the variation does not exceed 5 %.

An important variation period from May to October during which the radiation increase from 300 (May) to 360 kWh m$^{-2}$ (August) and decrease until 300 (October).

The regression formula between the atmospheric radiation and the screen temperature under clear conditions are :

R = 0.259 + 0.0049 T for daylight period
and
R = 0.235 + 0.0049 T for nightime period

T expressed in Celsius degrees and R in kW.h m$^{-2}$
The difference between day and nightime can be explained by three hypothesis : instrumental error, different vertical shape of the temperature during night and daytime, different water vapour concentration.

2.4.2 - a comparative study of the various formula used to calculate the atmospheric radiation when direct measurements are missing.

Heighteen formula have been checked on hourly and daily values. The conclusions of this study recommand the use of the following formula for the sky temperature :

Hourly values - Clear sky conditions - Daytime

$$T_s = T_a (0.741 + 0.0062 T_w)^{0.25} \qquad \text{from Berdahl}$$

Hourly values - Clear sky conditions - Nightime

$$T_s = T_a (0.741 + 0.0062 T_w)^{0.25} + 4.5$$

Hourly values - Cloudy sky - Daytime

$$T_s = T_a \left[ (1 - 0.056 \, n)(0.741 + 0.0062 T_w) + 0.056 \, n \right]^{0.25}$$
from Clarke

Hourly values - cloudy sky - Nightime

$$T_s = T_a \left[ (1 - 0.056 \, n)(0.741 + 0.0062 T_w) + 0.056 \, n \right]^{0.25} + 4.5$$

Daily values

$$T_s = T_a \left[ (1 - 0.056 \, n)(0.787 + 0.0028 T_w) + 0.056 \, n \right]^{0.25}$$

Where

$T_s$ = equivalent sky temperature

$T_a$ = air temperature (under a screen). For hourly values
  $T_a$ is the hourly air temperature, for daily values
  $T_a$ is the mean temperature of the day defined as the maximum
    plus the minimum air temperatures divided by two.
$T_w$ = dew point temperature
n  = cloudiness in tenths of the sky dome

Those formula give good results as well in Carpentras than in Trappes.

### 2.4.3 - sky temperature maps for France

Using the previous formula and the climatological data (Temperature, dew point and cloudiness) of the French meteorological station it was possible to establish the map of the sky temperature over the French territory.

This maps gives the values of the sky temperature for each month and for the whole year under clear sky conditions, under fully cloudy conditions and under natural observed conditions.

Figure 1 gives the yearly mean sky temperature under natural conditions.

## 2.5 - Atmospheric turbidity

the statistical study of the atmospheric turbidity data planned by the Direction de la Météorologie will be performed during the next months. At that time the data are on magnetic tapes and preliminary works have just begun. It is too soon at the present time to do some comments on this work.

## 3 - CONCLUSIONS

The most important part of the action 4.2 programme is now accomplished.

Interesting results have been outlined in the paragraph 2 of this paper. Those results will be used by the designers who have to take in account not only the energetical aspect of the solar devices but the surrounding conditions in which they operate and which can modified the efficiency ot the systems refered to laboratory experiments.

The action 4.2 studies can be applied for concentrating devices (circumsolar radiation), biomass and photovoltaic application (spectral measurement), low temperature applications (sky temperature, reflected radiation).

- - - -  Mountain areas

Figure 1 - Yearly mean sky temperature
in France

# STUDY OF I.R. RADIATION FROM THE SKY

CONTRACT NUMBER : ESF - 005 F

DURATION : 3 years (from 01/1980 to 06/1983)

TOTAL BUDGET : 662 098 FF - CEC contribution : 50%
(in national currency)

HEAD OF PROJECT : J.A. BEDEL

CONTRACTING ORGANIZATION : Direction de la Météorologie
ORGANIZATION                   73-77, rue de Sèvres
                                     F-92106 BOULOGNE

## SUMMARY

The first point consists in a statistical study of the atmospherical radiation data from Carpentras. The daily and monthly variations the good correlation between atmospheric radiation and other meteorological parameters are presented.

In the second point, after a bibliographical study, the different estimation formulae of the atmospheric radiation have been tested and compared, and we have selected the best formulae at the hourly and daily level for Carpentras and Trappes.

## I. PURPOSE AND REMINDERS

Every atmospheric volume - considered as a natural body, emits a long wave, radiation, in all directions. Through successive radiative transfers, a part of the radiated energy passes directly out to space and the remainder reaches the ground. This is the atmospheric radiation.

This radiation is permanent and plays a non negligible part in the natural energy exchanges at ground level. It is, in particular, an important element for assessing the achievements of solar energy sensors and more especially of passive systems, even though this radiation is involved for wavelengths more than 4 Nm.

This study aims at :
- assessing the significance of the atmospheric radiation, from the standpoint of energy, by means of a statistical analysis of a series of measurements made at Carpentras.

- reviewing, through a bibliographical research, the models, used for calculating atmospheric radiation.

- comparing the models to the series of the Carpentras data and also to a series of data obtained at Trappes.

It should be borne in mind that it is customary to modelize the atmosphere as a black body with a uniform temperature. To obtain the atmospheric radiation from the temperature referred to as the emittance temperature of the atmosphere, Stephan"s Law is applied :

$$R = \sigma T_c^{4}$$

R : atmospheric radiation in $W/m^2$

$T_c$: emittance temperature of the atmosphere in °K

$\sigma$: Stephan's Constant taken as $5.67 \ 10^{-8} \ W \ m^{-2} \ K^{-4}$

## II. DATA

The Meteorological Office carries out continuous, routine hourly measurements of total radiation at the Carpentras station. The measurements are carried out by means of a compensated pyrradiometer, developed at the Centre Technique du Matériel. Measurements of global solar radiation are also available.

To obtain the data concerning the incoming atmospheric radiation at the ground, we simply have to calculate the difference between the total radiation data and the global solar radiation data.

Since 1979 measurements of total radiation and global solar radiation have been taken by an automatic station at Trappes.

## III DATA ANALYSIS

The mean flux of the atmospheric radiation at Carpentras over the period 1976-1980 is 300 $Wm^{-2}$, i.e. on an average an energy of 2800 $J/cm^2$ /day (7.7 $kWh/m^2$/day). The daily average of global solar radiation at Carpentras - which is a site particularly suitable for solar energy - is 1500 $J/cm^2$/day (4 $kWh/m^2$/day).

A statistical analysis of the data has been performed and has put forward the following:

3.1) The monthly variation is important, more than 30 % (Fig. I, table I)

3.2) The daily variation is weak (Fig. II), about 10 %.

3.3) When there are clouds the incoming radiation at the ground is enchanced. For a similar screen temperature, the atmospheric radiation is an increasing function of cloudines (Fig. III). Besides, the lower the main cloud layer, more important is the incoming radiation at the ground. The average deviation between incoming radiation when the main cloud layer is high (more than 7000 m) and incoming radiation when the main cloud layer is low (less than 2400 m) is 44 W/m$^2$.

3.4) Under clear skies, there exists a very strong correlation between atmospheric radiation and screen temperature (Fig. VI). On an hourly basis, it is important to distinguish between night radiation and daylight radiation, this can be accounted for by a difference in vertical profiles of air temperature and water content, between night and daylight (Fig V). (Interpretations of Paltridge - Platt and Berdahl - Fromberg).

3.5) As was shown by Brunt in 1933, there exists a very good correlation between emittance (defined as the ratio of radiation emitted by the atmosphere to black-body radiation at screen temperature) and water content of the atmosphere, characterized either by water-vapour pressure or by dew-point temperature. The relation between emittance and the square root of the water-vapour pressure is almost linear (Fig. VI), as was suggested by Brunt.

## IV. ESTIMATION FORMULAE

There are various estimation formulae for the emittance temperature of the atmosphere. A bibliographical study enabled us to show that they are generally in the form of an estimation formula for clear skies, to which a correction formula is added, for cloudy skies. The correction formula is generally a function of cloudiness and of the value that would have been found for clear skies under the same conditions of temperature and moisture.

Table II shows the main estimation formulae of the emittance temperature of atmosphere for clear skies. It specifies the origin of the data utilized for their implementation and the date when the formula was established.

Table III gives the main correction formulae for cloudy skies.

## V. COMPARISON BETWEEN MODELS

The models in Tables II and III have been compared by means of the Carpentras data and the additional measurements taken at Trappes. We first sought to determine the best formula for clear skies, for Trappes and Carpentras on an hourly basis and on a daily basis. Then, from the best formula for clear skies, the different correction formulae have been tested. The comparisons concerned the computed and observed values of the emittance temperature of the sky.

### 5.1) Computation models for clear skies

On a daily basis, it seems the Clarke's formula No 2 yields the best results. For Carpentras, 421 clear sky days have been selected over the period 1976-1980. For these 421 days, the average of deviations between the observed emittance temperature and the computed emittance température is 0.16°K, and the standard deviation is 2.40°K, for Clarke's formula No 2. Goss and Brooks' formula also gives a fairly good result.

On an hourly basis, it is necessary to distinguish between night values and daylight values.

For daylight values Berdahl and Fromberg's formula No 1 yields good
results and so does Goss and Brooks' formula. To obtain night values,
it is suggested to add 4.5° to the daylight values. For the 862 hourly
observations made during clear skies, the average of deviations between
observed and calculated emittance temperatures is + 0.02° K and the
standard deviation is 3.69°K for Berdahl's formula No 1.

### 5.2) Correction formulae for cloudy skies

Among the seven correction formulae that have been kept (Table III),
the best one is Clarke's No 1.

On a daily basis, the average of deviations is 0.8°K and the standard
deviation is 2.88°K for the 1745 days kept for the purpose of comparison.

On an hourly basis, for 6983 observations, the average of deviations is
1.3°K and the standard deviation is 4.23°K.

### 5.3) Summary

For the calculation of emittance temperature it is suggested to retain
the following formulae.

On an hourly basis

daylight : $T_c = T_a [(1.0.056n) (0.741 + 0.0062 \ td) + 0.056n]^{0.25}$

night :    $T_c = T_a [(1.0.056n) (0.741 + 0.0062 \ td) + 0.056n]^{0.25} + 4.5$

On a daily basis

$$T_c = T_a [(1.0.056n) (0.787 + 0.0028 \ td) + 0.056n]^{0.25}$$

$T_c$ : emittance temperature of the atmosphere in °K

$T_a$ : screen - temperature in °K

td : dew-point temperature in °C

n  : cloudiness in tenths.

METEOROLOGIE (FRANCE) - SMM/CLIM/DEV      DOCUMENT ETABLI D APRES L ETAT DU FICHIER AU 92/05/14.

ALTITUDE : METRES
LATITUDE : 44.05 NORD (DEG.MIN)
LONGITUDE : 5.03 EST (DEG.MIN)

CARPENTRAS

COMPOSANTE VERTICALE DESCENDANTE DU RAYONNEMENT ATMOSPHERIQUE (INFRAROUGE)

DISTRIBUTION STATISTIQUE AU NIVEAU MENSUEL
PERIODE DU 1/1/1976 AU 31/12/1980

1.- DISTRIBUTION STATISTIQUE DES IRRADIATIONS QUOTIDIENNES - UNITE : JOULE/CM2/JOUR

| | JANV | FEVR | MARS | AVRI | MAI | JUIN | JUIL | AOUT | SEPT | OCTO | NOVE | DECE | ANNEE |
|---|---|---|---|---|---|---|---|---|---|---|---|---|---|
| NOMB. D OBS. | 152 | 140 | 146 | 140 | 142 | 140 | 145 | 149 | 145 | 146 | 146 | 153 | 1746 |

2.- FREQUENCES POUR MILLE DES CAS OU L IRRADIATION QUOTIDIENNE A DEPASSE LES SEUILS INDIQUES (EN JOULE/CM2/JOUR)

3.- DUREE MOYENNE PAR JOUR (EN HEURES X 10) DURANT LAQUELLE L ECLAIREMENT MOYEN HORAIRE A DEPASSE LES SEUILS INDIQUES (EN WATTS/M2)

## TABLE I

TABLE II - SELECTED ESTIMATION FORMULAE UNDER CLEAR SKIES

| Autor | Origine-Date | | Formula |
|---|---|---|---|
| Angstrom | USA | 1915 | $Tc=Ta (0.79 - 0.26 \exp(-0.052\ e))^{0.25}$ |
| Raiz | Allemagne | 1949 | $Tc=Ta( 0.82 - 0.25 \exp(-0.095\ e))^{0.25}$ |
| Chumakova I | Karadag | 1947 | $Tc=Ta( 0.90 - 0.18 \exp(-0.053\ e))^{0.25}$ |
| Brunt | Angleterre | 1932 | $Tc=Ta( 0.52 + 0.065\sqrt{e})^{0.25}$ |
| Goss & Brooks | USA | 1956 | $Tc=Ta( 0.66 + 0.039\sqrt{e})^{0.25}$ |
| Chumakova II | Karadag | 1947 | $Tc=Ta( 0.62 + 0.037\sqrt{e})^{0.25}$ |
| Boutaric | France | 1923 | $Tc=Ta( 0.50 + 0.042\sqrt{e})^{0.25}$ |
| Picha | USA | 1962 | $Tc=Ta( 0.50 + 0.062\sqrt{e})^{0.25}$ |
| Dogniaux | Belgique | 1980 | $Tc=\left(\frac{1}{\sigma}\sigma Ta^{4}(0.6+0.056\sqrt{e})+(\sigma Ta^{4}(0.135\ 0.128\sqrt{e})+43\sqrt{H})(1-\Sigma)\right)^{0.25}$ |
| Perrin de B | France | 1966 | $Tc=Ta - 35.8\ e^{-0.23}$ |
| Berdahl I | USA | 1979 | $Tc=Ta(0.741+ 0.0062\ td)^{0.25}$ |
| Berdahl II | USA | 1979 | $Tc=Ta(0.727+0.0060td)^{0.25}$ |
| Bliss | USA | 1966 | $Tc=Ta(0.8004 + 0.0039\ td)^{0.25}$ |
| Clarke I | USA | 1977 | $Tc=Ta(0.787 + 0.764\ Log(Td/273))^{0.25}$ |
| Clarke II | USA | 1977 | $Tc=Ta(0.787 + 0.0028\ td)^{0.25}$ |
| Swinbank | USA | 1963 | $Tc=Ta(9.366\ E\text{-}6\ Ta^{2})^{0.25}$ |
| Idso | USA | 1969 | $Tc=Ta( I-0.261\ \exp(-7.77\ E\text{-}4(273.16-Ta^{2})))^{0.25}$ |
| Monteith | Angleterre | 197? | $Tc= [(1/\sigma)(213+5.5\ Ta)]^{.25}$ |

Tc : emittance temperature of the atmosphere in °K
Ta : screen temperature in °K
e : water vapour pressure in mb
td : dew-point temperature in °C
Td : dew-point temperature in °K
$\Sigma$ : insolation fraction

TABLE III - CORRECTION FORMULAE

They concern the emittance of the atmosphere (the emittance temperature is obtained by applying the Stefan's law $Tc = Ta^{0.25}$), except for Mr Perrin's which directly concerns the radiation received at earth's surface.

| Autor | Formula |
|---|---|
| Chumakova | $\varepsilon_n=( I -0.0224\ n-0.004n^{2})\varepsilon_0 +0.024n +0.004n^{2}$ |
| Brunt | $\varepsilon_n= ( I - 0.09\ n)\varepsilon_0 +0.09\ n$ |
| Goss & Brooks | $\varepsilon_n= ( I - 0.07\ n)\varepsilon_0 +0.07\ n$ |
| Monteith | $\varepsilon_n= ( I -0.084n)\varepsilon_0 +0.084\ n$ |
| Clarke I | $\varepsilon'_n=( I - 0.056\ n)\varepsilon_0 +0.056\ n$ |
| Clarke II | $\varepsilon_n=(I-0.0224n+0.0035n^{2}-0.00028n^{3})\varepsilon_0 +0.0224n-0.0035n^{2}-0.00028n^{3}$ |
| Perrin de Brichambaut | $R_n= R_0 +70\ n$ |

$\varepsilon_n$ : emittance with a cloudiness if n tenths
$\varepsilon_0$ : emittance under clear sky
n: cloudiness in tenths
$R_n$ : atmospheric radiation with a cloudiness of n tenths in W/m²
$R_0$ : atmospheric radiation under clear sky in W/m²

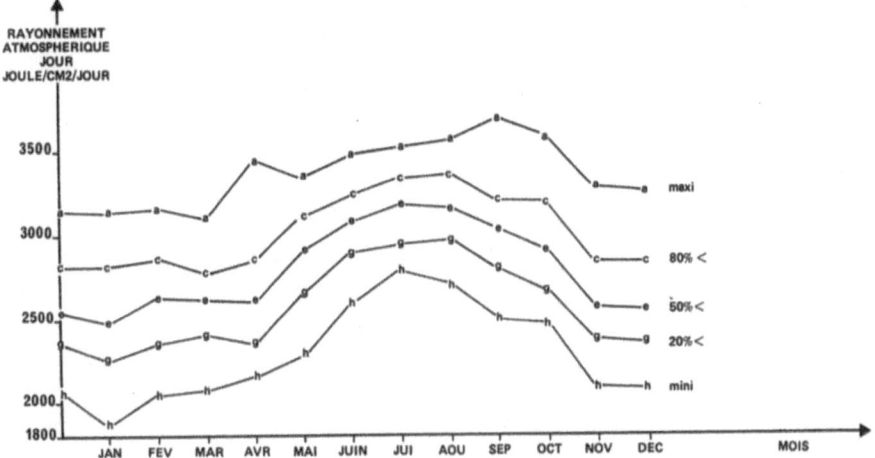

FIGURE I

Carpentras - Period 1976-1980

Annual variation of daily atmospheric radiation
The curve  a  gives, for each month, the maximum observed,
the curve  c  the superior quintain,
the curve  e  the medien,
the curve  g  the inferior quintain,
the curve  h  the minimum observed.

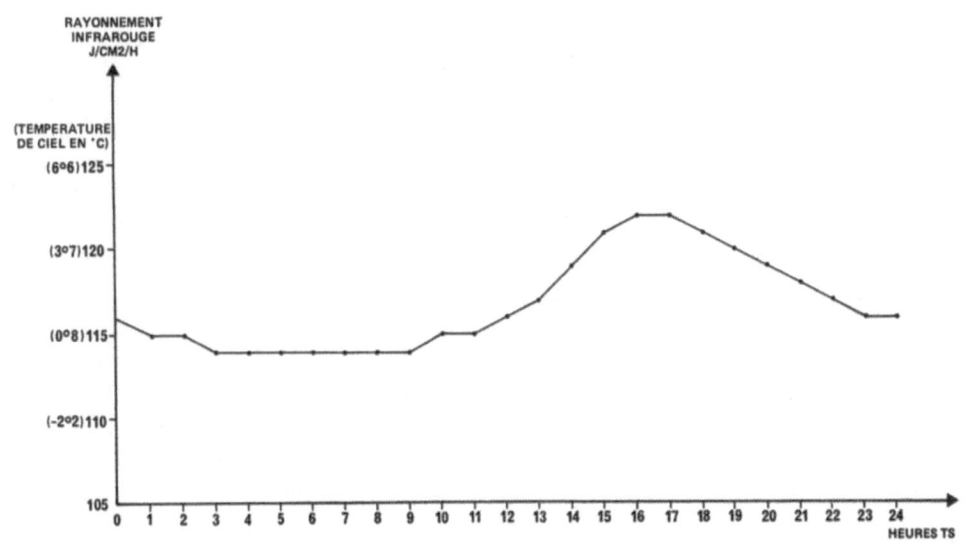

FIGURE II

Carpentras - Period 1976-1980

Daily variation of atmospheric radiation (annual mean).

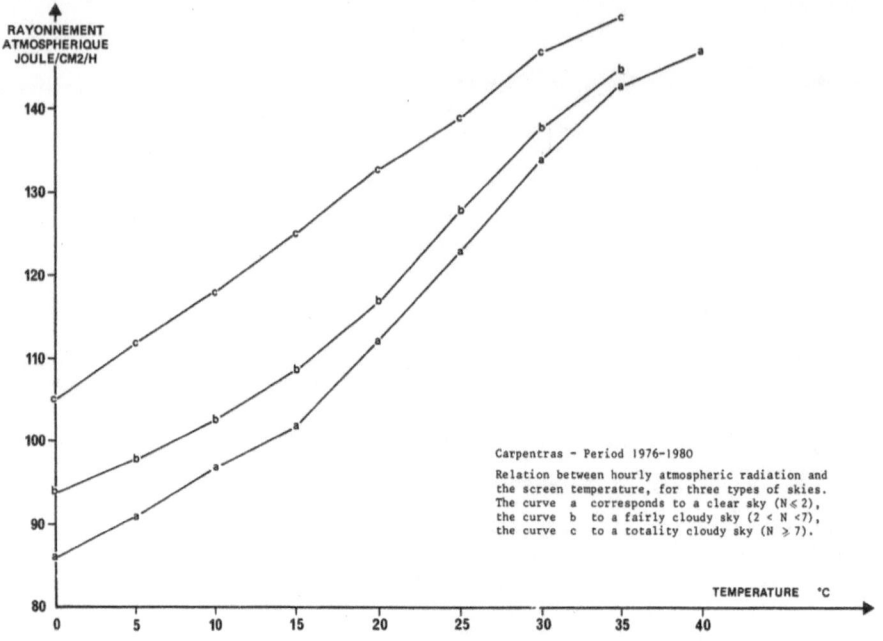

Carpentras - Period 1976-1980

Relation between hourly atmospheric radiation and
the screen temperature, for three types of skies.
The curve  a  corresponds to a clear sky (N ≤ 2),
the curve  b  to a fairly cloudy sky (2 < N < 7),
the curve  c  to a totality cloudy sky (N ≥ 7).

FIGURE III

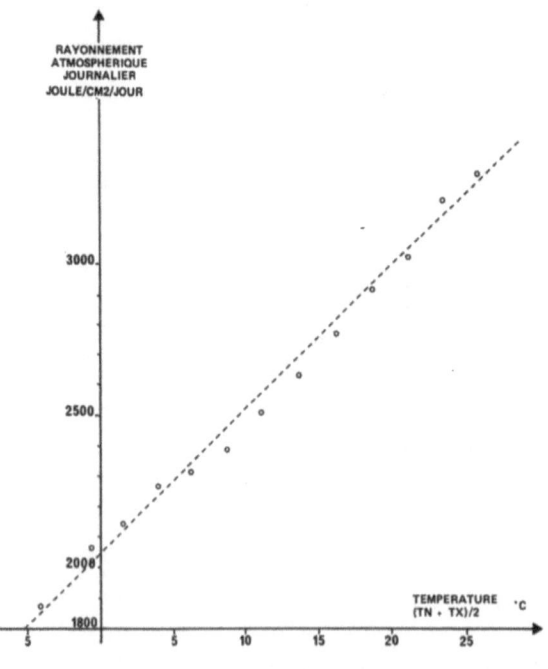

### FIGURE IV

Carpentras
Period 1976-1980

Relation between the daily
atmospheric radiation and
the mean screen temperature,
under clear skies (421 days).

The points give the means of
the daily atmospheric radiation for each class of temperature, which pitch is
2°,5C. The straight line is
calculated by least square
method.

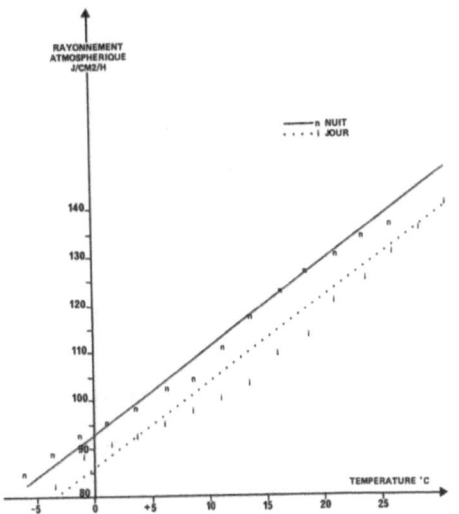

FIGURE V

Carpentras - Period 1976-80

Relation between hourly atmospheric radiation and screen temperature, for clear skies, with a distinction between night radiation and daylight radiation.

The points give the mean of the atmospheric radiation for each class of temperature, which pitch is 2°,5 C.

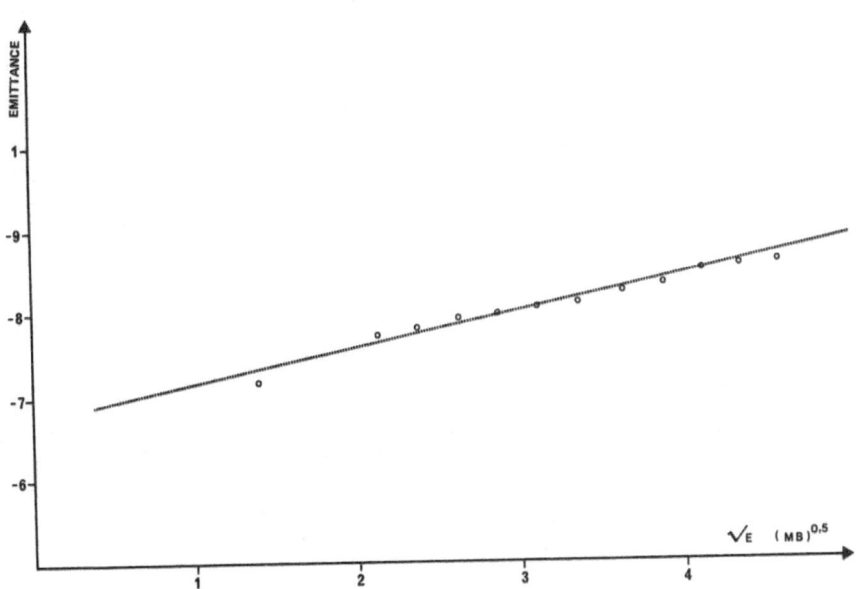

FIGURE VI - Carpentras - Period 1976-1980

Relation between emittance of the atmosphere and the square root of the water vapour pressure, for clear skies.

# THE FLUX DENSITY OF RADIATION ENERGY ORIGINATING FROM THE CIRCUMSOLAR SKY MEASURED AT GROUNDBASED STATIONS

Authors            : J. DÜDDER, R. EIDEN

Contract number : ESF-015-80 D(B)

Duration           : 35 months            1 February – 31 December 1982

Total budget     : DM 401.000,--        CEC contribution: DM 101.000,--

Head of project : Dr. R. Eiden, Lehrstuhl für Hydrologie,
                          Abt. Meteorologie, Universität Bayreuth

Contractor       : Johannes Gutenberg-Universität Mainz

Address           : Johannes Gutenberg-Universität Mainz
                          Saarstraße 21
                          P. O. Box 3980
                          D-6500 Mainz, FRG

## Summary

The aim of this project was the measurement of the radiant energy flux provided by the diffuse radiation of the ring shaped area of the sky around the unmasked sun disk. More than 200 measurements of the irradiance of this circumsolar cloudy or cloudfree sky up to an angular radius of 10° had been taken at Mainz (FRG), Deuselbach (FRG) and Adrano (Sicily). The measurements show that the irradiance of the circumsolar sky radiation may reach an amount of more then 30 % of the direct solar irradiance at the observation site. This means, optical systems combining the direct solar radiation and the circumsolar radiation may compensate - at least partly - the extinction losses of the direct solar radiation by the turbid and cloudy atmosphere. The most distinct compensation effect will be observed in the presence of thin cloud layers. A unique relationship between the irradiance of the circumsolar sky and the direct solar irradiance or an equivalent nonspectral turbidity factor could not be found.

# 1. Introduction

The gas molecules, aerosol and cloud particles, of which the atmosphere is composed, scatter and absorb the radiation emitted by the sun. Accordingly the direct sun radiation incident near the earth surface is reduced. A part of the scattered sun radiation will be received as diffuse sky light. The higher the reduction of the direct sun radiation by the atmospheric constituents, the higher is the amount of the diffuse sky radiation. So the decrease of the flux of radiant energy of the direct solar radiation at the earth surface is at least partly compensated by the diffuse sky radiation. The fraction of the radiant flux of diffuse radiation provided by a definite sky area increases with decreasing angular distance from the sun and with increasing radius and number concentration of the atmospheric scatterer especially in the size range of large particles (1 μm ≤ equivalent radius ≤ 30 μm). This means that the contribution to the diffuse radiant flux is highest from the sky areas close to the sun's limb and with cloud particles present in the atmosphere.

Taking into account this diffuse circumsolar radiation as an additional energy source, losses and variations of the direct solar radiation by changing turbidity and in the presence of thin, more or less homogenious cloud layers (sun disk still distinguishable) will be at least partly compensated and smoothed out. The aim of this project was the measurement of the circumsolar sky radiance and the determination of the radiant flux density provided by the circumsolar sky from the sun's limb up to an angular distance of 10° from the center of the sun. The results obtained as functions of this distance under changing atmospheric conditions and turbidities permit an evaluation of the reduction of the extinction losses of the direct solar radiation by the diffuse circumsolar sky radiation.

# 2. Description of Apparatus and Measuring Equipment

The photometer used is qualified for continuous operation and adapted for field operation. The photometer tubus is mounted parallaticly and follows automatically the sun. Simultaneously it swings and scans the sky above the sun tangential to the path line of the sun. So the angular distribution of the radiance of the diffuse sky radiation is measured continuously from $\varphi = 1°$ to $10°$ from the center of the sun.

The field of view of the photometer tubus is 0.5°. The tubus is constructed without using lenses or reflector systems. The error caused by the parasitic stray light within the tubus is suppressed by a diaphragm system. So the error can be kept below 1 % of the measured sky radiance $L(\varphi)$ even close to the sun.

The photometer detector system uses a silicon photovoltaic detector (UV 444 B, EG u. G) combined with a flat filter which levels the average photodiode sensivity to within ± 4 % between the wavelength 460 and 975 nm. The system is calibrated in absolute radiometric terms over the spectral range of 350 - 1.100 nm, taking into account also the endportions of the spectral sensivity range where it rolls of to zero. The calibration given by the manufacturer is controlled by a calibrated tungsten band lamp.

The turbidity factor T after Linke is measured by an auxiliary sun photometer with a photovoltaic detector too (UV 040 B, EG u. G). The factor T used here is an integral value covering the spectral range of 400 - 1030 nm, which is approximately identical with the spectral sensivity range of the photometer.

The radiant flux per unit area normal to the direction of the sun from the circumsolar sky - the radiant flux density or irradiance of interest - is obtained by integrating the primarily measured radiance $L(\varphi)$

$$E_{ci}(\varphi) = 2\pi \int_{0.5}^{\varphi} L(\varphi) \cos\varphi \sin\varphi d\varphi$$

assuming a rotational symmetry of the radiance $L(\varphi)$ with respect to the sun. The angular distance $\varphi$ defines the radius of the outer circle of the ring shaped circumsolar area that provides the irradiance $E_{ci}(\varphi)$.

## 3. Results and Comments

Measurements had been carried out at different sits:

Mainz       - polluted atmosphere
Deuselbach - clean continental atmosphere
Adrano     - clean maritime atmosphere.

As long as the circumsolar region of the sky was cloudless, the circumsolar irradiance determined showed statistical significant local differences depending on the turbidity. The highest values are found in the polluted, the lowest in the maritime atmosphere. All together the amount of the circumsolar irradiance is small and does hardly exceed 10 % of the irradiance of the direct normal solar radiation. The latter is only true for a highly polluted atmosphere and taking into account the circumsolar sky radiation up to an angular distance of $\varphi \approx 10°$.

The measurements taken so far during cloudy conditions didn't show any local dependance. The variation in cloud microstructure and thickness obscured all other atmospheric optical conditions. On the other side the radiant flux from the circumsolar sky influenced by clouds showed a significant different angular dependance compared with the cloudfree situation (Fig. 1). From the beginning, close to the sun, it started already with a considerable higher amount. But the increase with increasing angular distance was reduced. On the whole a cloudy atmosphere (sun disk not masked) yielded a higher circumsolar radiant energy flux than a cloudfree atmosphere with the same extinction.

This may demonstrated by an example: An atmosphere cloudy or not, characterized by a turbidity factor $T_L \approx 4.65$ and an air mass $m \approx 1.7$ produced in either case a direct normal solar irradiance $\dot{I} \approx 410$ W m$^{-2}$. In the cloudfree case the circumsolar irradiance was $E_{ci}$ ($\varphi = 2°$) $\approx$ 1.5 W m$^{-2}$ or $E_{ci}$ ($\varphi = 10°$) $\approx$ 19.0 W m$^{-2}$. In the cloudy case we got $E_{ci}$ ($\varphi = 2°$) $\approx$ 12.3 W m$^{-2}$ or $E_{ci}$ ($\varphi = 10°$) $\approx$ 35 W m$^{-2}$ (s. Fig. 1, curves No. 3 and 3c).

A unique relationship between the irradiance of the circumsolar sky and the direct solar irradiance or the equivalent nonspectral turbidity factor could not be observed. On the other side this seems unlikely, since particularly clouds change their droplet or ice particle sice distribution considerably with type and life-time. With the change of the size distribution the scattering pattern changes too, but not necessarily the total extinction.

The local frequency of the occurence of special clouds has not been studied within the frame of this project.

## 4. Conclusions

With increasing atmospheric turbidity the circumsolar radiation increases whereas the direct solar radiation decreases. The same total extinction of the direct solar radiation however may be caused by complete different structured atmospheres which leads to different radiant flux densities from the circumsolar sky. So a unique relationship between the direct solar irradiance or a nonspectral turbidity factor and the circumsolar irradiance could not be found. On the other side the highest values of circumsolar irradiance are supplied by a cloudy atmosphere. In general it can be stated that an optical system combining the direct solar radiation with the circumsolar radiation fluxes can compensate at least partly the extinction losses by the turbid atmosphere. The compensating effect increases with increasing turbidity. It is most effective in the presence of clouds and may exceed 30 % of the direct solar radiant flux density.

Fig. 1   Measured radiant flux $E_{ci}(\varphi)$ from the circumsolar sky per unit area (irradiance) normal to the direction of the sun as a function of the angular distance $\varphi$ of the center of the sun.
Measurements taken at Mainz and Deuselbach
Plotted are representative measurements with (index c) and without clouds (no index) but with similar air mass m, and turbidity factor $T_L$ or direct normal solar irradiance $\overset{.}{I}$   $(400 \leq \lambda \leq 1030 \text{ nm})$.

$1$ :m = 1.40; $T_L$=2.46; I=609 W m$^{-2}$
$1^c$: = 1.37;   $^L$=2.24;   =630 W m$^{-2}$

$2$ :m = 1.58; $T_L$=3.10; I=534 W m$^{-2}$
$2^c$: = 1.55;   $^L$=3.19;   =532 W m$^{-2}$

$3$ :m = 1.74; $T_L$=4.66; I=401 W m$^{-2}$
$3^c$: = 1.67;   $^L$=4.63;   =415 W m$^{-2}$

$4$ :m = 1.88; $T_L$=5.11; I=350 W m$^{-2}$
$4^c$: = 1.76;   $^L$=5.33;   =358 W m$^{-2}$

$5$ :m = 1.64; $T_L$=6.18; I=336 W m$^{-2}$
$5^c$: = 1.66;   $^L$=6.35;   =324 W m$^{-2}$

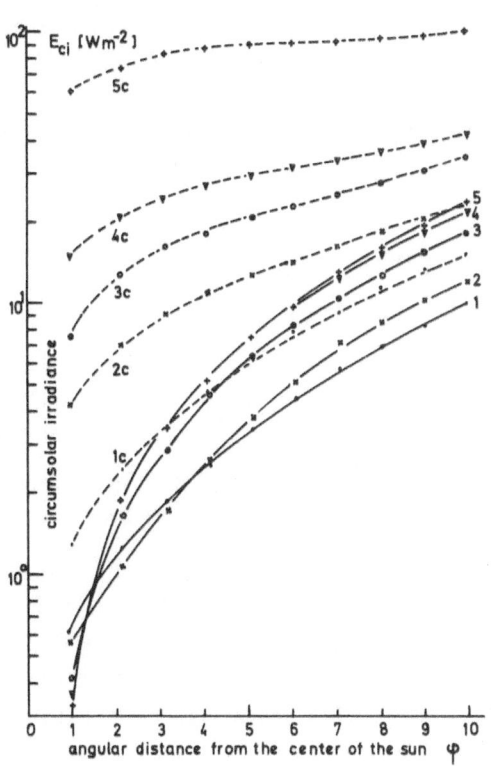

# SPECIAL MEASUREMENTS OF RADIATION IN HIGH ALTITUDE STATION

| | |
|---|---|
| Contract number | : ESF - 009 F |
| Duration | : 2 years (from 1981 to 1983) |
| Total budget (in national currency) | : 35.5 KFF - CEC contribution : 18 KFF |
| Head of project | : A. VIALARON (ask J.F. TRICAUD) |
| Authors | : J.F. TRICAUD, M. COURTEIX, Mme LE PHAT VINH, J. PY |
| Contracting organisation | : C.N.R.S. |
| Address | : Laboratoire d'énergétique solaire B.P. N° ODEILLO F-66120 FONT ROMEU |

## GOAL OF THE WORK

1) Installation for measurement of ground reflected radiation
2) Measurements of sushine duration for various thresholds of direct radiation
3) Extension of data logging system.

## STATUS AND FIRST RESULTS

Since October 1981 the data logging required for this work is running with a stored program computer.

- The measurement of ground reflected radiation for vertical south facing surface is made with à K d Z pyranometer on which a quarter of sphere put the sensor in the shadow for sun and sky radiation. In Table 1 we give an example of results for january 1982.

- The measurement of temperature is made with an "Analog devices" sensor "AD 590". Error on the measurement between - 25° and + 30°C is near the non-linearity error given by the manufacturer : ± 0,4°C. Data are available in the same presentation as Table 1.

- The data logging system has been completed, at present time forty Channels are available. A power supply is connected to protect the system.

- The measurement of sunshine duration is made for five thresholds of direct solar radiation (110,200,500,600 and 800 W.M-2). Each threshold is compare with the signal of the pyrheliometer; as soon as the required level is obtained the corresponding counting machine pass from 0 to 1 level. Twenty five measurements are made in a six minutes period. The mean value give the relative sunshine duration for the period.

For January 1982 some curves give examples of analysis of those data.

. The Fig.1 give the number of days/month where sunshine duration is above given thresholds of solar radiation.

. The Fig.2 give the number of hours/day where relative sunshine duration is above given threshold for the same three thresholds of direct solar radiation.

. At last Fig.3 also for the same thresholds of radiation give the number of set of consecutives days where duration is above given threshold.

## FUTURE WORK UNTIL THE END OF THE CONTRACT

Comparison of results of new measurements ground reflected radiation in ODEILLO with them of a low altitude station.

TABLE 1

ODEILLO    RELEVE MENSUEL DE RAYONNEMENT REFLECHI DU SOL SUR PLAN VERTICAL SUD ET D'INSOLATION

INSOLATION

| | | |
|---|---|---|
| LATITUDE : 42 29' N | !! PYRANOMETRE KIPP&ZONEN    NO:2208 | !! ANNEE: 1982 |
| LONGITUDE: 02 07' E | !! ENREGISTREUR HP85 | !! HELIOGRAPHE: PYRHELIOMETRE |
| ALTITUDE : 1580 M | !! INTEGRATEUR HP85 | !! AVEC SEUIL: 110W/M2 |
| | !! UNITE EMPLOYEE: J/CM2  RKM80 | !! UNITE EMPLOYEE: 1/10 HEURE |
| | | !! MOIS: JANVIER |

INTERVALLES HORAIRES EN TEMPS SOLAIRE VRAI — INTERVALLES HORAIRES EN TEMPS SOLAIRE VRAI — INSOLATION

| J | 0- -1 | 1- -2 | 2- -3 | 3- -4 | 4- -5 | 5- -6 | 6- -7 | 7- -8 | 8- -9 | 9- -10 | 10- -11 | 11- -12 | MAT | 12- -13 | 13- -14 | 14- -15 | 15- -16 | 16- -17 | 17- -18 | 18- -19 | 19- -20 | 20- -21 | 21- -22 | 22- -23 | 23- -24 | SOIR | TOT | H | S | 1 |
|---|---|---|---|---|---|---|---|---|---|---|---|---|---|---|---|---|---|---|---|---|---|---|---|---|---|---|---|---|---|---|
| 1 | 0 | 0 | 0 | 0 | 0 | 0 | 0 | 6 | 13 | 15 | 20 | 1 | 55 | 19 | 19 | 14 | 5 | 1 | 0 | 0 | 0 | 0 | 0 | 0 | 0 | 58 | 113 | 33 | 35 | 68 |
| 2 | 0 | 0 | 0 | 0 | 0 | 0 | 1 | 7 | 12 | 17 | 18 | 2 | 55 | 17 | 15 | 13 | 8 | 1 | 0 | 0 | 0 | 0 | 0 | 0 | 0 | 54 | 109 | 39 | 40 | 79 |
| 3 | 0 | 0 | 0 | 0 | 0 | 0 | 1 | 5 | 10 | 15 | 17 | 3 | 48 | 14 | 10 | 5 | 3 | 1 | 0 | 0 | 0 | 0 | 0 | 0 | 0 | 33 | 81 | 30 | 17 | 47 |
| 4 | 0 | 0 | 0 | 0 | 0 | 0 | 1 | 4 | 7 | 12 | 13 | 4 | 37 | 16 | 13 | 8 | 5 | 1 | 0 | 0 | 0 | 0 | 0 | 0 | 0 | 43 | 80 | 22 | 32 | 54 |
| 5 | 0 | 0 | 0 | 0 | 0 | 0 | 1 | 6 | 12 | 14 | 15 | 5 | 48 | 15 | 13 | 11 | 6 | 2 | 0 | 0 | 0 | 0 | 0 | 0 | 0 | 47 | 95 | 39 | 43 | 82 |
| 6 | 0 | 0 | 0 | 0 | 0 | 0 | 1 | 6 | 13 | 13 | 13 | 6 | 42 | 11 | 11 | 9 | 6 | 1 | 0 | 0 | 0 | 0 | 0 | 0 | 0 | 38 | 80 | 39 | 37 | 76 |
| 7 | 0 | 0 | 0 | 0 | 0 | 0 | 1 | 0 | 1 | 3 | 3 | 7 | 7 | 3 | 7 | 9 | 7 | 1 | 0 | 0 | 0 | 0 | 0 | 0 | 0 | 27 | 34 | 0 | 24 | 24 |
| 8 | 0 | 0 | 0 | 0 | 0 | 0 | 1 | 4 | 6 | 12 | 13 | 8 | 36 | 10 | 7 | 6 | 4 | 1 | 0 | 0 | 0 | 0 | 0 | 0 | 0 | 28 | 64 | 36 | 15 | 51 |
| 9 | 0 | 0 | 0 | 0 | 0 | 0 | 1 | 5 | 9 | 11 | 11 | 9 | 37 | 10 | 8 | 11 | 8 | 2 | 0 | 0 | 0 | 0 | 0 | 0 | 0 | 39 | 76 | 37 | 36 | 73 |
| 10 | 0 | 0 | 0 | 0 | 0 | 0 | 1 | 4 | 7 | 9 | 9 | 30 | 10 | 11 | 12 | 10 | 7 | 1 | 0 | 0 | 0 | 0 | 0 | 0 | 0 | 41 | 71 | 21 | 36 | 57 |
| TD | 0 | 0 | 0 | 0 | 0 | 0 | 8 | 47 | 86 | 122 | 132 | 395 | 126 | 115 | 96 | 59 | 12 | 0 | 0 | 0 | 0 | 0 | 0 | 0 | 408 | 803 | 295 | 314 | 609 |
| 11 | 0 | 0 | 0 | 0 | 0 | 0 | 1 | 4 | 12 | 12 | 13 | 42 | 8 | 8 | 8 | 6 | 1 | 0 | 0 | 0 | 0 | 0 | 0 | 0 | 32 | 74 | 22 | 20 | 42 |
| 12 | 0 | 0 | 0 | 0 | 0 | 0 | 0 | 6 | 10 | 12 | 13 | 42 | 14 | 13 | 12 | 7 | 2 | 0 | 0 | 0 | 0 | 0 | 0 | 0 | 48 | 89 | 38 | 39 | 77 |
| 13 | 0 | 0 | 0 | 0 | 0 | 0 | 0 | 7 | 11 | 11 | 14 | 43 | 14 | 14 | 9 | 6 | 2 | 0 | 0 | 0 | 0 | 0 | 0 | 0 | 42 | 85 | 38 | 37 | 75 |
| 14 | 0 | 0 | 0 | 0 | 0 | 0 | 0 | 3 | 3 | 4 | 3 | 13 | 5 | 3 | 3 | 3 | 1 | 0 | 0 | 0 | 0 | 0 | 0 | 0 | 15 | 28 | 0 | 0 | 0 |
| 15 | 0 | 0 | 0 | 0 | 0 | 0 | 0 | 3 | 3 | 3 | 1 | 10 | 1 | 1 | 1 | 1 | 0 | 0 | 0 | 0 | 0 | 0 | 0 | 0 | 4 | 14 | 0 | 0 | 0 |
| 16 | 0 | 0 | 0 | 0 | 0 | 0 | 1 | 4 | 4 | 8 | 9 | 26 | 8 | 9 | 4 | 4 | 1 | 0 | 0 | 0 | 0 | 0 | 0 | 0 | 26 | 52 | 0 | 0 | 0 |
| 17 | 0 | 0 | 0 | 0 | 0 | 0 | 2 | 5 | 9 | 13 | 20 | 49 | 18 | 11 | 7 | 4 | 1 | 0 | 0 | 0 | 0 | 0 | 0 | 0 | 41 | 90 | 1 | 0 | 1 |
| 18 | 0 | 0 | 0 | 0 | 0 | 0 | 2 | 5 | 0 | 0 | 0 | 6 | 0 | 0 | 0 | 0 | 0 | 0 | 0 | 0 | 0 | 0 | 0 | 0 | 2 | 8 | 0 | 0 | 0 |
| 19 | 0 | 0 | 0 | 0 | 0 | 0 | 4 | 26 | 20 | 21 | 16 | 87 | 15 | 24 | 17 | 9 | 2 | 0 | 0 | 0 | 0 | 0 | 0 | 0 | 67 | 154 | 27 | 30 | 57 |
| 20 | 0 | 0 | 0 | 0 | 0 | 0 | 2 | 10 | 17 | 21 | 22 | 72 | 22 | 20 | 16 | 12 | 3 | 0 | 0 | 0 | 0 | 0 | 0 | 0 | 73 | 145 | 41 | 44 | 85 |
| TD | 0 | 0 | 0 | 0 | 0 | 0 | 12 | 72 | 89 | 105 | 111 | 389 | 103 | 103 | 77 | 54 | 13 | 0 | 0 | 0 | 0 | 0 | 0 | 0 | 350 | 739 | 166 | 170 | 336 |
| 21 | 0 | 0 | 0 | 0 | 0 | 0 | 1 | 3 | 6 | 10 | 12 | 32 | 11 | 10 | 11 | 8 | 2 | 0 | 0 | 0 | 0 | 0 | 0 | 0 | 42 | 74 | 0 | 12 | 12 |
| 22 | 0 | 0 | 0 | 0 | 0 | 0 | 1 | 5 | 6 | 7 | 7 | 23 | 7 | 9 | 8 | 4 | 2 | 0 | 0 | 0 | 0 | 0 | 0 | 0 | 30 | 53 | 0 | 1 | 1 |
| 23 | 0 | 0 | 0 | 0 | 0 | 0 | 2 | 17 | 38 | 48 | 46 | 151 | 45 | 39 | 34 | 21 | 6 | 0 | 0 | 0 | 0 | 0 | 0 | 0 | 145 | 296 | 29 | 41 | 70 |
| 24 | 0 | 0 | 0 | 0 | 0 | 0 | 2 | 8 | 13 | 19 | 28 | 70 | 22 | 21 | 25 | 15 | 4 | 0 | 0 | 0 | 0 | 0 | 0 | 0 | 87 | 157 | 0 | 27 | 27 |
| 25 | 0 | 0 | 0 | 0 | 0 | 0 | 1 | 3 | 4 | 6 | 9 | 23 | 8 | 10 | 11 | 7 | 2 | 0 | 0 | 0 | 0 | 0 | 0 | 0 | 38 | 61 | 0 | 0 | 0 |
| 26 | 0 | 0 | 0 | 0 | 0 | 0 | 2 | 10 | 18 | 20 | 24 | 74 | 27 | 21 | 15 | 9 | 2 | 0 | 0 | 0 | 0 | 0 | 0 | 0 | 78 | 152 | 25 | 22 | 47 |
| 27 | 0 | 0 | 0 | 0 | 0 | 0 | 1 | 4 | 12 | 13 | 13 | 43 | 12 | 11 | 10 | 6 | 2 | 0 | 0 | 0 | 0 | 0 | 0 | 0 | 41 | 84 | 0 | 0 | 0 |
| 28 | 0 | 0 | 0 | 0 | 0 | 0 | 2 | 17 | 37 | 47 | 50 | 153 | 49 | 44 | 34 | 20 | 7 | 0 | 0 | 0 | 0 | 0 | 0 | 0 | 154 | 307 | 38 | 47 | 85 |
| 29 | 0 | 0 | 0 | 0 | 0 | 0 | 3 | 15 | 24 | 37 | 42 | 121 | 43 | 39 | 31 | 17 | 2 | 0 | 0 | 0 | 0 | 0 | 0 | 0 | 132 | 253 | 42 | 39 | 81 |
| 30 | 0 | 0 | 0 | 0 | 0 | 0 | 1 | 11 | 21 | 27 | 28 | 90 | 28 | 24 | 19 | 12 | 4 | 0 | 0 | 0 | 0 | 0 | 0 | 0 | 87 | 177 | 42 | 48 | 90 |
| 31 | 0 | 0 | 0 | 0 | 0 | 0 | 2 | 11 | 17 | 22 | 23 | 75 | 24 | 20 | 16 | 11 | 4 | 0 | 0 | 0 | 0 | 0 | 0 | 0 | 75 | 150 | 42 | 48 | 90 |
| TD | 0 | 0 | 0 | 0 | 0 | 0 | 20 | 103 | 197 | 253 | 282 | 855 | 276 | 252 | 214 | 130 | 37 | 0 | 0 | 0 | 0 | 0 | 0 | 0 | 909 | 1764 | 218 | 284 | 502 |
| TM | 0 | 0 | 0 | 0 | 0 | 0 | 40 | 222 | 372 | 480 | 525 | 1639 | 505 | 470 | 387 | 243 | 62 | 0 | 0 | 0 | 0 | 0 | 0 | 0 | 1667 | 3306 | 679 | 769 | 1448 |
| MM | 0 | 0 | 0 | 0 | 0 | 0 | 1 | 7 | 12 | 15 | 17 | 53 | 16 | 15 | 12 | 8 | 2 | 0 | 0 | 0 | 0 | 0 | 0 | 0 | 54 | 107 | 22 | 25 | 47 |

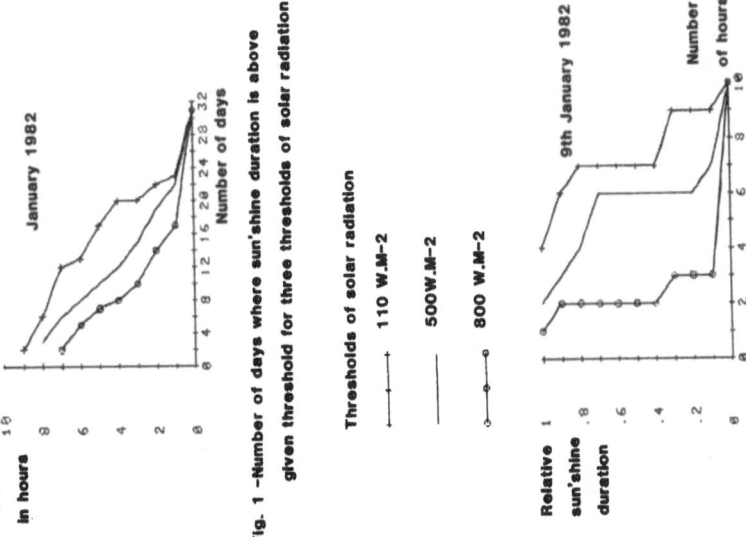

Fig. 3 –Number of set of consecutive days whose duration is above given threshold for three thresholds of solar radiation

Fig. 1 –Number of days where sun'shine duration is above given threshold for three thresholds of solar radiation

Thresholds of solar radiation

110 W.M-2

500W.M-2

800 W.M-2

Fig. 2 –For given day, number of hours where relative sun'shine duration is above given threshold for three thresholds of solar radiation

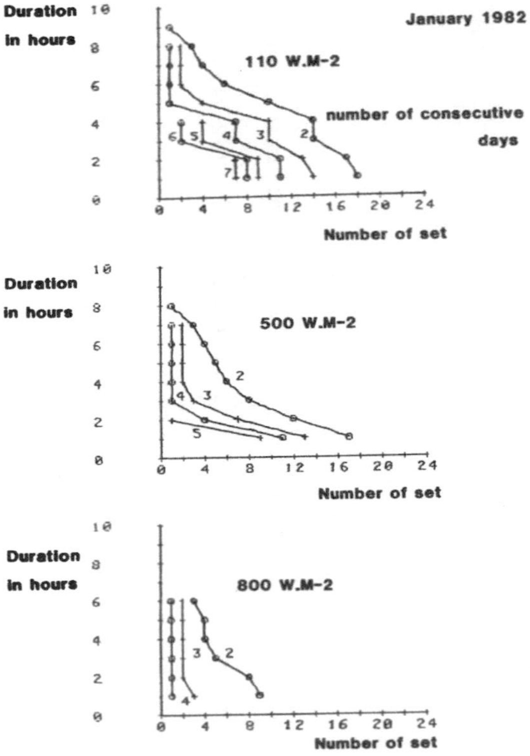

# INSTALLATION OF EQUIPMENT TO MEASURE THE SEPARATE COMPONENTS OF GLOBAL SOLAR RADIATION ON A VERTICAL SURFACE

Author              : E.J. Murphy

Contract number   : ESF-018-EIR(G)

Duration           : 36 months 1 January 1980 - 31 December 1982

Total budget       : IR£4,690   CEC contribution :   IR£2,345

Head of project   : E.J. Murphy

Contractor         : Irish Meteorological Service

Address            : Glasnevin Hill
                      Dublin 9
                      Ireland

## Summary

The purchase and installation of equipment to measure the separate components of global solar radiation on a vertical south-facing surface at Valentia Observatory was completed in April 1981 and data collection commenced on 1st May 1981. Participation in Action 4.2 was solely concerned with the purchase and installation of this equipment while the analysis of the recorded data is included under Action 3.2.

# 1. Project report

## 1.1 Introduction

Measurements of the direct, diffuse and ground-reflected components of solar radiation on a vertical south-facing surface at Valentia commenced on 1st May 1981. The preliminary data analysis (Action 3.2) was confined to the months of June and December 1981. Data for the 18 month period 1st May 1981 to 31st October 1982 will be utilised for the complete analysis and final report.

## 1.2 Measuring equipment

A Kipp and Zonen CM6 pyranometer and a BD8 Flatbed recorder have been installed to monitor the ground-reflected radiation.

Radiation on a vertical south-facing surface is monitored by two Kipp and Zonen CM6 pyranometers, one of which is screened by a 2 foot square honeycombed mat black metal surface to eliminate the ground reflected radiation. The 'screened' radiation is recorded on a Kipp and Zonen BD8 Flatbed recorder and an Eppley/Digitec electronic integrator/printer while the unshielded radiation is recorded by means of a Kipp and Zonen CC2 integrator.

Direct sun radiation at normal incidence is recorded using an Eppley pyrheliometer with a solar tracker and an Eppley/Digitec electronic integrator/printer. The component of direct radiation incident of the south-facing vertical surface is given by: $ICosECosA$.
where I is the measured direct radiation at normal incidence
E is the solar elevation and A is the solar azimuth.

The component of diffuse radiation incident on the south-facing vertical surface is obtained by subtracting $ICosECosA$ from the total radiation recorded by the screened pyranometer - (S).

The component of ground reflected radiation incident on the south-facing vertical surface is given by the difference between the radiation recorded by the screened and unscreened pyranometers.

## 1.3 Data analysis

The results of the data analysis for June and December 1981 may be found in the report for Action 3.2

Theoretically the radiation on a vertical south-facing surface is given by: $G_v$ = Direct$_v$ + Diffuse$_v$ + Reflected$_v$
$$= ICosECosA + \tfrac{1}{2}D + \tfrac{1}{2}\rho G$$
where $G_v$ = Global radiation on a south-facing vertical surface
$I$ = Direct radiation at normal incidence
$E$ = Solar elevation, $A$ = Solar azimuth, $\rho$ = Ground albedo
$D$ = Diffuse sky radiation on a horizontal surface
$G$ = Global radiation on a horizontal surface.

The theoretically expected values of the diffuse and ground reflected radiation are compared with the measured values. Some examples of the reflected values obtained for two days in both June and December are given hereunder

# Ground-reflected radiation (Reflected$_V$)

|        | 5 to 6 | 6 to 7 | 7 to 8 | 8 to 9 | 9 to 10 | 10 to 11 | 11 to 12 | 12 to 13 | 13 to 14 | 14 to 15 | 15 to 16 | 16 to 17 | 17 to 18 | 18 to 19 |
|--------|--------|--------|--------|--------|---------|----------|----------|----------|----------|----------|----------|----------|----------|----------|
| Measured values | 3 | 7 | 9 | 4 | 27 | 23 | 20 | 11 | 5 | 4 | 4 | 2 | 4 | |
| Theoretical values ($\frac{1}{2}\rho G$) | 3 | 6 | 10 | 6 | 18 | 21 | 20 | 10 | 6 | 4 | 4 | 3 | 3 | |

(1st June 1981 - sunny spells in morning but mainly overcast cloudy day)

|        | 5 to 6 | 6 to 7 | 7 to 8 | 8 to 9 | 9 to 10 | 10 to 11 | 11 to 12 | 12 to 13 | 13 to 14 | 14 to 15 | 15 to 16 | 16 to 17 | 17 to 18 | 18 to 19 |
|--------|--------|--------|--------|--------|---------|----------|----------|----------|----------|----------|----------|----------|----------|----------|
| Measured values | 9 | 16 | 30 | 40 | 43 | 41 | 35 | 28 | 21 | 13 | 8 | 5 | 14 | 10 |
| Theoretical values ($\frac{1}{2}\rho G$) | 7 | 13 | 20 | 23 | 27 | 30 | 31 | 31 | 31 | 27 | 25 | 20 | 17 | 9 |

(21st June 1981 - clear sunny day)

| | | | | | | 10 to 11 | 11 to 12 | 12 to 13 | 13 to 14 | 14 to 15 | 15 to 16 | | |
|--|--|--|--|--|--|----------|----------|----------|----------|----------|----------|--|--|
| Measured values | | | | | | 8 | 15 | 6 | 6 | 5 | 3 | | |
| Theoretical values ($\frac{1}{2}\rho G$) | | | | | | 3 | 7 | 6 | 5 | 4 | 3 | | |

(4th December 1981 - partly cloudy morning, overcast afternoon)

| | | | | | | 10 to 11 | 11 to 12 | 12 to 13 | 13 to 14 | 14 to 15 | 15 to 16 | 16 to 17 | 17 to 18 |
|--|--|--|--|--|--|----------|----------|----------|----------|----------|----------|----------|----------|
| Measured values | | | | | | 1 | 16 | 22 | 16 | 10 | 3 | 1 | 1 |
| Theoretical values ($\frac{1}{2}\rho G$) | | | | | | 1 | 3 | 8 | 10 | 10 | 9 | 4 | 1 |

(18th December 1981 - mainly clear and sunny)

An asymmetry between morning and afternoon values is quite apparent and is probably due to local factors. There is a similar asymmetry between the morning and afternoon values of the measured and theoretical diffuse components. This is discussed more fully in the Action 3.2 report.

ACTION 4.3


SATELLITE MEASUREMENTS


Action leaders's progress report
   J.M. MONGET, Ecole Nationale Supérieure des Mines de
   Paris, Centre de Télédétection et d'analyse des milieux
   naturels, Sophia Antipolis, Valbonne, France


Reports of action participants :


 - Determination of the global radiation and of cloudiness
   from satellite data


 - Satellite image analysis


 - Heliosat project : estimation of global radiation from
   satellite using a statistical approach

## ACTION 4.3 - SATELLITE MEASUREMENTS

Action Leader : J.M. MONGET
            Ecole Nationale Supérieure des Mines de Paris
            Centre de Télédétection et d'Analyse des Milieux Naturels
            Rue Claude Daunesse
            SOPHIA-ANTIPOLIS
            06565 VALBONNE Cédex (France)


Participants :

- Institut für Geophysik und Meteorologie
  Universität zu Köln - KÖLN
  Contracts : ESF-013-D and ESF-031-D
  E. RASCHKE - W. MÖSER

- Institut für Physikalishe Elektronik
  Universität zu Stuttgart - STUTTGART
  Contracts : ESF-016-80-D(B) and ESF-032-D
  E.R. REINHARDT - P. SCHWARZMANN

- Centre de Télédétection et d'Analyse des Milieux Naturels
  Ecole des Mines de Paris - SOPHIA-ANTIPOLIS
  Contract : ESF-008-F(G)
  J.M. MONGET - D. CANO


Task :

Data processing of meteorological satellite measurements in order to
produce an estimation and a cartography of the global radiation over the
geographic area covered by the EEC countries.

# 1. INTRODUCTION

This action is concerned with the estimation of global radiation using the radiometric observations made by the meteorological satellites.

The final goals are centered on the determination of hourly values and daily sums. The methods developed by the three groups in order to fulfil this goal are different in scope and in the set of satellite data set they handle.

This is why an intercomparison of results has been planned for the end of 1982.

# 2. ACHIEVEMENTS

The participants of Action 4.3 have now produced data processing schemes which allow to address the treatment of any kind of meteorological satellites whether they are :

- polar orbiting satellites of the NOAA-TIROS N series with observations transmitted in high resolution digital form (1 km) or low resolution analog form (4 km) ;

- geostationnary satellites like METEOSAT with digital data transmitted through the ESA-Darmstadt center or analog data retransmitted to low-cost SDUS receiving stations.

The methods permit the estimation of global radiation hourly values and daily sums on a regular grid over Europe. This grid has a size varying between 20 km and 5 km, depending of the proposed method.

The algorithms are relying on purely physical models (Köln) to statistical schemes working in time (Sophia-Antipolis) or in the two-dimensional space of the imagery using texture analysis (Stuttgart). But they all have to somehow estimate initial parameters by calibration with ground-truth measurements performed by pyranometers.

# 3. TEST OF PERFORMANCE

Test of the estimation accuracy of the three methods is conducted by reference to the data of the EEC pyranometer stations collected by Action 4.1 during the reference month of April 82.

The participants of Action 4.3 will provide for each of these stations, the following estimations during the period 24 to 30 April 1982 :

- Global hourly radiation values for three hourly intervals :

  . 1 : 08h20 to 09h20 GMT
  . 2 : 11h20 to 12h20 GMT
  . 3 : 14h20 to 15h20 GMT

- Global radiation daily sums (only for Köln and Stuttgart groups).

The evaluation of the accuracy of estimation will be conducted under the supervision of the Action Leader 4.1.

The final results will be presented on December 16, 1982.

4. <u>FUTURE WORK</u>

For the follow-on program, the participants recommend that actions should be taken in the following items :

- Use of satellite observations in the estimation of short-period variation statistics of solar radiation.

- Development of new ground-truth measurements directly compatible with satellite data (sky images, for example).

- Further research in the quantitative description of cloud structure.

## DETERMINATION OF THE GLOBAL RADIATION
## AND OF CLOUDINESS FROM SATELLITE DATA

Authors:            : W. Möser, E. Raschke

Contract number : ESF-o31-D(B)

Duration            : 8 months              o1 May 1982 - 31 December 1982

Total budget    : DM 114.2oo          CEC contribution: DM 57.1oo

Head of project : Prof. Dr. E. Raschke, Institut für Geophysik und
                       Meteorologie der Universität zu Köln

Contractor       : Universität zu Köln

Address            : Institut für Geophysik und Meteorologie der Universität
                       zu Köln, Kerpener Str. 13, D-5ooo Köln 41

## Summary

An operational method has been developed to determine the field of global
radiation at the surface from the imaging data of the geostationary
satellite METEOSAT.
Results of radiative transfer calculations in clear as well as in cloudy
atmospheres have been combined on a statistical basis with parameters
determined from these measurements. This provides a parameterization of
actual cloudiness which suffices in accuracy and is economical enough to
be run on a minicomputer.
Daily sums of global radiation measured during June 1979 by 19 pyranometers of the Deutscher Wetterdienst (DWD) radiation network have been
compared with the results derived from METEOSAT data of the same period
(up to six images per day: o6.23 GMT to 16.23 GMT, every 2 hours).
    The RMS difference for daily sums (using up to six satellite images
per day) reaches o.56 kWhm$^{-2}$ which is approximately 7 % of the corresponding daily sum under clear sky conditions. Using only up to three
images per day for the estimate of the daily sum we get a RMS difference
of o.7o kWhm$^{-2}$ which is approximately 9 % of the corresponding daily
sum under clear sky conditions.
Maps of global radiation have been derived for the area of Europe, the
Mediterranean Sea area and North Africa.

## 1.1 Introduction

The demand for more expanded use of available solar energy at ground attracted the interest of scientists to derive such informations also from satellite data, taken at least over areas where no or only few ground bound measurements are available.

Image data as obtained from the geostationary satellite METEOSAT provide us with sufficient information to derive estimates of global radiation in a wide range of time and space scales.

Basing on the results of EC-contract no. 3o2-77-ESD (concerning the theoretical treatment of the influence of clouds, aerosols and the condition of the atmosphere on the downward flux of global radiation at the surface as well as the upward flux of reflected radiation at the top of the atmosphere) and EC-contract no. ESF-o13-8o D(B) (concerning the evaluation of global radiation maps from METEOSAT image data) we have determined hourly mean values as well as daily sums of global radiation for the area of Europe, the Mediterranean Sea area and North Africa.

According to (1) who divides the methods recently developed in order to determine the solar irradiance at the surface into "statistical methods" (2), (3), (4) which use empirical relationships derived from fits between values of global radiation estimates from satellite data and those of nearby stations and "physical model methods" (5), (6), (7) which use the satellite data as indicators of parameters necessary for calculations of the global radiation at ground using radiative transfer models, our method might be classified as a "physical model method".

## 1.2 Description of the method

The method combines results of radiative transfer calculations basing on a modified version of a well calibrated two stream approximation (8) with parameters derived from the satellite data on a statistical basis. Ground truth data are needed only for an initial adjustment. In an operational phase only a few stations will be needed to monitor the quality of the results continuously.

A short description of the method and first results of ground truth comparisons (period: o1. - 15.June 1979) for hourly mean values are given in (9).

A more accurate description of the method and first results of ground truth comparisons on a monthly data basis (period: June 1979) for hourly mean values are given in (1o).

Meanwhile we have extended our model for the calculation of daily sums of global radiation:

$$M_G^{ds} = M_{Go}^{ds} \cdot \frac{\Sigma M_G^{hm}}{\Sigma M_{Go}^{hm}} \qquad (1)$$

where ds indicates daily sum and hm indicates hourly mean. $M_{Go}$ is the maximum value of $M_G$ under cloudless conditions. $M_{Go}^{ds}$ and $M_{Go}^{hm}$ have been derived from radiative transfer calculations whereas $M_G^{hm}$ has been derived from up to six METEOSAT images measured at approximately o6.23 GMT to 16.23 GMT, every two hours.

## 1.3 Results

For the time period of June 1979 ground truth comparisons have been done using pyranometer measurements from 19 stations of the radiation network

of the Deutscher Wetterdienst (DWD). In Fig. 1 daily sums of global radiation derived from up to 6 images per day have been compared with the
corresponding pyranometer measurements. The RMS difference reaches o.56
kWhm$^{-2}$ which is approximately 7 % of the corresponding daily sum under
clear sky conditions. Fig. 2 shows the influence temporal averaging has
to the RMS difference $\sigma_{xy}$. We can see an exponential decrease of $\sigma_{xy}$ with
a remaining systematic difference of o.22 kWhm$^{-2}$ for the monthly mean at
19 pyranometer stations (curve 1). Fig. 3 to Fig. 6 show the distribution
of global radiation for the area of Europe, the Mediterranean Sea area
and North Africa for different scales of temporal averaging. In Fig. 3
and Fig. 4 the coastlines and the station dots change from black to white
at 5 kWhm$^{-2}$ (in Fig. 5 at 6 kWhm$^{-2}$, in Fig. 6 at 2 kWhm$^{-2}$).

## 1.4 Analysis of results and comments

To get an optimal adjustment of the estimates of daily sums of global
radiation as derived from the satellite image data to those as measured
at the surface from pyranometers it would have been appropriate to use
satellite image data measured every hour (up to twelve images per day instead of up to six images per day). But as to be seen in Fig. 2 even the
daily sums gained from only up to three images per day, every four hours,
give a reasonable RMS difference for single day values of o.7 kWhm$^{-2}$ and
averaging over n days (n$\geq$ 5) reduces the difference between the curves
to about o.o5 kWhm$^{-2}$. This is why we think that for the calculation of
weekly or monthly averaged global radiation maps 4-5 images per day,
measured every three hours, might be sufficient whereas for the calculation of single day values at least six images per day, measured every two
hours, will be needed.
The global radiation maps shown in Fig. 3 to Fig. 6 give some impression
of how the spatial variability of the radiation field decreases using
different periods of temporal averaging. The lower values in Fig. 5 for
the monthly mean over ground (France, Ireland, Italy) compared to nearby
situated sea areas are due to convective cloudiness especially in the
afternoon whereas the low values of the North Spanish coast are due to the
orographic influence of the Spanish Highlands.

## 2. Conclusions

A method has been developed to estimate the global radiation at the surface
from image data of the geostationary satellite METEOSAT.
     Although actual cloudiness has been parameterized in a simple way
first ground truth comparisons have shown that our method is able to re-
produce the field of global radiation in its temporal and spatial variation with an accuracy sufficient for most users.
     The method is fully developed and ready for operational use.
     We are now enlarging the geographical region of our ground truth com-
parisons to the area of whole Europe. In addition we have started processing image data of the period April 1982.

## References

(1) ELLINGSON, R.G.: On estimating the radiation budget at the surface
    using satellite-observed radiance. - Proceedings of a workshop on
    applications of existing satellite data to the study of the ocean sur-
    face energetics, 19.-21.11.198o. - University of Wisconsin (Mad.),
    61-72 (1981)
(2) HANSON, K.: A new estimate of solar irradiance at the earth's surface
    on zonal and global scales. - J.Geophys.Res., 81, 4435-4443 (1976)

(3) ELLIS, J.S. and VONDER HAAR, T.H.: Solar radiation reaching the ground determined from meteorological satellite data. - Proceedings of the third conference on atmospheric radiation, - Davis (Cal.), 187-189 (1978)

(4) TARPLEY, J.D.: Estimating incident solar radiation at the surface from geostationary satellite data. - J.Appl.Met., 18, 1172-1181 (1979)

(5) RASCHKE, E. and PREUSS, H.-J.: The determination of the solar radiation budget at the earth's surface from satellite measurements. - Meteorol.Rdsch., 32, 18-28 (1979)

(6) GAUTIER, C. et al.: A simple physical model to estimate incident solar radiation at the surface from GOES satellite data. - J.Appl.Met., 19, 1oo5-1o12 (198o)

(7) GAUTIER, C.: Mesoscale insolation variability derived from satellite data. - J.Appl.Met., 21, 51-58 (1982)

(8) KERSCHGENS, M.: Berechnungen des solaren Strahlungstransportes in Atmosphäre und Ozean mit Hilfe einer Zweistrommethode. - Mitteilungen aus dem Institut für Geophysik und Meteorologie der Universität zu Köln (1978)

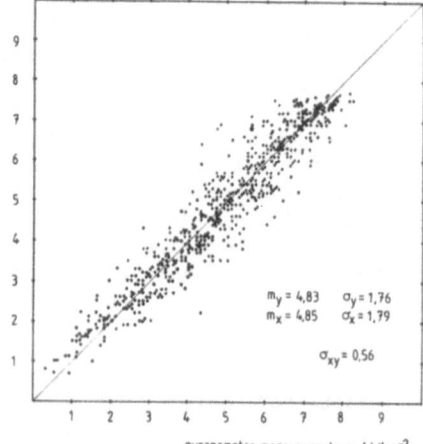

N = 573
r = 0,95

Fig.1: Satellite estimates of global radiation compared with
pyranometer measurements
(daily sums):
Up to 6 images per day,
measured at 6.23 GMT to
16.23 GMT, every 2 hours,
have been used.
m: mean of global radiation
$\sigma$: standard deviation of
global radiation
$\sigma_{xy}$: RMS difference of
global radiation
r: correlation coefficient

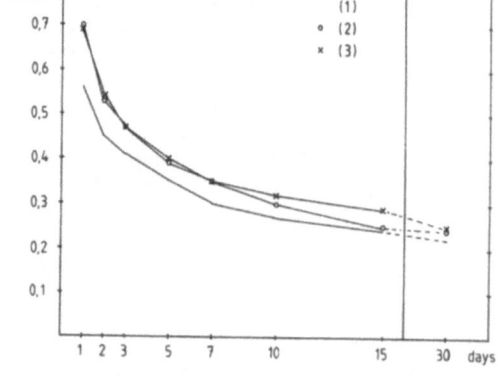

Fig.2: Satellite estimates of global
radiation compared with pyranometer measurements:
RMS difference of global radiation (daily sums averaged
over n days).
(1) up to 6 images per day,
measured at 6.23 GMT to
16.23 GMT, every 2 hours,
(2) up to 3 images per day,
measured at 8.23 GMT to
16.23 GMT, every 4 hours,
(3) up to 3 images per day,
measured at 6.23 GMT to
14.23 GMT, every 4 hours,
have been used.

(9)  MÖSER, W. and RASCHKE, E.: Determination of global radiation and of
     cloudiness from METEOSAT imaging data. - Annalen der Meteorologie
     (Neue Folge) Nr. 18, 161-163, Offenbach a.M., Selbstverlag des Deut-
     schen Wetterdienstes (1982)

(1o) MÖSER, W. and RASCHKE, E.: Mapping of global radiation and of cloudi-
     ness from METEOSAT image data. - accepted for publication in the
     Meteorol.Rdsch.

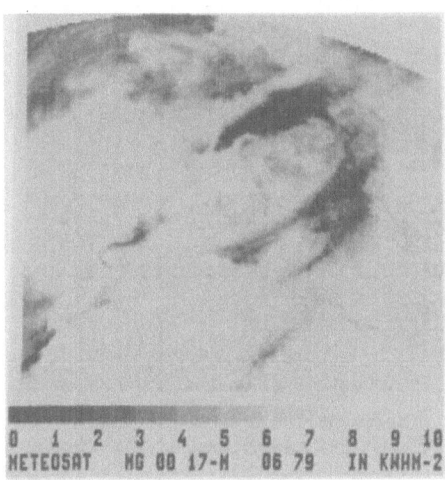

Fig.3: Global radiation map:
        Single day value of daily
        sums (17.6.1979).

Fig.4: Global radiation map:
        5-day mean of daily
        sums (16.-20.6.1979).

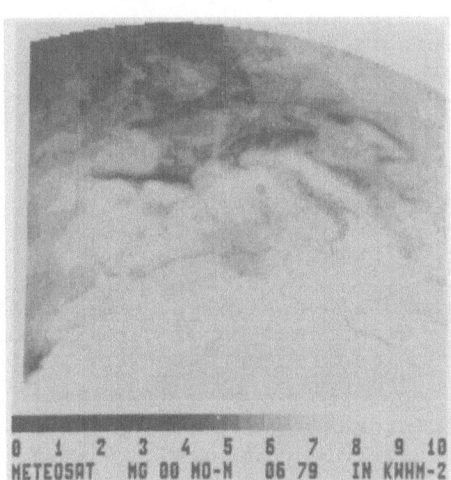

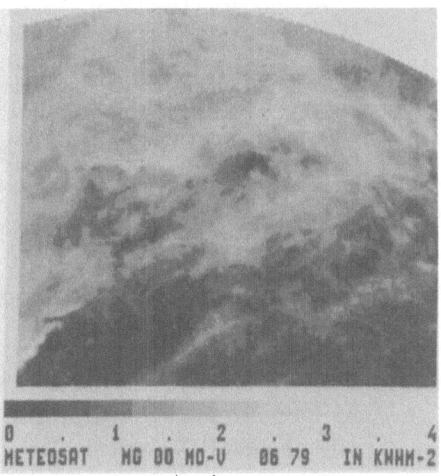

Fig.5: Global radiation map:
        Monthly mean of daily
        sums (June 1979).

Fig.6: Global radiation map:
        Monthly standard deviation
        of daily sums (June 1979).

# SATELLITE IMAGE ANALYSIS

| | |
|---|---|
| Authors | : E.R.Reinhardt, P.Schwarzmann, R.Lenz |
| Contract number | : ESF-016-80-D(B) |
| Duration | : 36 months   1 Jan.1980 - 31 Dec.1982 |
| Total budget | : DM 554 520   CEC contribution 249 534 |
| Head of project | : E.R.Reinhardt |
| Contractor | : University of Stuttgart |
| Address | : Institut fuer Physikalische Elektronik<br>Boeblingerstr. 70<br>D7000 Stuttgart - 1 |

## Summary

A method for the determination of global radiation from multispectral satellite images has been developed and applied to about 85 NOAA6 and NOAA7 scenes of Europe. The mean characteristics of the analysis procedure are:
- The analysis procedure is based on statistical methods using selftraining algorithms
- The image evaluation process is based on the analysis of local neighbourhood relations in satellite images
- High processing speed is achieved by processor oriented algorithms
- Analysis process is adaptable to different types of satellite images
- Analysis procedure can be transfered to operational systems.

Preliminary results demonstate impressively the efficiency of the procedure for the determination of global radiation values. The correlation coefficients between estimated and ground truth data are better than 0.9.

By-products of the analysis process are sunshine duration and quantitative description of cloudiness.

All results can be presented as maps.

## 1.1 Introduction.

Application of solar energy in a reasonable scale depends on data of the availability of global radiation. It is obvious that satellite image data are excellent data sources with high spatial and temporal resolution.

Regarding the large number of meteorologic parameters relevant for radiation transfer in the atmosphere, statistic methods are adequate for such problems.

## 1.2 Strategy.

The principle of the analysis strategy is outlined in fig. 1. From an economic as well as from a technical point of view, the guidline for the development of the analysis algorithms can be summarized as follows:

The analysis procedure of the different subtasks should be based on same  processing
modules ( processor oriented method).
The strategy does not take into account
explicit albedo information.

## 1.3 Equipment.

The processing system used during the development phase of the algorithms and for the experiments is a highly flexible image processing system. It consists of a minicomputer ( as host ) with standard periphery devices, special fast hardware processors and image memories. The components are oriented to TV-frame processing. Components of this system are transferable to an operational system.

## 1.4 Results.

Experiments performed so far are based on NOAA6 and NOAA7 satellite image data of April 82 received by operational stations in FRG (DFVLR Oberpfaffenhofen) and UK (University of Dundee).

The results are presented in the form of correlation coefficients between calculated values from the satellite images and measured values of 39 meteorologic network stations in UK (Meteorological Office) and FRG (Deutscher Wetterdienst).

From each satellite scene global radiation sums for 1,3 and 5 hours have been computed. These sums are the base for the calculation of daily sums (normally at least one pass in the morning, afternoon and evening is available for every geographic location in the EC).

The results are summarized in table I :

| Global radiation Interval | 1 hour | 3 hours | 5 hours | daily sum |
|---|---|---|---|---|
| Correlation coefficient | 0.91 | 0.93 | 0.91 | 0.93 |

Table I. Correlation coefficients.

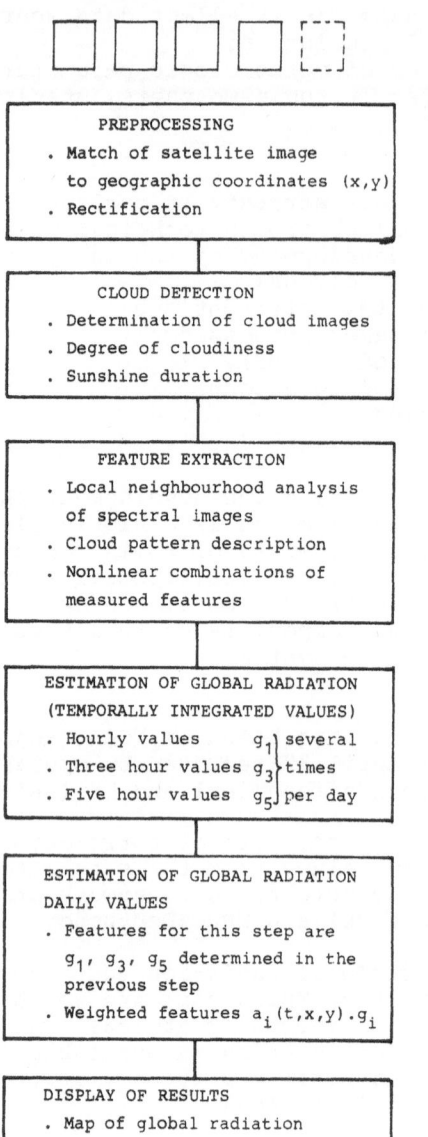

Spectral satellite images

PREPROCESSING
. Match of satellite image
  to geographic coordinates (x,y)
. Rectification

CLOUD DETECTION
. Determination of cloud images
. Degree of cloudiness
. Sunshine duration

Classifier for cloud detection
(point by point), determined in
training phase on the base of
data bank containing visually
classified cloud images.

FEATURE EXTRACTION
. Local neighbourhood analysis
  of spectral images
. Cloud pattern description
. Nonlinear combinations of
  measured features

Composition of relevant feature
set; determined in training phase
by experiments with classified
data sets.

ESTIMATION OF GLOBAL RADIATION
(TEMPORALLY INTEGRATED VALUES)
. Hourly values        $g_1$ ⎤ several
. Three hour values $g_3$ ⎬ times
. Five hour values  $g_5$ ⎦ per day

Estimation function determined
in training phase based on data
bank containing ground truth data
of global radiation (hourly values).

ESTIMATION OF GLOBAL RADIATION
DAILY VALUES
. Features for this step are
  $g_1$, $g_3$, $g_5$ determined in the
  previous step
. Weighted features $a_i(t,x,y) \cdot g_i$

Estimation function determined
in training phase based on data
of global radiation (daily sums).

DISPLAY OF RESULTS
. Map of global radiation
. Map of cloudiness
. Map of sunshine duration

Fig. 1. Flowchart of procedure for
estimation of global radiation.

The correlation values are based on about 800 samples each.
For illustration calculated sums are plotted against measured sums. In fig. 2-6 the x-axis corresponds to the measured values and the y-axis to the calculated global radiation data.

## 2 Conclusion.
The results demonstrate that global radiation data can be determined from satellite images with high accuracy. By-products of the process are temporally and spatially resolved values of the meteorologic data

- Degree of cloudiness
- Sunshine duration
- Cloud structure
- Temporal variances of cloudiness etc.

All results can be presented in the form of maps.
A concept for an operational system is based on hardware and software components developed so far.

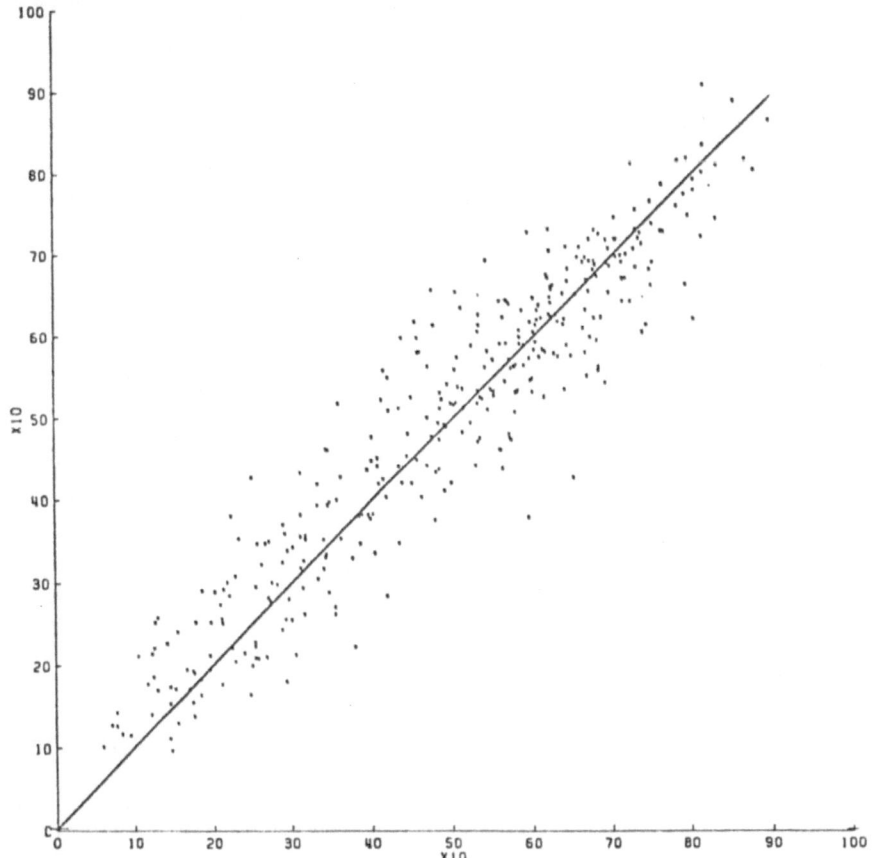

Fig.2. Correlation of global radiation (3 hours).

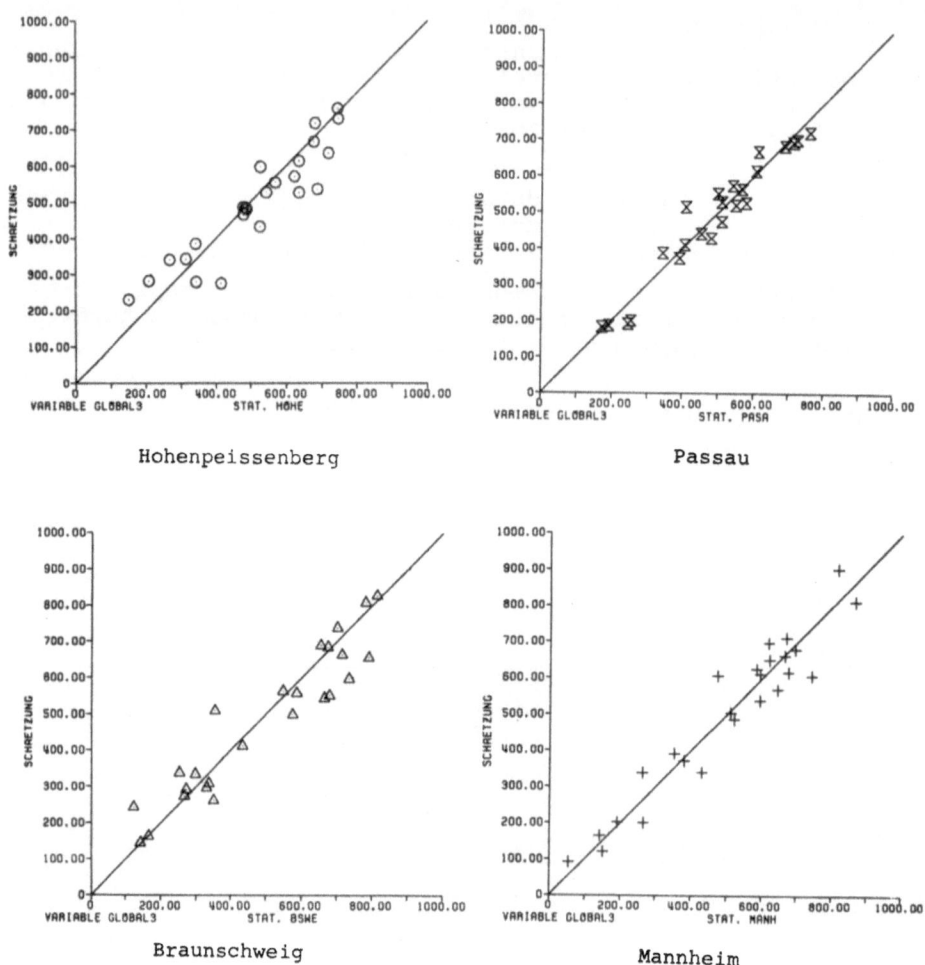

Fig.3-6. Correlation (3 hour values).

# HELIOSAT PROJECT : ESTIMATION OF GLOBAL RADIATION
## FROM SATELLITE USING A STATISTICAL APPROACH

Authors          : D. Cano, J.M. Monget

Contract number : ESF-008-F(G)

Duration         : 42 months            01 January 1980 - 30 June 1983

Total budget     : 914.450 FF           CEC contribution : 457.225 FF

Head of project : J.M. Monget, Director of Centre de Télédétection et
                   d'Analyse des Milieux Naturels

Contractor       : A.R.M.I.N.E.S.

Address          : Ecole des Mines, Rue Claude Daunesse,
                   Sophia-Antipolis, 06565 Valbonne Cédex (France)

## Abstract

The global objective of the HELIOSAT project is to develop an operational system for estimating the global hourly radiation from the geostationnary satellite METEOSAT, using a low-cost SDUS receiving station.

The estimation method is based on the determination of a cloud cover
index from satellite. Linear regression of this index with ground pyranometer measurements provides coefficients which are used for prediction.

Results are presented for a test period 5 to 21 May 1979, using the
ground data provided by Météorologie Nationale for 27 stations over France.
Linear regression coefficients are determined using the period 5 to 12 May.

Comparison between predicted and measured hourly global radiation is
provided for period 12 to 21 May, and three hour intervals (8 - 9h GMT,
10 - 11h GMT, and 12 - 13h GMT). The RMS difference is computed at pixel
level (5 x 7 km), it is on the average of 42 $J/cm^2$ (0.12 $kWh/m^2$) ranging
from 14 $J/cm^2$ (0.04 $kWh/m^2$) to 73 $J/cm^2$ (0.20 $kWh/m^2$) depending on the
station.

# 1. Introduction

The global objective of the HELIOSAT project is to develop an operational system for estimating the global radiation at the surface from the images of the geostationnary satellite METEOSAT retransmitted to the earth using the WEFAX diffusion system.

The main factor for the determination of the global radiation from satellite, is to estimate the intensity of the cloud cover over a given point. This approach is the basis of a two step process which has been designed during the present work. For each pixel (i, j) one first computes a cloud cover index which is then used as a basis for a statistical estimation of the global radiation.

The determination of the cloud cover index has been reported previously (reports ref. CTAMN/80/R/08 and CTAMN/81/R/03). We will present the first results in this paper regarding the period 5 to 21 May 1979 for which 47 images were processed.

Following this determination, a statistical method has been derived for the computation of the global radiation and applied to the same set of data.

# 2. Evaluation of the cloud cover index

## 2.1. Definition

A cloud cover index $(n_{ij}^t)$ is defined at a given time t, as a function of the characteristic albedo of the earth $(\rho_{ij}^S)$, the apparent albedo at the same point as measured by the satellite $(\rho_{ij}^t)$ and the average albedo of the cloud tops $(\rho^{NU})$. It is computed for each pixel using the following formula :

$$n_{ij}^t = (\rho_{ij}^t - \rho_{ij}^S) \,/\, (\rho^{NU} - \rho_{ij}^S)$$

The characteristic albedo of the Europe area was determined in previous work (Cf. report CTAMN/81/R/03) for each 5 × 7 km pixel using an iterative algorithm for the detection and elimination of cloudy areas. The computation of cloud albedo $(\rho^{NU})$ is performed using an inverse algorithm retaining all the cloudy areas followed by the construction of an albedo histogram which modal value is retained as an estimation of $\rho^{NU}$.

## 2.2. Use of thermal infra-red over snow covered areas

The previous method does not perform accurately when the albedo of the earth surface is of the same order of magnitude as the albedo of the clouds. Under the latitude of Europe, this is specifically the case of snowy areas.

In this case, an alternate cloud index is defined using the radiance measured by the satellite in the thermal infra-red :

$$n'_{ij}^t = (R_{ij}^t - R^{SN}) \,/\, (R^{NU} - R^{SN})$$

In this formula, $R^{NU}$ is the radiance of the clouds, and $R^{SN}$ is the radiance of cloud-freee snow areas, both determined using image histogram in the thermal infra-red.

## 2.3. Sequential updating processing scheme

The characteristic albedo map of Europe ($\rho_{ij}^S$) for the period 5 - 21 May 1979 was first separated into two     categories using its histogram thus defining two classes : pixels with high albedo (snow or everlasting cloud cover), and pixels with normal ground albedo.

Each of the 47 images were then compared to this characteristic map in order to sequentially update the surface albedo in order to account for long term seasonal variations which had not been accounted for in the characteristic albedo determination.

## 2.4. Experiment

The algorithm was applied to 47 images (VIS and IR) for the period 5 to 21 May 1979, belonging to the following daily hour intervals :
- 8 to  9 h - Slot 16 (15 images)
- 10 to 11 h - Slot 20 (16 images)
- 12 to 13 h - Slot 24 (16 images)

The characteristic albedo image (Fig. 1) was produced in 35 minutes of computing time and the cloud index compution spent 1.5 minute per image (VIS and IR). Fig. 2 illustrates the results for the 8 - 9h interval on 10 May 1979.

## 3. Statistical relationship between cloud cover index and ground measured transmission coefficient

Contrary to the methodology used by other authors [C. Gauthier (1), E. Rashke (2) and J. Amada (3)], we are using physical models of the transmission of radiation through the atmosphere, we have chosen to apply a purely statistical method in order to estimate the global hourly radiation from satellite measurements.

We have successfully developed linear relationships between the satellite determined cloud cover index and the ground measured transmission coefficient. This simple and efficient approach is based on the work of Centre d'Energétique de l'Ecole Nationale Supérieure des Mines de Paris (4) and correlations determined by Pastre (5) between insolation duration and cloud cover index.

### 3.1. Definitions and interpretation method

The global radiation atmospheric transmission coefficient $K_{ij}$ is sometimes also called global irradiation fraction, determined as the ratio of global ground irradiance on an horizontal surface ($G_{ij}$) to the mean horizontal irradiance outside the atmosphere ($G_{ij}^0$) :

$$K_{ij} = G_{ij} / G_{ij}^0$$

This quantity varies typically from 0.2 to 0.8.

If one interprets the cloud cover index ($n_{ij}^t$) as the percentage of surface of the surface of the pixel (i, j) covered by clouds, the global ground irradiance at time t is expressed as a linear combination :

$$G_{ij}^t = n_{ij}^t \, G_{ij}^1 + (1 - n_{ij}^t) \, G_{ij}^2$$

where $G_{ij}^1$ and $G_{ij}^2$ are the global ground irradiance for respectively cloud covered and clear skys.

For each of these extreme conditions, one can define a transmission coefficient, respectively $K^1$ and $K^2$, which is supposed to be constant during the same hour. This realistic hypothesis leads to the following linear relation :

$$ K_{ij}^t = n_{ij}^t \ (K_{iJ}^2 - K_{ij}^1) + K_{ij}^1 $$

## 3.2. Verification of linear relation between $n_{ij}^t$ and $K_{ij}^t$

The Météorologie Nationale provided us with pyranometer ground hourly global radiation measurements for each of 27 stations listed on Table 1.

These horizontal measurements were reduced to transmission coefficients and compared to the closest 5 x 7 km pixel in the image with a location error better than one pixel (5 km).

The correlation coefficient almost everywhere superior to 80 % thus indicating that the hypothesis of linear relationship can be relied upon (fig. 3).

## 4. Determination of hourly ground radiation from satellite

### 4.1. Method of estimation

The previous step provides for each pixel (i, j) and hour interval t, the coefficients $a_{ij}$ and $b_{ij}$, defining a linear regression between the transmission coefficient $K_{ij}^t$ and the cloud cover index $n_{ij}^t$ determined from satellite.

The hypothesis is that this relation is stable in time or slow varying at the seasonal level. For a new image, it is thus possible to estimate the transmission coefficient :

$$ \hat{K}_{ij}^t = a_{ij} \ n_{ij}^t + b_{ij} $$

from which is computed the estimation of the hourly global radiation :

$$ \hat{G}_{ij}^t = \hat{K}_{ij}^t \ G_{ij}^{0t} $$

### 4.2. Verification and performance of estimation

The test period of May 1979 was split into two parts for verification purposes :
- training period  :  5 to 12 May 1979
- estimated period : 13 to 21 May 1979

Ground truth measurements of global hourly radiation were used during the training period in order to determine the coefficients $a_{ij}$ and $b_{ij}$ for the 27 stations over France. The coefficients were then used for the prediction of global hourly radiation during the estimated period for a pixel size 5 x 7 km.

For verification, estimated and measured values were then compared for the period 13 to 21 May 1979 using a linear regression.

The correlation coefficient between measured and estimated is superior to 80 %, except for four stations (Carpentras, Pau, Montpellier, Marignane) and the average R.M.S. error is 42 J/cm² (0.12 kWh/m²) (Table 2). Results are presented for the station of Nice and Strasbourg (Fig. 4).

## 5. Conclusion

The hypothesis of linear correlation between the ground measured transmission coefficient and the METEOSAT computed cloud cover index, is now verified for the 27 pyranometer stations of France.

Two problems will be adressed in the future :
- Stability in time of the regression coefficients will be studied for slow seasonal variations ;
- The spatial correlation between these coefficients will be determined in order to develop a method of interpolation between the stations.

The use of the regression coefficients will be verified for the test period of April 1982 for which SDUS METEOSAT data were received from 18 to 30 April.

## References

(1) GAUTHIER C., DIAK MASSE : A simple physical model to estimate incident solar radiation at the surface from GOES satellite data. Journal of Applied Meteorology, vol. 19, n° 8, August 1979.
(2) RASHKE E., PREUSS : The determination of the solar radiation budget at the earth's surface from satellite measurements. Meteorology Rdsch 32, pp. 18-28, February 1979.
(3) AMADO, DELORME Ch., DELORME Cl., RABERANTO : Méthodologie d'utilisation des images du satellite METEOSAT pour la détermination du gisement solaire. Rapport COMES-GDTA n° 81.877 GS.7, Août 1981.
(4) Anonyme : Courbes de fréquences cumulées de l'irradiation solaire globale horaire reçue par une surface plane. Rapport du Centre d'Energétique de l'Ecole Nationale Supérieure des Mines de Paris, Juillet 1979.
(5) PASTRE : Développement d'une méthode de détermination du rayonnement solaire global à partir des données METEOSAT. Preprint, revue La Météorologie VI, série n° 24, Mars 1981.

## Acknowledgements

This work is also conducted under the sponsorship of AFME (COMES), grants n° 80/06/002 and 82/11/006/2256.

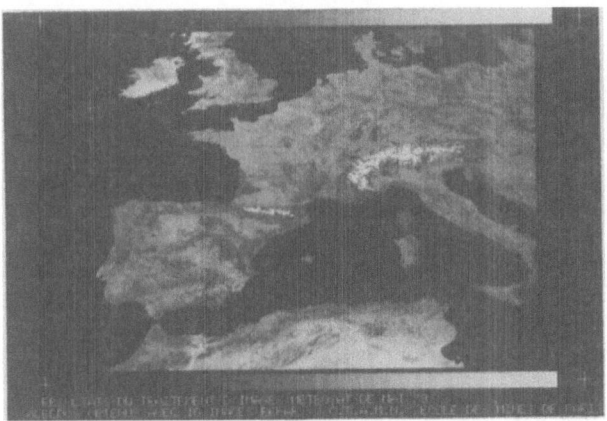

Fig. 1 :

Characteristic albedo map of Europe obtained with METEOSAT for period 5 - 21 May 1979.

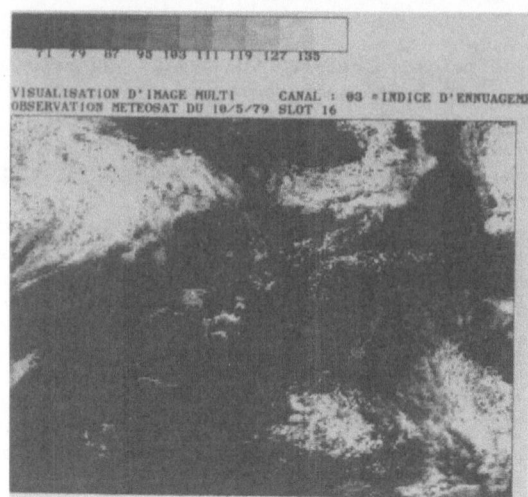

Fig. 2 :

Cloud cover index map
obtained from satellite for
the 8 - 9 h GMT hour
interval on 10 May 1979

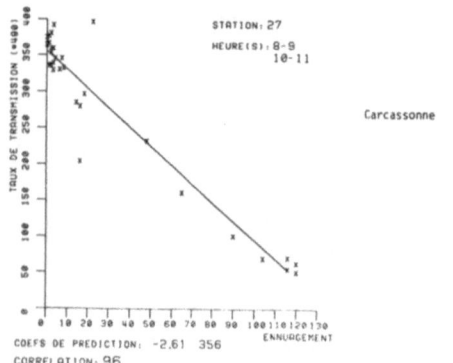

Fig. 3 :

Example of linear regression
between satellite cloud index
and pyranometer measured
transmission coefficient.

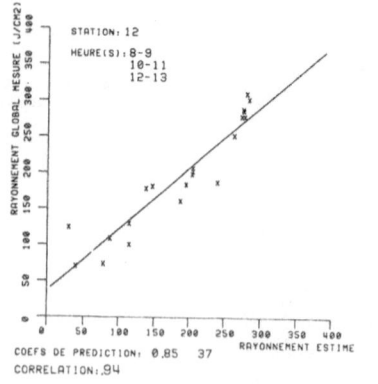
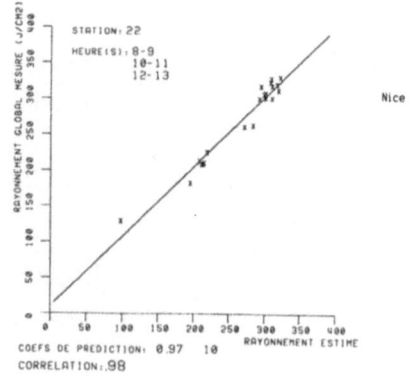

Fig. 4 : Examples of relation between satellite
derived and ground measured hourly
global radiations.

| | |
|---|---|
| 1 - AJACCIO | 15 - TOURS |
| 2 - NANCY | 16 - ILES DU LEVANT |
| 3 - MILLAU | 17 - MUCOR |
| 4 - PARIS | 18 - LIMOGES |
| 5 - SAINT-QUENTIN | 19 - MARIGNANE |
| 6 - CLERMONT-FERRAND | 20 - AGEN |
| 7 - CARPENTRAS | 21 - BORDEAUX |
| 8 - REIMS | 22 - NICE |
| 9 - DIJON | 23 - LA ROCHELLE |
| 10 - PAU | 24 - AUXERRE |
| 11 - RENNES | 25 - BISCAROSSE |
| 12 - STRASBOURG | 26 - CAEN |
| 13 - MONTPELLIER | 27 - CARCASSONNE |
| 14 - TRAPPES | 28 - Toutes stations confondues |

*Tab. 1:   Station codes of the Météorologie Nationale ground pyranometer net*

| | r | a | b | Δ |
|---|---|---|---|---|
| 1 | CORRELATION:.906 | COEF.: 1.200 | -46.244 | ERREUR: 14.295 |
| 2 | CORRELATION:.968 | COEF.: 1.040 | -6.071 | ERREUR: 27.146 |
| 3 | CORRELATION:.978 | COEF.: .955 | 2.180 | ERREUR: 24.547 |
| 4 | CORRELATION:.922 | COEF.: .859 | 40.761 | ERREUR: 34.694 |
| 5 | CORRELATION:.938 | COEF.: .955 | 2.916 | ERREUR: 28.138 |
| 6 | CORRELATION:.918 | COEF.: 1.028 | 8.602 | ERREUR: 45.625 |
| 7 | CORRELATION:.715 | COEF.: .462 | 115.900 | ERREUR: 55.353 |
| 8 | CORRELATION:.956 | COEF.: .919 | 27.939 | ERREUR: 26.388 |
| 9 | CORRELATION:.895 | COEF.: 1.041 | -24.708 | ERREUR: 50.140 |
| 10 | CORRELATION:.777 | COEF.: .955 | -4.755 | ERREUR: 69.171 |
| 11 | CORRELATION:.914 | COEF.: .968 | 19.831 | ERREUR: 36.841 |
| 12 | CORRELATION:.961 | COEF.: 1.138 | -40.769 | ERREUR: 24.786 |
| 13 | CORRELATION:.713 | COEF.: 1.108 | -80.907 | ERREUR: 72.916 |
| 14 | CORRELATION:.964 | COEF.: .942 | 38.926 | ERREUR: 26.276 |
| 15 | CORRELATION:.851 | COEF.: 1.230 | -61.943 | ERREUR: 44.812 |
| 16 | CORRELATION:.966 | COEF.: .959 | 5.212 | ERREUR: 22.324 |
| 17 | CORRELATION:.943 | COEF.: .955 | -10.024 | ERREUR: 36.300 |
| 18 | CORRELATION:.853 | COEF.: .955 | -31.211 | ERREUR: 56.402 |
| 19 | CORRELATION:.484 | COEF.: .562 | 91.889 | ERREUR: 72.069 |
| 20 | CORRELATION:.863 | COEF.: 1.012 | -29.283 | ERREUR: 55.104 |
| 21 | CORRELATION:.962 | COEF.: 1.131 | -48.760 | ERREUR: 28.225 |
| 22 | CORRELATION:.805 | COEF.: .913 | -2.537 | ERREUR: 60.487 |
| 23 | CORRELATION:.919 | COEF.: 1.043 | -11.162 | ERREUR: 41.450 |
| 24 | CORRELATION:.816 | COEF.: .837 | 7.272 | ERREUR: 57.247 |
| 25 | CORRELATION:.928 | COEF.: .910 | -.287 | ERREUR: 36.015 |
| 26 | CORRELATION:.806 | COEF.: 1.131 | -13.197 | ERREUR: 58.465 |
| 27 | CORRELATION:.961 | COEF.: .957 | -15.957 | ERREUR: 28.251 |
| 28 | CORRELATION:.871 | COEF.: .916 | 6.707 | ERREUR: 50.659 |

*Tab. 2:
Statistical results of comparison between satellite derived and ground measured global hourly radiations using the 27 stations of Météorologie Nationale. Prediction was performed on 13 to 21 May 1979 period using 5 to 12 May 1979 period for training (1 kWh/m² = 360 J/cm²).*

r   = Coefficient de corrélation
a, b = Paramètres de la droite d'estimation (rayonnement mesuré/
      rayonnement estimé)
Δ   = Variance de l'erreur (J/cm²/h)

# LIST OF PARTICIPANTS

AYDINLI, S.

Institut für Lichttechnik der
Technischen Universität Berlin
Einsteinufer 19
D - 1000 BERLIN 10
Tel. 314-3489 (-2401 -2536)
Telex 184262 tubln d

BEDEL, J.-A.

Service Météorologique Métropolitain
Direction de la Météorologie
2, avenue Rapp
F - 75340 PARIS CEDEX 07
Tel. 555 95 02
Telex

BOURGES, B.

ARMINES/Ecole des Mines de Paris
Centre d'Energétique
60, Boulevard St. Michel
F - 75272 PARIS CEDEX 06
Tel. (1) 329 21 05
Telex

BUIS, H.

Commission of the European Communities
D.G. Science, Research and Development
200, rue de la Loi - SDM 3/35
B - 1049 BRUSSELS
Tel. 235 84 44
Telex 21877 comeu b

CENA, V.

ADES srl
23, via Giubbonari
I - 00186 ROME
Tel. (06) 65 98 62
Telex

DEHNE, K.

Deutscher Wetterdienst
Meteorologisches Observatorium Hamburg
Frahmredder 95
D - 2000 HAMBURG 65
Tel. (040) 601 79 24
Telex 2162912 dwsa d

DOGNIAUX, R.

Institut royal météorologique de Belgique
3, avenue Circulaire
B - 1180 BRUXELLES
Tel. 374 67 88
Telex meteor bru 21315

EIDEN, R.                   Universität Bayreuth, Inst. für
                            Geowissenschaften, Abt. für Meteorologie
                            Universitätsstrasse
                            D - 8580 BAYREUTH
                            Tel. (0921) 55 22 50
                            Telex

EIDORFF, S.                 Technical University of Denmark
                            Thermal Insulation Laboratory
                            Building 118
                            DK - 2800 LYNGBY
                            Tel. 88 35 11
                            Telex 37529 dthdia dk

FRYDENDAHL, K.              Danish Meteorological Institute
                            Lyngbyvej 100
                            DK - 2100 KØBENHAVN
                            Tel. (01) 29 21 00
                            Telex

GANDINO, C.                 Commission of the European Communities
                            Joint Research Center
                            Meteorological Observatory
                            I - 21020 ISPRA (Varese)
                            Tel. (0332) 78 02 71
                            Telex 380042/380058 eur i

HALLGREEN, L.               Teknologisk Institut
                            Gregersensvej
                            DK - 2630 TAASTRUP
                            Tel. (02) 99 66 11
                            Telex

JENKINS, G.                 UK Meteorological Office
                            MET.O.IC
                            Eastern Road
                            UK - BRACKNELL, Berkshire RG12 2SZ
                            Tel. (0344) 202 42 x 2538
                            Telex

KASTEN, F.                  Deutscher Wetterdienst
                            Meteorologisches Observatorium Hamburg
                            Frahmredder 95
                            D - 2000 HAMBURG 65
                            Tel. (40) 601 79 24
                            Telex 2162912 dwsa d

LASNIER, F.                 ARMINES/Ecole des Mines de Paris
                            Centre d'Energétique
                            60. Bd. St. Michel
                            F - 75006 PARIS
                            Tel. (1) 329 21 05
                            Telex

LAVAGNINI, A.         Istituto di Fisica dell'Atmosfere
                      P.le L. Sturzo 31
                      I - 00144 ROME
                      Tel. (06) 591 09 41
                      Telex 614344 atmos


LEMOINE, M.           Institut royal météorologique
                      3, avenue Circulaire
                      B - 1180 BRUXELLES
                      Tel. 374 67 87
                      Telex 21315 meteor b


LUND, H.              Technical University of Denmark
                      Thermal Insulation Laboratory
                      Building 118
                      DK - 2800 LYNGBY
                      Tel. (02) 88 35 11 x 5286
                      Telex 37529 dthdia dk


McGREGOR, J.          Solar Energy Unit, Dept.of Mechanical
                      Engineering, University College Cardiff
                      Newport Road
                      UK - CARDIFF, Wales
                      Tel. (0222) 44 211 x 7029
                      Telex


McWILLIAMS, S.        Irish Meteorological Service
                      Valentia Observatory
                      IRL - CAHIRCEVEEN, Co. Kerry
                      Tel. Cahirceveen 27
                      Telex 26912 mtva ei


MOESER, W.            Institut für Geophysik und Meteorologie
                      der Universität zu Köln
                      Kerpenerstr. 13
                      D - 5000 KOELN 41
                      Tel. (0221) 470 36 82
                      Telex


MONGET, J.-M.         Centre de Télédétection et Analyse des
                      Milieux naturels - Ecole des Mines
                      Sophia-Antipolis
                      F - 06560 VALBONNE
                      Tel. (93) 33 05 58
                      Telex


MURPHY, E.            Irish Meteorologicql Service
                      Valentia Observatory
                      IRL - CAHIRCEVEEN, Co. Kerry
                      Tel. Cahirceveen 27
                      Telex 26912 mtva ei

NICOLAY, D.                  Commission of the European Communities
                            D.G. Information Market and Information
                            B.P. 1907 - JMO B4/072
                            L - 1019 LUXEMBOURG
                            Tel. 43011 x 2946
                            Telex 3423/3446 comeur lu

PAGE, J.                     University of Sheffield
                            Dept. of Building Science
                            UK - SHEFFIELD, S10 2TN
                            Tel. (0742) 78 555
                            Telex 54348 ulshef g

PALZ, W.                     Commission of the European Communities
                            D.G. Science, Research and Development
                            200, rue de la Loi - SDM 3/19
                            B - 1049 BRUSSELS
                            Tel. 235 69 22 - 234 71 24
                            Telex 21877 comeu b

PLAZY, J.-L.                 Agence Française pour la Maîtrise de l'Energie
                            Commissariat à l'Energie solaire
                            Sophia Antipolis
                            F - 06565 VALBONNE
                            Tel. (93) 74 79 79
                            Telex comessa 461357 f

SCHWARZMANN, P.              University of Stuttgart
                            Inst. für Physikalische Elektronik
                            Boeblingerstrasse 70
                            D - 7000 STUTTGART
                            Tel. (0711) 665 370
                            Telex

SLOB, W.                     The Royal Netherlands Meteorological Institute
                            Wilhelminalaan 10
                            NL - 3730 AE DE BILT
                            Tel. (030) 76 69 11
                            Telex

STEEMERS, T.                 Commission of the European Communities
                            D.G. Science, Research and Development
                            200, rue de la Loi
                            B - 1049 BRUSSELS
                            Tel. (02) 235 68 78
                            Telex 21877 comeu b

TRICAUD, J.-F.               CNRS
                            Laboratoire d'Energétique Solaire
                            B.P. 5 - Odeillo
                            F - 66120 FONT-ROMEU
                            Tel. (68) 30 10 24
                            Telex odeilab 500167 f

VALKO, P.                  Swiss Meteorological Institute
                           Krähbühlstr. 58
                           CH - 8044 ZUERICH
                           Tel. (01) 252 67 20
                           Telex


VAN PAASSEN, A.H.C.        Delft University of Technology
                           Mekelweg 2
                           NL - DELFT
                           Tel. (015) 78 66 75
                           Telex